Acoustic Design of Schools

School Buildings and Design Unit
Department for Education and Skills

London : TSO

DfES Project Team

Richard Daniels	Building Services Engineer, School Buildings & Design Unit
Alex Freemantle	Architect, Formerly of School Buildings & Design Unit
Mukund Patel	Head of School Buildings & Design Unit

Acknowledgements

DfES would like to thank the following:

Editors:

Bridget Shield	London South Bank University
Carl Hopkins	BRE Acoustics, Building Research Establishment Ltd (BRE)

Principal authors:

Carl Hopkins & Robin Hall	BRE Acoustics, Building Research Establishment Ltd (BRE)
Adrian James	Adrian James Acoustics Ltd
Raf Orlowski & Sam Wise	Arup Acoustics
David Canning	City University, London

Other authors and advisors:

Stephen Chiles	University of Bath
David Dennis	London Borough of Newham
Nigel Cogger	The English Cogger Partnership
John Miller & Theodoros Niaounakis	Bickerdike Allen and Partners
Les Fothergill	Building Regulations Division, Office of the Deputy Prime Minister
Guy Shackle	Barron and Smith Architects
Julie Dockrell	Institute of Education, University of London
Mindy Hadi	Building Research Establishment Ltd (BRE)
Matthew Ling	Formerly of Building Research Establishment Ltd (BRE)
Russell Brett	British Association of Teachers of the Deaf
Richard Vaughan	National Deaf Children's Society
Roz Comins	Voice Care Network UK
Thomas Wulfrank	Arup Acoustics
David Coley & Andrew Mitchell	Centre for Energy and the Environment, Exeter University.
Derek Poole	Formerly of University of Wales, College of Cardiff
John Lloyd & Tom Cecil	Faber Maunsell
Peter Brailey	Hawksmoor Engineering Ltd.
Andrew Parkin	RW Gregory LLP
Terry Payne	Monodraught Ltd
Wayne Aston	Passivent
Tim Spencer	Rockwool Rockfon Ltd
David Whittingham	Formerly of Ecophon Ltd

Photographer:	Philip Locker, Photo Graphic Design, Bolton
Design & Mac file:	Malcolm Ward, Malcolm Studio, Croydon

ISBN 0 11 271105 7

Contents

Contents

Introduction

Building Bulletin 93 aims to:

- provide a regulatory framework for the acoustic design of schools in support of the Building Regulations
- give supporting advice and recommendations for planning and design of schools
- provide a comprehensive guide for architects, acousticians, building control officers, building services engineers, clients, and others involved in the design of new school buildings.

The constructional standards for acoustics for new school buildings, as given in Section 1 of this document, are required to be achieved under the Building Regulations. This represents a significant tightening of the regulation of acoustic design in schools, to reflect a general recognition, supported by research, that teaching and learning are acoustically demanding activities. In particular, there is a consensus that low ambient noise levels are required, particularly in view of the requirements of the Special Educational Needs and Disability Act 2001[1] for integration of children with special needs in mainstream schools.

Unfortunately, a large number of classrooms in the UK currently suffer from poor acoustics. The most serious acoustic problems are due to noise transfer between rooms and/or excessive reverberation in rooms. There are many reasons for the poor acoustics, for example:

- The acoustics of the stock of old Victorian schools are often unsuitable for modern teaching methods.
- Modern constructions do not always provide adequate sound insulation and may need special treatment.
- Open-plan, or semi open-plan layouts, designed to accommodate a number of different activities, are areas where background noise and sound intrusion often cause problems.
- The acoustics of multi-purpose rooms, such as halls, have to be suitable for a variety of activities, for example music (which requires a long reverberation time) and speech (which requires shorter reverberation times).
- Many activities, such as music and design technology lessons, can be noisy and will cause problems if there is inadequate sound insulation between areas for these activities and those requiring quieter conditions.

Poor acoustic conditions in the classroom increase the strain on teachers' voices as most teachers find it difficult to cope with high noise levels. This often leads to voice problems due to prolonged use of the voice and the need to shout to keep control. Recent surveys in the UK and elsewhere show that teachers form a disproportionate percentage of voice clinic patients.

Historically, there have been a number of factors preventing good acoustic design and this Building Bulletin addresses these issues.

- Before 2003, Part E of the Building Regulations did not apply to schools. It now includes schools within its scope.
- Although the constructional standards for schools previously quoted Building Bulletin 87[2] as the standard for acoustics in schools, many designers were unaware of the requirements of BB87 and the standards were rarely enforced. These standards have been updated to reflect current research and the relevant requirements of the Disability Discrimination Act, and are included in the compliance section, Section 1, of this bulletin.
- The pressure on finances has meant in the past that acoustics came low on the list of design priorities. The acoustic design will now have a higher priority as it will be subject to building control

1. Now incorporated as Section IV of the Disability Discrimination Act[1]

approval procedures.

• There has been little guidance available in the past on how to achieve the right balance of acoustics in the complex and dynamic environment of a school. Architects and designers have had a difficult time finding relevant information to help them choose the correct target values of appropriate parameters.

Overall, Building Bulletin 93 recommends a structured approach to acoustic design at each stage of the planning and design process, as shown in the table below.

A structured approach to acoustic design at each stage of the planning and design process	
Feasibility/sketch design	■ Selection of the site ■ Noise survey to establish external noise levels ■ Orientation of buildings ■ Massing and form of the buildings ■ Consideration of need for external noise barriers using the buildings, fences and screens and landscape features ■ Preliminary calculation of sound insulation provided by building envelope including the effect of ventilation openings
Detailed design	■ Determine appropriate noise levels and reverberation times for the various activities and room types ■ Consider the special educational needs of the pupils ■ Consider the design of music, drama and other specialist spaces separately from that of normal classrooms as the design criteria are very different ■ Provide the necessary façade sound insulation whilst providing adequate ventilation, particularly in the case of spaces such as classrooms and science laboratories which require high ventilation rates ■ Architectural/acoustic zoning: plan the disposition of 'quiet' and 'noisy' spaces, separating them wherever possible by distance, external areas or neutral 'buffer' spaces such as storerooms or corridors ■ Consider sound insulation separately from other aspects of room acoustics using walls, floors and partitions to provide adequate sound insulation ■ Design the acoustics of the rooms by considering their volume and shape, and the acoustic properties of their surfaces ■ Specify the acoustic performance of doors, windows and ventilation openings ■ Specify any amplification systems
Building Control approval	■ Submit plans, including specific details of the acoustic design, for approval by Building Control Body

SCOPE of Building Bulletin 93

Section 1 of Building Bulletin 93 supersedes Section A of Building Bulletin 87[2] as the constructional standard for acoustics for new school buildings.

In addition, Part E of the Building Regulations includes schools within its scope and Approved Document E[3] gives the following guidance: *"In the Secretary of State's view the normal way of satisfying Requirement E4 will be to meet the values for sound insulation, reverberation time and internal ambient noise which are given in Section 1 of Building Bulletin 93 'The Acoustic Design of Schools', produced by DfES."*

The requirements of Section 1 came into force on 1st July 2003, at the same time as those contained in the new Approved Document Part E[3], in support of the Building Regulations.

Requirement E4 from Part E of Schedule 1 to The Building Regulations 2000 (as amended) states that:

"Each room or other space in a school building shall be designed and constructed in such a way that it has the acoustic conditions and the insulation against disturbance by noise appropriate to its intended use."

The Education (School Premises) Regulations 1999, SI 1999 No.2 [4] which applies to both new and existing school buildings, contains a similar statement:

"Each room or other space in a school building shall have the acoustic conditions and the insulation against disturbance by noise appropriate to its normal use."

Compliance with the acoustic performance standards specified in Section 1 will satisfy both regulations for new schools.

Although Building Regulations do not apply to all alteration and refurbishment work, it is desirable that such work should consider acoustics and incorporate upgrading of the acoustics as appropriate. (In the case of existing buildings, the Building Regulations apply only to 'material alterations' as defined in Regulations 3 and 4.) Although it would be uneconomic to upgrade all existing school buildings to the same standards as new school buildings, where there is a need for upgrading the acoustic performance of an existing building or when refurbishment is happening for other reasons, then the designer should aim to meet the acoustic performance given in Section 1 of BB93 to satisfy the School Premises Regulations and the Disability Discrimination Act.

The exemption of Local Education Authority (LEA) maintained schools from the Building Regulations has ended. New school buildings, including extensions to existing school buildings and new schools formed by change of use of other buildings, are now included in the Building Regulations and may be subject to detailed design checks and on-site inspections by Building Control Bodies.

The Building Regulations and hence the requirements of BB93 only apply in England and Wales. They apply to both LEA maintained schools and independent schools.

Temporary buildings are exempt from the Building Regulations. Temporary buildings are defined in Schedule 2 to the Building Regulations as those which are not intended to remain in place for longer than 28 days. What are commonly called temporary buildings in schools are classed as prefabricated buildings and are normally subject to the same Building Regulations requirements as other types of building. Additional guidance is given in Clause 0.6 of Approved Document E[3]. A building that is created by dismantling, transporting and re-erecting the sub-assemblies on the same premises, or is constructed from sub-assemblies obtained from other premises or from stock manufactured before 1st July 2003, would normally be considered to meet the requirements for schools if it satisfies the relevant provisions relating to acoustic standards set out in the 1997 edition of Building Bulletin 87[2].

The extension of Part E of Schedule 1 to the Building Regulations 2000 (as amended by SI 2002/2871) to schools applies to teaching and learning spaces.

Therefore the performance standards in the tables in Section 1 are required for compliance with Part E for all teaching and learning spaces. Part E of the Building Regulations is not intended to cover the acoustic conditions in administration and ancillary spaces not used for teaching and learning except in as far as they affect conditions in neighbouring teaching and learning spaces. Therefore consideration needs to be given to adjoining areas, such as corridors, which might have doors, ventilators, or glazing separating them from a teaching or learning space. The performance standards given in the tables for administration and ancillary spaces are for guidance only.

Rooms used for nursery and adult/community education within school complexes are also covered by Part E. Part E does not apply to nursery schools which are not part of a school, sixth form colleges which have not been established as schools, and Universities or Colleges of Further and Higher Education[2]. However, many of the acoustic specifications are desirable and can be used as a guide to the design of these buildings. The standards are particularly appropriate for nursery schools as figures are quoted for nursery spaces within primary schools.

The Disability Discrimination Act 1995[1], as amended by the Special Educational Needs and Disability Act 2001, places a duty on all schools and LEAs to plan to increase over time the accessibility of schools for disabled pupils and to implement their plans. Schools and LEAs are required to provide:

- increased access for disabled pupils to the school curriculum. This covers teaching and learning and the wider curriculum of the school such as after-school clubs, leisure and cultural activities.
- improved access to the physical environment of schools, including physical aids to assist education. This includes acoustic improvements and aids for hearing impaired pupils.

When alterations affect the acoustics of a space then improvement of the acoustics to promote better access for children with special needs, including hearing impairments, should be considered. Approved Document M, Access and facilities for disabled people, in support of the Building Regulations[5] includes requirements for access for children with special needs. See also BS 8300: 2001 Design of buildings and their approaches to meet the needs of disabled people[6].

2. Part E of the Building Regulations quotes the definition of school given in Section 4 of the 1996 Education Act. In the case of sixth form colleges Section 4 of the 1996 Act should be read in conjunction with Section 2 of the same Act, in particular subsections (2), (2A) and (4) which deal with the definition of secondary education.

If a sixth form college is established as a school under the 1998 School Standards and Framework Act then it will be classed as a school under Section 4 of the 1996 Education Act and Part E of the Building Regulations on acoustics will apply. Only one sixth form college has been established in this way up until now.

Therefore, most sixth form colleges are institutions in the Further Education sector and not schools, and Part E of the Building Regulations will not apply.

In the case of a new sixth form college it will be necessary to contact the LEA to enquire if the sixth form college has been established as a school or as an Institute of Further Education.

Overview of contents of Building Bulletin 93

Section 1: Specification of Acoustic Performance consists of three parts.

Section 1.1 gives the performance standards for new school buildings to comply with the Building Regulations. These provide a good minimum standard for school design. However, on occasion higher standards will be necessary.

Section 1.2 sets out the preferred means of demonstrating compliance to the Building Control Body.

Section 1.3 gives the tests recommended to be conducted as part of the building contract.

Section 2: Noise Control describes how to conduct a site survey and to plan the school to control noise. It also includes recommendations on maximum external noise levels applying to playing fields, recreational areas and areas used for formal and informal outdoor teaching. External levels are not covered by Building Regulations but are taken into consideration in planning decisions by local authorities[7].

Section 3: Sound Insulation gives detailed guidance on constructions to meet the performance standards for sound insulation specified in Section 1.1.

Section 4: The Design of Rooms for Speech and Section 5: The Design of Rooms for Music give guidance on various aspects of acoustic design relevant to schools.

Section 6: Acoustic Design and Equipment for Pupils with Special Hearing Requirements discusses design appropriate for pupils with hearing impairments and special hearing requirements. It discusses the necessary acoustic performance of spaces and describes the range of aids available to help these pupils.

Section 7 contains 10 case studies illustrating some of the most important aspects of acoustic design of schools.

Appendix 1 defines the basic concepts and technical terms used in the Bulletin.

Appendices 2 and 3 describe the basic principles of room acoustics and sound insulation.

Appendices 4 to 7 give examples of calculations of sound insulation, reverberation time and absorption.

Appendix 8 gives equipment specifications for sound field systems to guide those who need to specify this type of equipment.

Appendix 9 gives an overview of the Noise at Work Regulations as they relate to teachers.

Appendix 10 gives an example of a submission for approval by a Building Control Body.

The DfES acoustics website www.teachernet.gov.uk/acoustics contains further reference material which expands on the source material for acousticians and designers. For example, it links to a spreadsheet which can be used to calculate the sound insulation of the building envelope and the reverberation time of internal rooms. The website will be regularly updated with new information, discussion papers and case studies. The website also contains complete downloads of Building Bulletin 93.

References

[1] Disability Discrimination Act (1995) Part IV
www.hmso.gov.uk
[2] Building Bulletin 87, Guidelines for Environmental Design in Schools (Revision of Design Note 17), The Stationery Office, 1997. ISBN 011 271013 1. (Now superseded by 2003 version of BB87, which excludes acoustics, and is available on www.teachernet.gov.uk/energy)
[3] Approved Document E, Resistance to the passage of sound. Stationery Office, 2003. ISBN 0 11 753 642 3.
www.odpm.gov.uk
[4] Statutory Instruments: 1999: 2. *The Education (School Premises) Regulations 1999: Education, England & Wales.* London: The Stationery Office. ISBN 0110803310.
www.legislation.hmso.gov.uk/si/si1999/19990002.htm
[5] Approved Document M, Access and facilities for disabled people, in support of the Building Regulations, The Stationery Office, 1999 ISBN 0 11 753469. To be replaced shortly by Approved Document M, Access to and use of buildings.
www.odpm.gov.uk
[6] BS 8300: 2001 Design of buildings and their approaches to meet the needs of disabled people, Code of Practice.
[7] PPG 24, Planning Policy Guidance: Planning and Noise, Department of the Environment, The Stationery Office, September 1994. To be replaced by revised Planning Policy documents.

Specification of acoustic performance

SECTION 1

Section 1 of Building Bulletin 93 sets the performance standards for the acoustics of new buildings, and describes the normal means of demonstrating compliance with The Building Regulations.

Contents

The normal way of satisfying Requirement E4 of The Building Regulations is to demonstrate that the performance standards in Section 1.1, as appropriate, have been met.

Section 1.2 sets out the preferred means for demonstrating compliance of the design to the Building Control Body.

Section 1.3 describes acoustic tests that can be used to demonstrate compliance with the performance standards in Section 1.1. It is strongly recommended that the client require acoustic testing to be carried out as part of the building contract, because testing of the completed construction is the best practical means of ensuring that it achieves the design intent.

In all but the simplest of projects it is advisable to appoint a suitably qualified acoustic consultant[1] at an early stage of the project, before the outline design has been decided. This will prevent simple mistakes which can be costly to design out at a later stage. An acoustic consultant will normally be needed to check the design details and on-site construction, and to carry out acoustic tests to confirm that the building achieves the required acoustic performance.

1 The primary professional body for acoustics in the UK is the Institute of Acoustics. An experienced professional acoustician who is competent to be responsible for the acoustic design of school buildings would normally be a corporate member of the Institute of Acoustics.

1 Specification of acoustic performance

1.1 Performance standards

The overall objective of the performance standards in Section 1.1 is to provide acoustic conditions in schools that (a) facilitate clear communication of speech between teacher and student, and between students, and (b) do not interfere with study activities.

Performance standards on the following topics are specified in this section to achieve this objective:

- indoor ambient noise levels
- airborne sound insulation between spaces
- airborne sound insulation between corridors or stairwells and other spaces
- impact sound insulation of floors
- reverberation in teaching and study spaces
- sound absorption in corridors, entrance halls and stairwells
- speech intelligibility in open-plan spaces.

All spaces should meet the performance standards defined in Tables 1.1, 1.2, 1.3, 1.4 and 1.5 for indoor ambient noise level, airborne and impact sound insulation, and reverberation time. Open-plan spaces should additionally meet the performance standard for speech intelligibility in Table 1.6.

The notes accompanying Tables 1.1, 1.2, 1.3 and 1.5 contain additional guidance that should be considered when designing the spaces to meet the performance standards in these tables. Although good practice, this guidance will not be enforced under the Building Regulations.

1.1.1 Indoor ambient noise levels in unoccupied spaces

The objective is to provide suitable indoor ambient noise levels (a) for clear communication of speech between teacher and student, and between students and (b) for study activities.

The indoor ambient noise level includes noise contributions from:

- external sources outside the school premises (including, but not limited to, noise from road, rail and air traffic, industrial and commercial premises)
- building services (eg ventilation system, plant, etc). If a room is naturally ventilated, the ventilators or windows should be assumed to be open as required to provide adequate ventilation. If a room is mechanically ventilated, the plant should be assumed to be running at its maximum operating duty.

The indoor ambient noise level excludes noise contributions from:

- teaching activities within the school premises, including noise from staff, students and equipment within the building or in the playground. Noise transmitted from adjacent spaces is addressed by the airborne and impact sound insulation requirements.
- equipment used in the space (eg machine tools, CNC machines, dust and fume extract equipment, compressors, computers, overhead projectors, fume cupboards). However, these noise sources should be considered in the design process.
- rain noise. However, it is essential that

NOTES ON TABLE 1.1

1 Research indicates that teaching can be disrupted by individual noisy events such as aircraft flyovers, even where the noise level is below the limits in Table 1.1. For rooms identified in Table 1.1 having limits of 35 dB or less the noise level should not regularly exceed 55 dB $L_{A1,30min}$.

2 Acoustic considerations of open-plan areas are complex and are discussed in Section 1.1.7 and Section 4.

3 Studios require specialised acoustic environments and the noise limits for these will vary with the size, intended use and type of room. In some cases noise limits below 30 dB L_{Aeq} may be required, and separate limits for different types of noise may be appropriate; specialist advice should be sought.

4 Halls are often multi-functional spaces (especially in primary schools) used for activities such as dining, PE, drama, music, assembly, and performing plays and concerts. In such multi-functional spaces the designer should design to the lowest indoor ambient noise level for which the space is likely to be used. For large halls used for formal drama and music performance lower noise levels than those in Table 1.1 are preferable, and levels of 25 dB $L_{Aeq,30min}$ may be appropriate. In these cases specialist advice should be sought.

Type of room	Room classification for the purpose of airborne sound insulation in Table 1.2		Upper limit for the indoor ambient noise level $L_{Aeq,30min}$ (dB)
	Activity noise (Source room)	Noise tolerance (Receiving room)	
Nursery school playrooms	High	Low	35[1]
Nursery school quiet rooms	Low	Low	35[1]
Primary school: classrooms, class bases, general teaching areas, small group rooms	Average	Low	35[1]
Secondary school: classrooms, general teaching areas, seminar rooms, tutorial rooms, language laboratories	Average	Low	35[1]
Open-plan[2]			
Teaching areas	Average	Medium	40
Resource areas	Average	Medium	40
Music			
Music classroom	Very high	Low	35[1]
Small practice/group room	Very high	Low	35[1]
Ensemble room	Very high	Very low	30[1]
Performance/recital room	Very high	Very low	30[1]
Recording studio[3]	Very high	Very low	30[1]
Control room for recording	High	Low	35[1]
Lecture rooms			
Small (fewer than 50 people)	Average	Low	35[1]
Large (more than 50 people)	Average	Very low	30[1]
Classrooms designed specifically for use by hearing impaired students (including speech therapy rooms)	Average	Very low	30[1]
Study room (individual study, withdrawal, remedial work, teacher preparation)	Low	Low	35[1]
Libraries			
Quiet study areas	Low	Low	35[1]
Resource areas	Average	Medium	40
Science laboratories	Average	Medium	40
Drama studios	High	Very low	30[1]
Design and Technology			
• Resistant materials, CADCAM areas	High	High	40
• Electronics/control, textiles, food, graphics, design/resource areas	Average	Medium	40
Art rooms	Average	Medium	40
Assembly halls[4], multi-purpose halls[4] (drama, PE, audio/visual presentations, assembly, occasional music)	High	Low	35[1]
Audio-visual, video conference rooms	Average	Low	35[1]
Atria, circulation spaces used by students	Average	Medium	45
Indoor sports hall	High	Medium	40
Dance studio	High	Medium	40
Gymnasium	High	Medium	40
Swimming pool	High	High	50
Interviewing/counselling rooms, medical rooms	Low	Low	35[1]
Dining rooms	High	High	45
Ancillary spaces Kitchens*	High	High	50
Offices*, staff rooms*	Average	Medium	40
Corridors*, stairwells*	Average - High	High	45
Coats and changing areas*	High	High	45
Toilets*	Average	High	50

** Part E of Schedule 1 to the Building Regulations 2000 (as amended by SI 2002/2871) applies to teaching and learning spaces and is not intended to cover administration and ancillary spaces (see under Scope in the Introduction). For these areas the performance standards are for guidance only.*

Table 1.1: Performance standards for indoor ambient noise levels – upper limits for the indoor ambient noise level, $L_{Aeq,30min}$

Table 1.2: Performance standards for airborne sound insulation between spaces – minimum weighted BB93 standardized level difference, $D_{nT\,(T_{mf,max}),w}$

Minimum $D_{nT\,(T_{mf,max}),w}$ (dB)		**Activity noise in source room** *(see Table 1.1)*			
		Low	Average	High	Very high
Noise tolerance in receiving room *(see Table 1.1)*	High	30	35	45	55
	Medium	35	40	50	55
	Low	40	45	55	55
	Very low	45	50	55	60

NOTES ON TABLE 1.2

1 Each value in the table is the minimum required to comply with the Building Regulations. A value of 55 dB $D_{nT\,(T_{mf,max}),w}$ between two music practice rooms will not mean that the music will be inaudible between the rooms; in many cases, particularly if brass or percussion instruments are played, a higher value is desirable.

2 Where values greater than 55 dB $D_{nT\,(T_{mf,max}),w}$ are required it is advisable to separate the rooms using acoustically less sensitive areas such as corridors and storerooms. Where this is not possible, high performance constructions are likely to be required and specialist advice should be sought.

3 It is recommended that music rooms should not be placed adjacent to design and technology spaces or art rooms.

4 These values of $D_{nT\,(T_{mf,max}),w}$ include the effect of glazing, doors and other weaknesses in the partition. In general, normal (non-acoustic) doors provide much less sound insulation than the surrounding walls and reduce the overall $D_{nT\,(T_{mf,max}),w}$ of the wall considerably, particularly for values above 35 dB $D_{nT\,(T_{mf,max}),w}$. Therefore, doors should not generally be installed in partitions between rooms requiring values above 35 dB $D_{nT\,(T_{mf,max}),w}$ unless acoustic doors, door lobbies, or double doors with an airspace are used. This is not normally a problem as rooms are usually accessed via corridors or circulation spaces so that there are at least two doors between noise-sensitive rooms. For more guidance see Section 3.

this noise is considered in the design of lightweight roofs and roof lights as it can significantly increase the indoor ambient noise level (see the design guidance in Section 3.1.1). It is intended that a performance standard for rain noise will be introduced in a future edition of BB93. To satisfy this edition of BB93 it should be demonstrated to the Building Control Body that the roof has been designed to minimise rain noise (see Section 1.2).

Table 1.1 contains the required upper limits for the indoor ambient noise levels for each type of unoccupied space. The noise levels in Table 1.1 are specified in terms of $L_{Aeq,30min}$. This is an average noise level over 30 minutes, as explained in Appendix 1. The specified levels refer to the highest equivalent continuous A-weighted sound pressure level, $L_{Aeq,30min}$, likely to occur during normal teaching hours. The levels due to external sources will depend on weather conditions (eg wind direction) and local activities. High noise levels due to exceptional events may be disregarded.

The indoor ambient noise levels in Table 1.1 apply to finished but unoccupied and unfurnished spaces.

Tonal and intermittent noises are generally more disruptive than other types of noise at the same level. Noise from plant, machinery and equipment in noise–sensitive rooms should therefore be constant in nature and should not contain any significant tonal or intermittent characteristics. Noise from building services which is discontinuous, tonal, or impulsive (ie noise which can be distracting) should be reduced to a level at least 5 dB below the specified maximum.

In rooms with very low noise tolerance, including music rooms, studios and rooms used for formal music and drama performance, any audible intermittent noise source of this type is likely to cause problems and specialist advice should be sought.

1.1.2 Airborne sound insulation between spaces

The objective is to attenuate airborne sound transmitted between spaces through walls and floors.

Table 1.2 contains the required minimum airborne sound insulation values between rooms. These values are defined by the activity noise in the source room and the noise tolerance in the receiving room. The activity noise and noise tolerance for each type of room are given in Table 1.1. The airborne sound insulation is quoted in terms of the weighted BB93 standardized level difference, $D_{nT(T_{mf,max}),w}$, between two rooms.

The BB93 standardized level difference, $D_{nT(T_{mf,max})}$, is the level difference, in decibels, corresponding to a BB93 reference value of the reverberation time in the receiving room:

$$D_{nT(T_{mf,max})} = D + 10 \lg \frac{T}{T_{mf,max}} \text{ dB}$$

where

D is the level difference (dB)

T is the reverberation time in the receiving room (s)

$T_{mf,max}$ is the reference reverberation time equal to the upper limit of the reverberation time, T_{mf}, given in Table 1.5 for the type of receiving room. This reference reverberation time shall be used for all frequency bands.

The BB93 standardized level difference, $D_{nT(T_{mf,max}),w}$, is measured in accordance with BS EN ISO 140-4:1998[1] in octave or one-third octave bands, the results are weighted and expressed as a single-number quantity, $D_{nT(T_{mf,max}),w}$, in accordance with BS EN ISO 717-1:1997[2].

The prediction and measurement of $D_{nT(T_{mf,max}),w}$ between two rooms must be carried out in both directions as its value depends upon the volume of the receiving room, see the example below.

1.1.3 Airborne sound insulation between circulation spaces and other spaces used by students

The objective is to attenuate airborne sound transmitted between circulation spaces (eg corridors, stairwells) and other spaces used by students.

Table 1.3 contains the required minimum airborne sound insulation for the separating wall construction, any doorset in the wall and any ventilators in the wall. The airborne sound insulation for walls and doorsets is quoted in terms

Example to determine the performance standards for airborne sound insulation between a music classroom and a secondary school general teaching area.

***From* the music classroom (source room) *to* the general teaching area (receiving room):**
Table 1.1 shows that music classrooms have 'very high' activity levels and that general teaching areas have 'low' tolerance. Table 1.2 shows that at least 55 dB $D_{nT(0.8s),w}$ is required.

***From* the general teaching area (source room) *to* the music classroom (receiving room):**
Table 1.1 shows that general teaching areas have 'average' activity levels and that music classrooms have 'low' tolerance. Table 1.2 shows that at least 45 dB $D_{nT(1.0s),w}$ is required.

In this example the requirement to control noise from the music classroom to the general teaching area is more stringent.

The construction should be designed to achieve at least 55 dB $D_{nT(0.8s),w}$ from the music classroom (source room) to the general teaching area (receiving room), and at least 45 dB $D_{nT(1.0s),w}$ from the general teaching area (source room) to the music classroom (receiving room).

Table 1.3: Performance standards for airborne sound insulation between circulation spaces and other spaces used by students – minimum sound reduction index, R_w and minimum $D_{n,e,w}$ – 10lg$N$ (laboratory measurements)

Type of space used by students	Minimum R_w (dB)		Minimum $D_{n,e,w} - 10\lg N$ (dB)
	Wall including any glazing	Doorset[1]	
All spaces except music rooms	40	30	39
Music rooms[2]	45	35	45[3]

NOTES ON TABLE 1.3

1 The R_w ratings are for the doorset alone. Manufacturers sometimes provide doorset sound insulation data as a combined rating for the wall and doorset where the R_w refers to the performance of an ≈10 m^2 high-performance wall containing the doorset. This is not appropriate as it gives higher figures than the R_w of the doorset itself. However, with knowledge of the wall and doorset areas the R_w of the doorset can be calculated from these test results.

2 Special design advice is recommended.

3 Wherever possible, ventilators should not be installed between music rooms and circulation spaces.

of the weighted sound reduction index, R_w, which is measured in the laboratory. The airborne sound insulation for ventilators is quoted in terms of the weighted element-normalized level difference, $D_{n,e,w}$. The performance standard for ventilators is quoted in terms of $D_{n,e,w} - 10\lg N$ where N is the number of ventilators with airborne sound insulation $D_{n,e,w}$.

The weighted sound reduction index is measured in accordance with BS EN ISO 140-3:1995[3] and rated in accordance with BS EN ISO 717-1:1997[2].

The weighted element-normalized level difference is measured in accordance with BS EN 20140-10:1992[4] and rated in accordance with BS EN ISO 717-1:1997[2].

Table 1.3 excludes:

- service corridors adjacent to spaces that are not used by students
- lobby corridors leading only to spaces used by students that have a high tolerance to noise as defined in Table 1.1.

The performance standard is set using a laboratory measurement because of the difficulty in accurately measuring the airborne sound insulation between rooms and corridors, or rooms and stairwells in the field. Therefore it is crucial that the airborne sound insulation of the wall and/or doorset is not compromised by flanking sound transmission, eg sound transmission across the junction between the ceiling and the corridor wall (see guidance in Section 3.10.3).

1.1.4 Impact sound insulation of floors

The objective is to attenuate impact sound (eg footsteps) transmitted into spaces via the floor.

Table 1.4 contains the recommended maximum weighted BB93 standardized impact sound pressure level, $L'_{nT(T_{mf,max}),w}$, for receiving rooms of different types and uses.

The BB93 standardized impact sound pressure level, $L'_{nT(T_{mf,max})}$, is the impact sound pressure level in decibels corresponding to a BB93 reference value of the reverberation time in the receiving room:

$$L'_{nT(T_{mf,max})} = L_i - 10 \lg \frac{T}{T_{mf,max}} \text{ dB}$$

where

L_i is the impact sound pressure level (dB)

T is the reverberation time in the receiving room (s)

$T_{mf,max}$ is the reference reverberation time equal to the upper limit of the reverberation time, T_{mf}, given in Table 1.5 for the type of receiving room. This reference reverberation time shall be used for all frequency bands.

The BB93 standardized impact sound pressure level, $L'_{nT(T_{mf,max})}$, is measured in accordance with BS EN ISO 140-7:1998[5] in octave or one-third octave bands, the results are weighted and expressed as a single-number quantity, $L'_{nT\,(T_{mf,max}),w}$, in accordance with BS EN ISO 717-2:1997[6].

Impact sound insulation should be designed and measured for floors without a soft covering (eg carpet, foam backed vinyl) except in the case of concrete structural floor bases where the soft covering is an integral part of the floor.

1.1.5 Reverberation in teaching and study spaces

The objective is to provide suitable reverberation times for (a) clear communication of speech between teacher and student, and between students, in teaching and study spaces and (b) music teaching and performance.

Table 1.5 contains the required mid-frequency reverberation times for rooms which are finished but unoccupied and unfurnished. The reverberation time is quoted in terms of the mid-frequency reverberation time, T_{mf}, the arithmetic average of the reverberation times in the 500 Hz, 1 kHz and 2 kHz octave bands.

Sound absorption from pinboards and noticeboards can change when they are covered up or painted. Absorption coefficients for pinboards and noticeboards used in design calculations should be for fully covered or painted boards, as appropriate. If these data are not available then the absorption coefficient for the board area used in the design calculation should be the absorption coefficient of the wall to which the board is attached.

Table 1.4: Performance standards for impact sound insulation of floors – maximum weighted BB93 standardized impact sound pressure level $L'_{nT\,(T_{mf,max}),w}$

Type of room (receiving room)		Maximum weighted BB93 standardized impact sound pressure level $L'_{nT\,(T_{mf,max}),w}$ (dB)
Nursery school playrooms		65
Nursery school quiet rooms		60
Primary school: classrooms, class bases, general teaching areas, small group rooms		60
Secondary school: classrooms, general teaching areas, seminar rooms, tutorial rooms, language laboratories		60
Open-plan		
Teaching areas		60
Resource areas		60
Music		
Music classroom		55
Small practice/group room		55
Ensemble room		55
Performance/recital room		55
Recording studio		55
Control room for recording		55
Lecture rooms		
Small (fewer than 50 people)		60
Large (more than 50 people)		55
Classrooms designed specifically for use by hearing impaired students (including speech therapy rooms)		55
Study room (individual study, withdrawal, remedial work, teacher preparation)		60
Libraries		60
Science laboratories		65
Drama studios		55
Design and Technology		
• Resistant materials, CADCAM areas		65
• Electronics/control, textiles, food, graphics, design/resource areas		60
Art rooms		60
Assembly halls, multi-purpose halls (drama, PE, audio/visual presentations, assembly, occasional music)		60
Audio-visual, video conference rooms		60
Atria, circulation spaces used by students		65
Indoor sports hall		65
Gymnasium		65
Dance studio		60
Swimming pool		65
Interviewing/counselling rooms, medical rooms		60
Dining rooms		65
Ancillary spaces	Kitchens*	65
	Offices*, staff rooms*	65
	Corridors*, stairwells*	65
	Coats and changing areas*	65
	Toilets*	65

** Part E of Schedule 1 to the Building Regulations 2000 (as amended by SI 2002/2871) applies to teaching and learning spaces and is not intended to cover administration and ancillary spaces (see under Scope in the Introduction). For these areas the performance standards are for guidance only.*

Type of room	T_{mf}[1] (seconds)
Nursery school playrooms	<0.6
Nursery school quiet rooms	<0.6
Primary school: classrooms, class bases, general teaching areas, small group rooms	<0.6
Secondary school: classrooms, general teaching areas, seminar rooms, tutorial rooms, language laboratories	<0.8
Open-plan	
Teaching areas	<0.8
Resource areas	<1.0
Music	
Music classroom	<1.0
Small practice/group room	<0.8
Ensemble room	0.6 - 1.2
Performance/recital room[3]	1.0 - 1.5
Recording studio	0.6 - 1.2
Control room for recording	<0.5
Lecture rooms[3]	
Small (fewer than 50 people)	<0.8
Large (more than 50 people)	<1.0
Classrooms designed specifically for use by hearing impaired students (including speech therapy rooms)	<0.4
Study room (individual study, withdrawal, remedial work, teacher preparation)	<0.8
Libraries	<1.0
Science laboratories	<0.8
Drama studios	<1.0
Design and Technology	
• Resistant materials, CADCAM areas	<0.8
• Electronics/control, textiles, food, graphics, design/resource areas	<0.8
Art rooms	<0.8
Assembly halls, multi-purpose halls (drama, PE, audio/visual presentations, assembly, occasional music)[2,3]	0.8 - 1.2
Audio-visual, video conference rooms	<0.8
Atria, circulation spaces used by students	<1.5
Indoor sports hall	<1.5
Gymnasium	<1.5
Dance studio	<1.2
Swimming pool	<2.0
Interviewing/counselling rooms, medical rooms	<0.8
Dining rooms	<1.0
Ancillary spaces	
Kitchens*	<1.5
Offices*, staff rooms*	<1.0
Corridors, stairwells	See Section 1.1.6
Coats and changing areas*	<1.5
Toilets*	<1.5

Table 1.5: Performance standards for reverberation in teaching and study spaces – mid-frequency reverberation time, T_{mf}, in finished but unoccupied and unfurnished rooms

** Part E of Schedule 1 to the Building Regulations 2000 (as amended by SI 2002/2871) applies to teaching and learning spaces and is not intended to cover administration and ancillary spaces (see under Scope in the Introduction). For these areas the performance standards are for guidance only.*

1.1.6 Sound absorption in corridors, entrance halls and stairwells

The objective is to absorb sound in corridors, entrance halls and stairwells so that it does not interfere with teaching and study activities in adjacent rooms.

The requirement is to provide additional sound absorption in corridors, entrance halls and stairwells. The amount of additional absorption should be calculated according to Approved Document E[7], Section 7. This describes two calculation methods, A and B, for controlling reverberation in the common internal parts of domestic buildings. One of these methods should be used to determine the amount of absorption required in corridors, entrance halls and stairwells in schools. (See sample calculations using calculation methods A and B in Appendix 7.)

Sound absorption from pinboards and noticeboards can change when they are covered up or painted. Absorption coefficients for pinboards and noticeboards used in design calculations should be for fully covered or painted boards, as appropriate. If these data are not available then the absorption

NOTES ON TABLE 1.5

1 Common materials often absorb most sound at high frequencies. Therefore reverberation times will tend to be longer at low frequencies than at high frequencies. In rooms used primarily for speech, the reverberation times in the 125 Hz and 250 Hz octave bands may gradually increase with decreasing frequency to values not more than 30% above T_{mf}.

2 For very large halls and auditoria, and for halls designed primarily for unamplified music rather than speech, designing solely in terms of reverberation time may not be appropriate and specialist advice should be sought. In large rooms used primarily for music, it may be appropriate for the reverberation times in the 125 Hz and 250 Hz octave bands to gradually increase with decreasing frequency to values up to 50% above T_{mf}. For more guidance see Section 5.

3 Assembly halls, multi-purpose halls, lecture rooms and music performance/recital rooms may be considered as unfurnished when they contain permanent fixed seating. Where retractable (bleacher) seating is fitted, the performance standards apply to the space with the seating retracted.

coefficient for the board area used in the design calculation should be the absorption coefficient of the wall to which the board is attached.

1.1.7 Speech intelligibility in open-plan spaces

The objective is to provide clear communication of speech between teacher and student, and between students, in open-plan teaching and study spaces.

For enclosed teaching and study spaces it is possible to achieve good speech intelligibility through specification of the indoor ambient noise level, sound insulation and reverberation time. Open-plan spaces require extra specification as they are significantly more complex acoustic spaces. The main issue is that the noise from different groups of people functioning independently in the space significantly increases the background noise level, thus decreasing speech intelligibility.

Open-plan spaces are generally designed for high flexibility in terms of the layout of teaching and study spaces. In addition, the layout is rarely finalised before the school is operational. This increases the complexity of assessing speech intelligibility in the open-plan space. Therefore, at an early stage in the design, the designer should establish the expected open-plan layout and activity plan with the client.

The open-plan layout should include:

- the positions at which the teacher will give oral presentations to groups of students
- the seating plan for the students and teachers in each learning base
- the learning base areas.

The activity plan should include:

- the number of teachers giving oral presentations to groups of students at any one time
- the number of students engaged in discussion at any one time
- the number of people walking through the open-plan space (eg along corridors and walkways) during teaching and study periods
- any machinery (eg engraving machines, CNC machines, dust and fume extract equipment, computers, printers, AVA) operating in the open-plan space.

The expected open-plan layout and activity plan should be agreed as the basis on which compliance with BB93 can be demonstrated to the Building Control Body.

The activity plan should be used to establish the overall noise level due to the combination of the indoor ambient noise level, all activities in the open-plan space (including teaching and study), and transmitted noise from adjacent spaces. A computer prediction model should be used to calculate the Speech Transmission Index (STI)[8] in the open-plan space, using the overall noise level as the background noise level. Other methods of estimating STI may also be applicable.

The performance standard for speech intelligibility in open-plan spaces is described in terms of the Speech Transmission Index in Table 1.6. The calculated value of STI should be between 0.60 and 1.00, which gives an STI rating of either 'good' or 'excellent'. This performance standard applies to speech transmitted from teacher to student, student to teacher and student to student.

The performance standard in Table 1.6 is intended to ensure that open-plan spaces in schools are only built when suited to the activity plan and layout. With some activity plans, room layouts and open-plan designs it will not be possible to achieve this performance standard. At this point in the design process the decision to introduce an open-plan space into the school should be thoroughly re-assessed. If, after re-assessment, there is still a need for the open-plan space, then the inclusion of operable walls between learning bases should be considered. These operable walls will form classrooms and be subject to the airborne sound insulation requirements in Table 1.2. It is not appropriate to simply adjust the activity

Room type	Speech Transmission Index (STI)
Open-plan teaching and study spaces	>0.60

Table 1.6: Performance standard for speech intelligibility in open-plan spaces – Speech Transmission Index (STI)

plan until the performance standard for speech intelligibility is met.

Computer prediction software capable of simulating an impulse response should be used to create a three-dimensional geometric model of the space, comprising surfaces with scattering coefficients and individually assigned absorption coefficients for each frequency band. The model should allow for the location and orientation of single and multiple sources with user-defined sound power levels and directivity. (See guidance on computer prediction models on the DfES acoustics website www.teachernet.gov.uk/acoustics.)

Assumptions to be made in the assessment of speech intelligibility are:

- For students, when seated, the head height (for listening or speaking) is 0.8 m for nursery schools, 1.0 m for primary schools and 1.2 m for secondary schools.
- For students, when standing, the head height (for listening or speaking) is 1.0 m for nursery schools, 1.2 m for primary schools and 1.65 m for secondary schools.
- For teachers, when seated, the head height (for listening or speaking) is 1.2 m.
- For teachers, when standing, the head height (for listening or speaking) is 1.65 m.
- The background noise level is the overall noise level due to all activities (including teaching and study) in the open-plan space.

1.2 Demonstrating compliance to the Building Control Body

The preferred means of demonstrating compliance to the Building Control Body is to submit a set of plans, construction details, material specifications, and calculations, as appropriate for each area of the school which is covered by Requirement E4 of the Building Regulations.

The plans should identify:

- the highest estimate for the indoor ambient noise level, $L_{Aeq,30min}$, in each space and the appropriate upper limit from Table 1.1
- the estimated weighted BB93 standardized level difference, $D_{nT\,(T_{mf,max}),w}$, between spaces and the appropriate minimum value from Table 1.2
- the proposed values of R_w for partition walls and for doors, $D_{n,e,w} - 10\lg N$ for ventilators between circulation spaces and other spaces used by students, and the appropriate minimum values from Table 1.3
- the estimated weighted BB93 standardized impact sound pressure level, $L'_{nT\,(T_{mf,max}),w}$, of floors above spaces and the appropriate maximum values from Table 1.4
- the estimated value of mid-frequency reverberation time T_{mf} in each space and the appropriate range of values from Table 1.5
- the proposed absorption treatments in corridors, entrance halls and stairwells
- for open-plan spaces, the estimated range of STI values for speech communication from teacher to student, student to teacher and student to student.

The supporting information should include:

- construction details and material specifications for the external building envelope
- construction details and material specifications for all wall and floor constructions, including all flanking details
- calculations of the sound insulation $D_{nT\,(T_{mf,max}),w}$ and $L'_{nT\,(T_{mf,max}),w}$
- calculations of reverberation times in teaching and study spaces
- calculations of the absorption area to be applied in corridors, entrance halls and stairwells
- measurements and/or calculations demonstrating how rain noise has been controlled
- sound insulation test reports (laboratory and/or field)
- sound absorption test reports (laboratory)
- activity plan and layout for open-plan spaces.

An example of a submission to a Building Control Body, with explanatory notes, is contained in Appendix 10.

1.2.1 Alternative performance standards

In some circumstances alternative performance standards may be appropriate for specific areas within individual schools for particular educational, environmental or health and safety reasons. In these cases, the following information should be provided to the Building Control Body:

- a written report by a specialist acoustic consultant, clearly identifying (a) all areas of non-compliance with BB93 performance standards (b) the proposed alternative performance standards and (c) the technical basis upon which these alternative performance standards have been chosen
- written confirmation from the educational provider (eg school or Local Education Authority) of areas of non-compliance, together with the justification for the need and suitability of the alternative performance standards in each space.

1.3 Demonstrating compliance to the client

To ensure that the performance standards are met, it is recommended that the client should include a requirement for acoustic testing in the building contract.

The design calculations submitted to the Building Control Body demonstrate only that the construction has the potential to meet the performance standards in Section 1.1. In practice, the performance of the construction is strongly influenced by workmanship on site. If the design calculations and detailing are correct, the most likely causes of failure to meet the performance standards will be poor workmanship, product substitution and design changes on site. Therefore, acoustic testing is recommended.

The DfES acoustics website (www.teachernet.gov.uk/acoustics) will be used to encourage manufacturers and others to disseminate acoustic test results alongside construction details for constructions that consistently satisfy the performance standards.

1.3.1 Timetabling of acoustic testing

Timetabling of acoustic testing is important because any test that results in a failure to satisfy the performance standards will require remedial work to rectify the failure and potential design changes to other parts of the building. For this reason it is desirable, where possible, to complete a sample set of rooms in the school for advance testing.

1.3.2 Remedial treatments

Where the cause of failure is attributed to the construction, other rooms that have not been tested may also fail to meet the performance standards. Therefore, remedial treatment may be needed in rooms other than those in which the tests were conducted. The efficacy of any remedial treatment should be assessed through additional testing.

1.3.3 Indoor ambient noise levels in unoccupied spaces

To demonstrate compliance with the values in Table 1.1, measurements of indoor ambient noise levels should be taken in at least one in four rooms intended for teaching and/or study purposes, and should include rooms on the noisiest façade. These rooms should be finished and unoccupied but may be either furnished or unfurnished. Measurements should be made when external noise levels are representative of conditions during normal school operation.

During measurements, the following should apply:

- Building services (eg ventilation system, plant) should be in use during the measurement period.
- For mechanically ventilated rooms, the plant should be running at its maximum design duty.
- For naturally ventilated rooms, the ventilators or windows should be open as required to provide adequate ventilation.
- There should be no more than one person present in the room. (The values in Table 1.1 allow for one person to be present in the room during the test)
- There should be dry weather conditions outside.

Measurements of $L_{Aeq,T}$ should be made at least 1 m from any surface of the room and at 1.2 m above floor level in at least three positions that are normally occupied during teaching or study periods. A sound level meter complying with BS EN 60804:2001 (IEC 60804:2001)[9] should be used. Further information on noise measurement techniques is available in the Association of Noise Consultants Guidelines on Noise Measurement in Buildings[10].

Where there is negligible change in noise level over a teaching period, measurements of $L_{Aeq,T}$ over a time period much shorter than 30 minutes (eg $L_{Aeq,5min}$) can give a good indication of whether the performance standard in terms of $L_{Aeq,30min}$ is likely to be met. However, if there are significant variations in noise level, for example due to intermittent noise events such as aircraft or railways, measurements should be taken over a typical 30 minute period in the school day.

1.3.4 Airborne sound insulation between spaces

To demonstrate compliance with the values in Table 1.2, measurements of airborne sound insulation should be taken between vertically and horizontally adjacent rooms where the receiving room is intended for teaching and/or study purposes. At least one in four rooms intended for teaching and study purposes should be tested. Measurements should be taken in the direction with the more stringent airborne sound insulation requirement.

During measurements, the ventilators or windows should be open as required to provide adequate ventilation in both the source room and the receiving room.

Measurements should be made in accordance with BS EN ISO 140-4:1998[1] and the additional guidance in Approved Document E[7] Annex B, paragraphs B2.3 – B2.8. Performance should be rated in accordance with BS EN ISO 717-1:1997[2].

1.3.5 Airborne sound insulation between circulation spaces and other spaces used by students

It is not intended that field measurements should be taken between circulation spaces and other spaces used by students. Laboratory data for the wall, doorsets (if any) and ventilators (if any) should be presented as evidence of compliance with the values in Table 1.3.

1.3.6 Impact sound insulation

To demonstrate compliance with the values in Table 1.4, measurements of impact sound insulation should be taken between vertically adjacent rooms, where the receiving room is intended for teaching and study purposes. At least one in four teaching/study rooms below a separating floor should be tested.

Measurements should be made in accordance with BS EN ISO 140-7:1998[5]. Performance should be rated in accordance with BS EN ISO 717-2:1997[6].

Impact sound insulation should be measured on floors without a soft covering (eg carpet, foam backed vinyl), except in the case of concrete structural floor bases where the soft covering is an integral part of the floor.

1.3.7 Reverberation in teaching and study spaces

To demonstrate compliance with the values in Table 1.5, measurements of reverberation time should be taken in at least one in four rooms intended for teaching and study purposes.

One person may be present in the room during the measurement.

Depending upon the completion sequence for spaces within the school, it may be possible to reduce the measurement effort by utilising measurements of reverberation time that are required as part of airborne or impact sound insulation measurements. For this reason, two measurement methods, described below, are proposed for the measurement of reverberation time. For the purpose of demonstrating compliance, either method can be used to assess whether the performance standards have been met. If one method demonstrates compliance with the performance standard and the other demonstrates failure, then the performance standard should be considered to have been met.

Measurement method 1: Measurements should be made in accordance with either low coverage or normal coverage measurements described in BS EN ISO 3382:2000[11].

Measurement method 2: Reverberation time measurements should be made in accordance with BS EN ISO 140-4:1998[1] (airborne sound insulation) or BS EN ISO 140-7:1998[5] (impact sound insulation) in octave bands.

1.3.8 Sound absorption in corridors, entrance halls and stairwells

It is not intended that field measurements of reverberation time should be taken in corridors, entrance halls and stairwells.

1.3.9 Speech intelligibility in open-plan spaces

To demonstrate compliance with the values in Table 1.6, measurements of the Speech Transmission Index (STI) should be taken in at least one in ten student positions in the open-plan spaces.

Measurements should be made in accordance with BS EN 60268-16:1998[8].

Measurements should be made using the following heights for listening or speaking:

- to represent seated students, a head height of 0.8 m for nursery schools, 1.0 m for primary schools and 1.2 m for secondary schools
- to represent standing students, a head height of 1.0 m for nursery schools, 1.2 m for primary schools and 1.65 m for secondary schools
- to represent seated teachers, a head height of 1.2 m
- to represent standing teachers, a head height of 1.65 m.

Simulation of the estimated occupancy noise should be carried out in the STI measurement. This noise level will have been established at the design stage (see Section 1.1.7) and is defined as the noise level due to the combination of the indoor ambient noise level, all activities in the open-plan space (including teaching and study), and transmitted noise from adjacent spaces.

References

[1] BS EN ISO 140-4:1998 Acoustics – Measurement of sound insulation in buildings and of building elements. Part 4. Field measurements of airborne sound insulation between rooms.

[2] BS EN ISO 717-1:1997 Acoustics – Rating of sound insulation in buildings and of building elements. Part 1. Airborne sound insulation.

[3] BS EN ISO 140-3:1995 Acoustics – Measurement of sound insulation in buildings and of building elements. Part 3. Laboratory measurement of airborne sound insulation of building elements.

[4] BS EN 20140-10:1992 Acoustics – Measurement of sound insulation in buildings and of building elements. Part 10. Laboratory measurement of airborne sound insulation of small building elements.

[5] BS EN ISO 140-7:1998 Acoustics – Measurement of sound insulation in buildings and of building elements. Part 7. Field measurements of impact sound insulation of floors.

[6] BS EN ISO 717-2:1997 Acoustics – Rating of sound insulation in buildings and of building elements. Part 2. Impact sound insulation.

[7] Approved Document E – Resistance to the passage of sound. Stationery Office 2003. ISBN 0 11 753 642 3. www.odpm.gov.uk

[8] BS EN 60268-16:1998 Sound system equipment – Part 16: Objective rating of speech intelligibility by speech transmission index.

[9] BS EN 60804:2001 (IEC 60804:2001) Integrating-averaging sound level meters.

[10] Guidelines on Noise Measurement in Buildings, Part 1: 1997 Noise from Building Services and Part 2: 1998 Noise from External Sources. Association of Noise Consultants.

[11] BS EN ISO 3382:2000 Acoustics – Measurement of the reverberation time of rooms with reference to other acoustical parameters.

Noise control

SECTION 2

Section 2 gives recommendations and guidance concerning noise control, starting with the choice of a site and the control of external noise. Local government planning policy will be influenced by the recommendations on maximum external noise levels in playing fields and other external areas used by the school. Section 2 also includes discussion of the means of controlling indoor ambient noise.

2.1 Choosing a site

The acoustic design of a school starts with the selection of the site, a noise survey of the site and planning the layout of the school buildings.

Economic sites for new schools with easy access to transport often suffer from traffic noise and pollution. In the past, schools have sometimes been built on sites which would not normally have been considered suitable for housing. This has been in part because schools have not always been recognised as requiring particularly high environmental standards, and in part because there has been less formal control or regulation of noise levels in schools than for housing.

Where school sites are adjacent to busy roads they will require the use of intelligent design, zoning, noise screening and, if necessary, sound insulating building envelopes together with mechanical ventilation or acoustically designed passive ventilation.

Many of the acoustic problems in existing schools derive directly from the school's location in a noisy area. For existing schools, noise from road traffic is a common problem, but in some areas noise from railways and aircraft is intrusive[1]. Noise from industrial and leisure sources is a less frequent problem and can normally be dealt with at source by the Local Authority using their powers under the Environmental Pollution Act.

2.2 Recommendations for external noise levels outside school buildings

Although Requirement E4 does not apply to external noise, the following recommendations are considered good practice for providing good acoustic conditions outside school buildings.

For new schools, 60 dB $L_{Aeq,30min}$ should be regarded as an upper limit for external noise at the boundary of external premises used for formal and informal outdoor teaching, and recreational areas.

Under some circumstances it is possible to meet the specified indoor ambient noise levels on sites where external noise levels are as high as 70 dB $L_{Aeq,30min}$ but this will require considerable building envelope sound insulation, screening or barriers.

Noise levels in unoccupied playgrounds, playing fields and other outdoor areas should not exceed 55 dB $L_{Aeq,30min}$ and there should be at least one area suitable for outdoor teaching activities where noise levels are below 50 dB $L_{Aeq,30min}$. If this is not possible due to a lack of suitably quiet sites, acoustic screening should be used to reduce noise levels in these areas as much as practicable, and an assessment of predicted noise levels and of options for reducing these should be carried out.

Playgrounds, outdoor recreation areas and playing fields are generally considered to be of relatively low sensitivity to noise, and indeed playing fields may be used as buffer zones to separate school buildings from busy roads where necessary. However, where outdoor areas are used for teaching, eg. sports lessons, outdoor ambient noise levels have a significant impact on communication in an environment which is already acoustically less favourable than most classrooms. Ideally, noise levels on unoccupied playing fields used for teaching sport should not exceed 50 dB $L_{Aeq,30min}$. If this is not possible at all locations, there should be at least one area at which noise levels are below 50 dB $L_{Aeq,30min}$ so that some outdoor teaching is possible.

Acoustic screening from fences, walls or buildings may be used to protect playgrounds from noise. At positions near the screen, traffic noise can be reduced by up to 10 dB(A).

All external noise levels in this section apply to measurements made at approximately head height and at least 3 m from any reflecting surface other than the ground.

2.3 Noise survey

Figure 2.1 shows typical external and internal sources of noise which can affect noise levels inside a school.

In order to satisfy the limits for the indoor ambient noise levels in Table 1.1, it is necessary to know the external noise level so that the building envelope can be designed with the appropriate sound insulation.

The external noise level should be established by carrying out a noise measurement survey. (Note that a brief survey is advisable even if the site appears to be quiet, in case there are noisy events at certain times of the day.) The measurements should be taken during typical school hours and include noisy events (eg road traffic at peak hours, worst case runway usage in the case of airports, etc). The measurements must also take account of the weather conditions. For long-distance propagation of noise, the measured level is affected by wind gradients, temperature gradients and turbulence. With wind, the noise level is generally increased downwind or reduced upwind. (Note that temperature inversions can radically change noise propagation, but tend to occur only at night-time, outside school hours.)

A noise measurement survey must include octave or one-third octave frequency band levels. This is because the attenuation of sound, for example by a sound insulating wall or noise barrier, depends upon the frequency of sound. In general materials and barriers are less effective at controlling low frequency noise than mid and high frequency noise. Although overall noise levels and performance standards can be quoted as overall A-weighted levels, calculations must be carried out in octave or one-third octave bands (see Appendix 1) and the results converted into overall A-weighted levels.

In addition to the noise measurement

Figure 2.1: Typical sources of noise

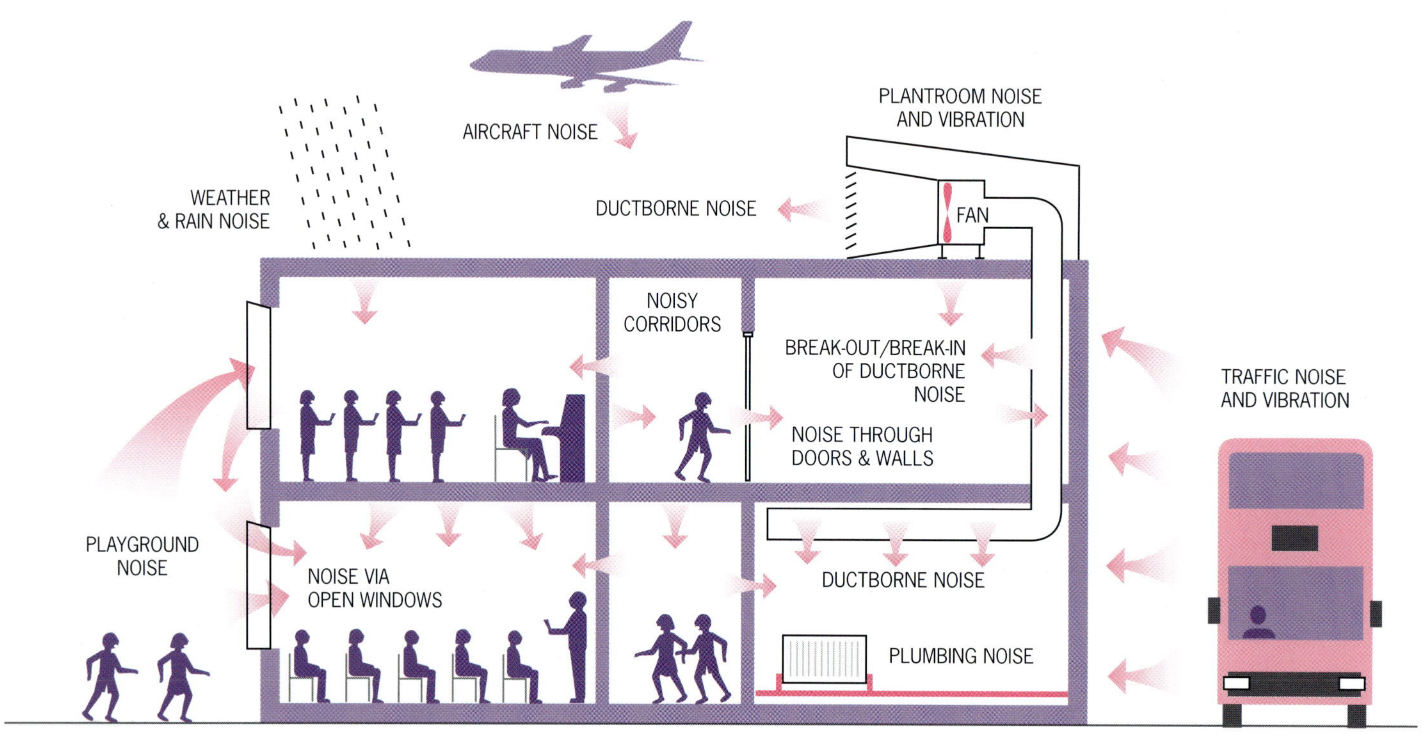

survey, consideration should be given to predicting the potential increases in noise levels due to future developments (eg increases in traffic flows, new transport schemes, changes in flight paths). The local highway authority should be able to advise on whether significant changes in road traffic noise are expected in the future. This is likely to be relevant for developments near new or recently improved roads. Where road traffic noise levels are likely to increase, it is reasonable to base the sound insulation requirements on the best estimate of noise levels in 15 years time. Similar information is likely to be available from railway operators, and airports. The prediction[2,3] of future external noise levels should be carried out by an acoustic consultant.

If the noise measurement survey shows that the ambient external noise levels on the site are below 45 dB $L_{Aeq,30min}$, and prediction work shows that they will remain below 45 dB $L_{Aeq,30min}$ in the future, no special measures are likely to be necessary to protect the buildings or playing fields from external noise.

2.4 Road and rail noise

Sources of road and rail noise require individual assessment because of their characteristics.

Road traffic noise is a function of traffic flow, percentage of heavy goods vehicles, traffic speed gradient (rate of acceleration), road surface and propagation path of the noise.

Rail noise is a function of train type, number, speed, rail type and propagation path of the noise.

In general it is advisable to locate a school at least 100 m away from busy roads and railways, but in towns and cities this is often not possible. However, the use of distance alone is a relatively ineffective way to reduce noise. Simple rules of thumb are that the noise level from a busy road increases by 3 dB(A) for a doubling of the traffic flow and decreases by 3 dB(A) for a doubling of distance from the road (over hard ground).

2.5 Aircraft noise

Where a school is to be located in an area affected by aircraft noise, special measures are necessary and an acoustic consultant should be appointed.

2.6 Vibration

Railways, plant and heavy vehicles close to a school can lead to vibration within the school buildings. This vibration can re-radiate as audible noise, even when the vibration itself is not perceptible as shaking in the building. The propagation of vibration depends on ground conditions but in general when planning a new school building it is advisable for the noise survey to include vibration measurements when there is a railway within 30 m of a building, or a road with significant HGV traffic within 20 m. In these cases airborne noise is also likely to be a problem.

Figure 2.2: Attenuation by a noise barrier as a function of path difference

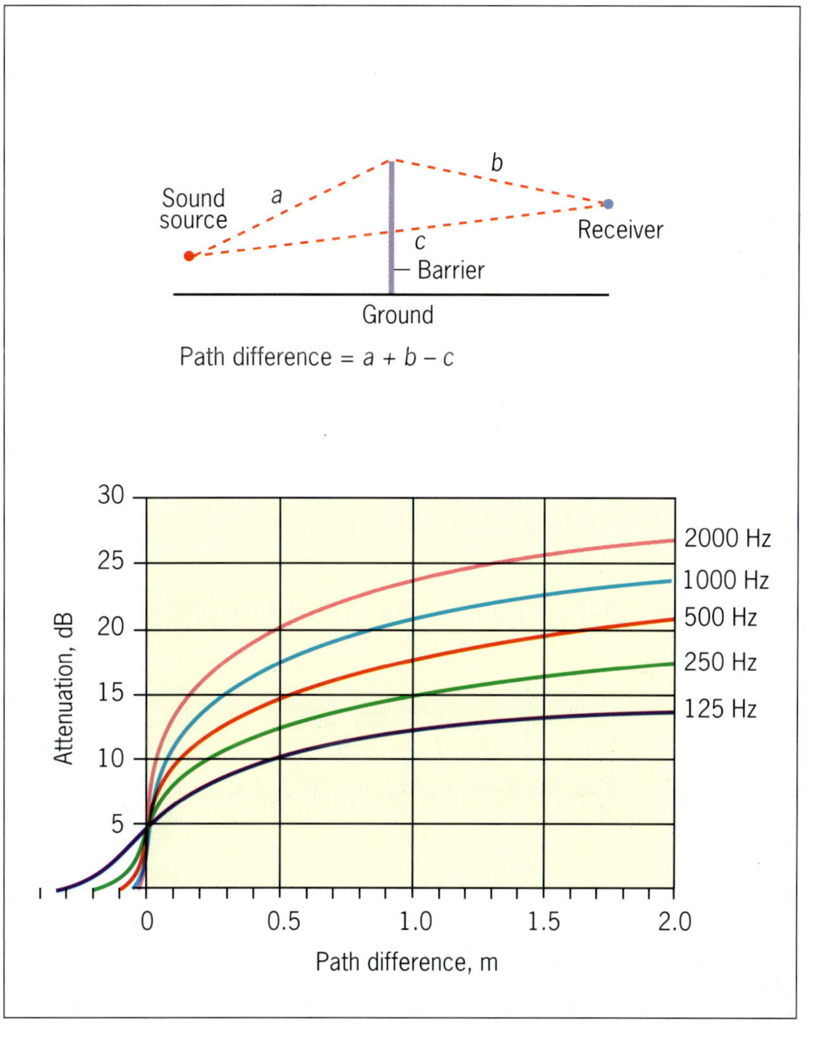

2.7 Noise barriers

Noise barriers are much more effective than distance in reducing noise from road or rail traffic. In its simplest form a noise barrier can be a continuous close-boarded wooden fence, with a mass of not less than 12 kg/m^2. There is relatively little point in increasing the weight of the barrier beyond this because a significant proportion of the noise passes over the top (or round the ends) of the barrier.

The attenuation of a barrier is a function of the path difference, that is the extra distance that the sound has to travel to pass over the top of the barrier, see Figure 2.2. Barriers are less effective at reducing low frequency noise than mid and high frequency noise. Hence, to calculate the effectiveness of a noise barrier it is necessary to know the source noise levels in octave or one-third octave bands (see Appendix 1).

Hedges or single trees (or rows of trees) do not in themselves make effective noise barriers. A common and effective solution is a wooden fence to act as a noise barrier, located within a band of trees to create an acceptable visual effect.

Barriers can also be formed by other buildings or by landscaping using earth bunds, see Figure 2.3. The path difference, and hence the attenuation, will be affected by whether the road or railway is in a cutting or on an embankment.

2.8 Noise from schools to surrounding areas

Noise from schools to the surrounding area can also be a problem, and consideration should be given to nearby residential and other noise-sensitive developments which could be disturbed by noise from playgrounds, playing fields, music rooms and halls used for events such as after school concerts and discos. The local planning authority will normally consider this in assessing any planning application for new schools or extensions to existing premises.

The effect of playground noise on children inside parts of the school near the playground should also be considered as part of the design.

2.9 Planning and layout

Among the most common problems found in schools is noise transfer between rooms. To a large extent this can be designed out without resort to very high performance sound insulating walls or floors, but by good planning and zoning of the building at the earliest stages of design. At this stage it is possible to identify noise-sensitive areas and to

Figure 2.3: Traffic noise barriers

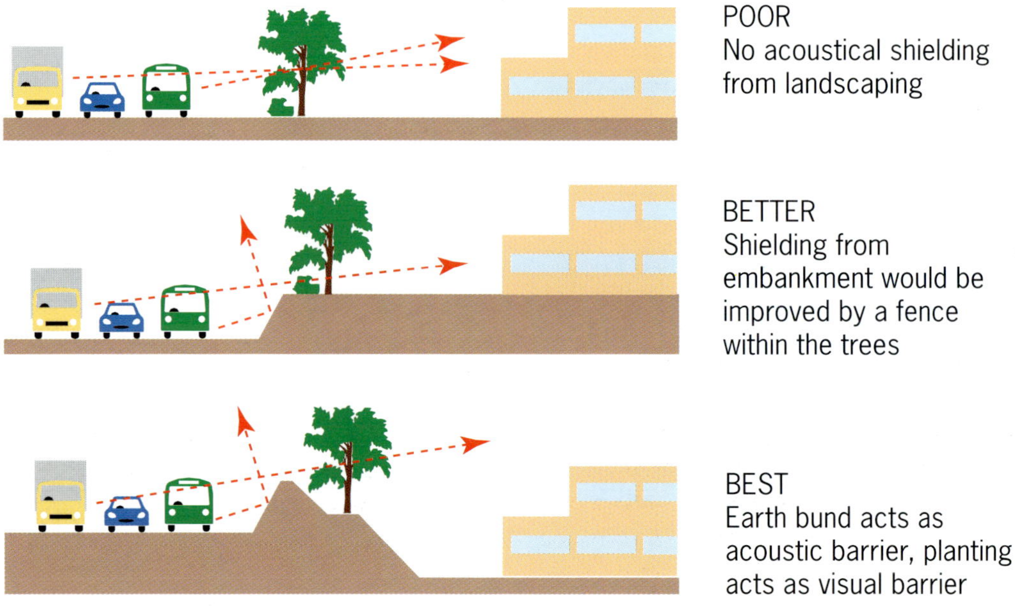

separate these from noisy areas using buffer zones such as storerooms, corridors or less sensitive rooms, or by locating buildings a suitable distance apart. See Figure 2.4 for an example of room layout in a music department using buffer zones.

When considering external noise such as that from roads, it is sensible to locate noise-sensitive rooms, such as classrooms, away from the source.

Tables 1.1 and 1.2 give the required maximum indoor ambient noise values and the minimum sound insulation values between rooms. The performance standards in these tables should be used in the early planning stages of a project to determine (a) the layout of the school (b) the constructions needed to provide sound insulation and (c) the compatibility of school activities in adjacent rooms.

2.10 Limiting indoor ambient noise levels

The total indoor ambient noise level is determined by combining the noise levels from all the known sources. The indoor ambient noise level due to external sources such as traffic must be added to the noise from mechanical ventilation, heating systems, lighting and other building services. Unless care is taken, these individual sources can be loud enough to cause disturbance, particularly in spaces where low indoor ambient noise levels are required.

It should be noted that noise levels in dB or dB(A) cannot be simply added together. For example, two noise levels of 40 dB(A) when combined will produce a level of 43 dB(A). The addition of noise levels is explained in Appendix 1.

2.11 Impact noise

Impact noise within a space from footfalls on balconies, stairs and circulation routes, or from movement of furniture or other class activities, can be a significant distraction to teaching and learning.

Carpets and other soft yet resilient floor finishes such as resilient backed vinyl or rubber type flooring materials can be useful in limiting this impact noise within a space. However, carpets may be difficult to clean and are sometimes not used because of their effect on indoor air quality and resultant health implications.

Resilient feet can also be fitted to furniture to reduce impact noise within a space.

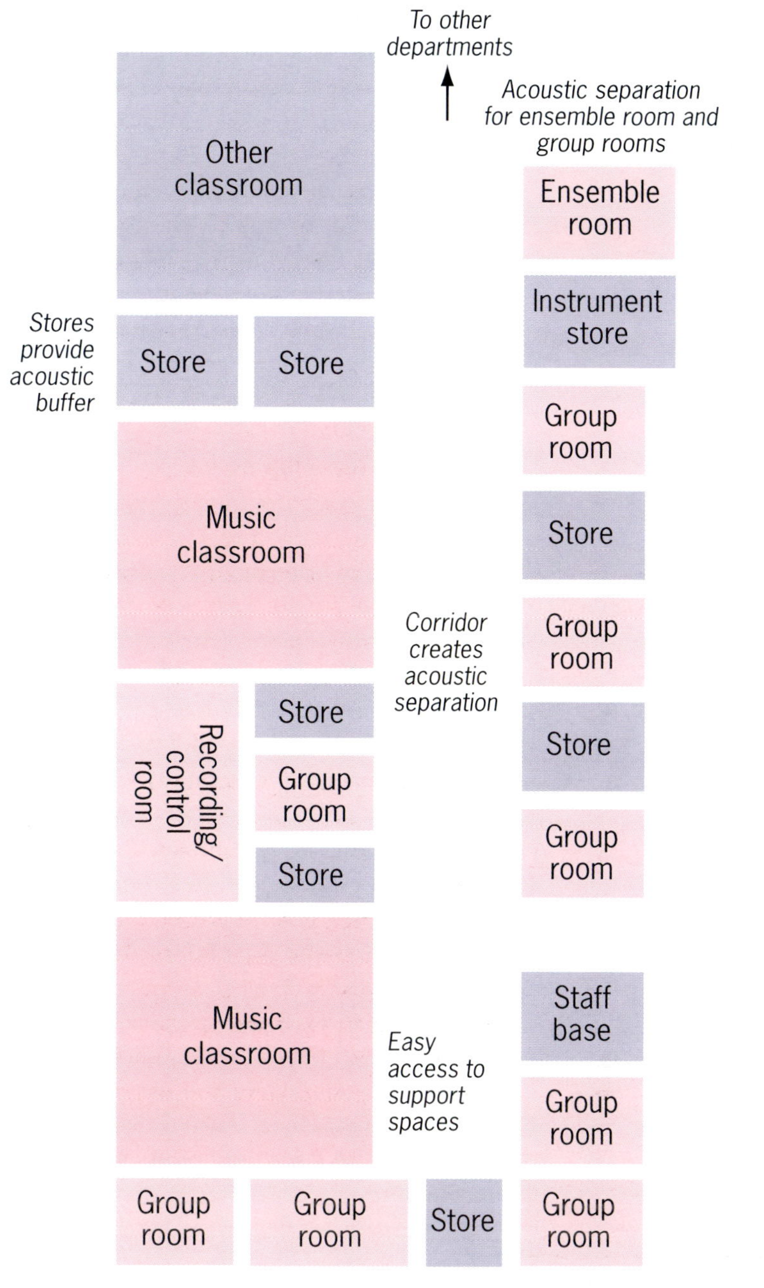

Fig 2.4: Planning acoustic 'buffer zones'

2.12 Corridors, entrance halls and stairwells

Noise in corridors, entrance halls and stairwells can cause disturbance to neighbouring classrooms and other teaching spaces. It is therefore important that reverberation in corridors, entrance halls and stairwells is kept as low as possible in order to minimise noise levels in these areas. The requirement is to provide sound absorption in accordance with Section 1.1.6. To satisfy this requirement, corridors outside classrooms typically need acoustically absorbent ceilings and/or wall finishes. Carpets and other soft floor finishes can also help to reduce reverberation and the noise from footfalls. However, as discussed in Section 2.11, the use of carpets may not be appropriate in all schools.

2.13 Masking noise

The audibility and intrusiveness of noise from other areas (break-in noise) is a function of both the level of the break-in noise and the noise level in the room under consideration (the receiving room). If the ambient noise level in the receiving room is unnecessarily low, break-in noise will be more audible. Hence where rooms are mechanically ventilated, the noise from the ventilation system can be used to mask the noise from activities in neighbouring rooms. In these cases ventilation noise should be not more than 5 dB below the maximum ambient noise levels listed in Table 1.1. For this type of masking to work it is important to ensure that the ventilation noise follows a specific masking noise curve and has no tonal or intermittent characteristics. Specialist acoustic advice is required before using building services noise for masking.

Other possible sources of masking noise are fan convectors, electric lighting circuits, and constant levels of road traffic noise, for example from distant arterial roads. However it should be noted that the noise from some sources (eg fans and other mechanical equipment) may cause annoyance to individuals, particularly hearing impaired people, in some circumstances. Also, some building services systems may only operate at certain times of the year.

2.14 Low frequency noise and hearing impaired pupils

Many hearing impaired pupils make use of low frequencies below 500 Hz to obtain information from speech. Therefore, for hearing impaired pupils to be included in classes with pupils having normal hearing, special care should be taken to minimise low frequency indoor ambient noise levels. Given the prevalence of infections leading to temporary hearing loss, it is advisable to minimise low frequency indoor ambient noise levels in all classrooms, especially those used by younger pupils.

The indoor ambient noise levels in Table 1.1 are given in terms of $L_{Aeq,30min}$ which is an A-weighted noise level. This is a convenient and widely-used parameter but is not a good indicator of low frequency noise. To assess indoor noise there are other rating systems in use which address low frequency noise but these are beyond the scope of this document. In cases where low frequency noise is likely to be a problem, specialist advice from an acoustics consultant should be sought. Such cases include schools exposed to high levels of external noise (in excess of 60 dB $L_{Aeq,30min}$, see Section 2.2), where sound insulation may reduce high frequency noise while leaving comparatively high levels of low frequency noise.

More information is given in CIBSE Guide B5 Noise and Vibration Control for HVAC.[4]

References

[1] B Shield, J Dockrell, R Jeffery and I Tachmatzidis. The effects of noise on the attainments and cognitive performance of primary school children. Department of Health, 2002.

[2] Calculation of Road Traffic Noise (CRTN), Department of Transport, The Stationery Office, 1988.

[3] Calculation of Railway Noise (CRN), (Supplement 1), Department of Transport, The Stationery Office, 1995.

[4] CIBSE Guide B5, Noise and vibration control for HVAC, CIBSE, 2002 ISBN 1 903287 2 51.

Sound insulation

SECTION 3

General principles of sound insulation and typical constructions are discussed in this section. Space does not allow all details for each type of construction to be shown. Many such details are illustrated and discussed in greater detail in *Approved Document E*[1]. Further guidance and illustrations are also available in *Sound Control for Homes*[2] and in manufacturers' literature for proprietary materials and systems.

3.1 Roofs

The sound insulation of a pitched roof depends upon the mass of the ceiling and the roof layers and the presence of a sound absorbing material in the roof space. Mineral wool, used as thermal insulation in the ceiling void, will also provide some acoustic absorption, which will have a small effect on the overall sound insulation of a roof. A denser specification of mineral wool, as commonly used for acoustic insulation, would have a greater effect on the overall sound insulation of the roof.

Where it is necessary to ventilate the roof space, it is advisable to make any necessary improvements to the sound insulation by increasing the mass of the ceiling layer, which should be airtight. Recessed light fittings can make this difficult and sometimes it is better to place the sound insulating material below the roof covering and to extend partition walls up to the roof layer (providing adequate ventilation can be maintained).

3.1.1 Rain noise

The impact noise from rain on the roof can significantly increase the indoor noise level; in some cases the noise level inside a school due to rain can be as high as 70 dB(A).

Although rain noise is excluded from the definition of indoor ambient noise in Section 1.1, it is a potentially important noise source which must be considered at an early point in the roof design to minimise disturbance inside the school.

Excessive noise from rain on the roof can occur in spaces (eg sports halls, assembly halls) where the roof is made from profiled metal cladding and there is no sealed roof space below the roof to attenuate the noise before it radiates into the space below. With profiled metal cladding, the two main treatments that should be used in combination to provide sufficient resistance to impact sound from rain on the roof are:

- damping of the profiled metal cladding (eg using commercial damping materials)
- independent ceilings (eg two sheets of 10 kg/m^2 board material such as plasterboard, each supported on its own frame and isolated from the profiled metal cladding, with absorptive material such as mineral fibre included in the cavity.)

Profiled metal cladding used without a damping material and without an independent ceiling is unlikely to provide sufficient resistance to impact sound from rain on the roof. A suitable system that could be used in schools is shown in Figure 3.1. The performance of such a system was measured by McLoughlin et al[3].

Prediction models are available to predict the noise radiated from a single sheet of material; however, a single sheet will not provide sufficient attenuation of impact noise from rain. Suitable lightweight roof constructions that do provide sufficient attenuation will consist of many layers. For these multi-layer roof constructions, laboratory measured data for the entire roof construction is needed. At the time of writing, a new laboratory

Figure 3.1: Profiled metal clad roof incorporating acoustic damping

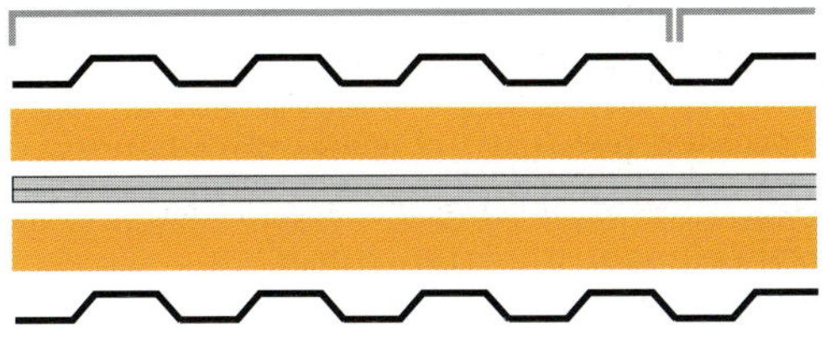

measurement standard for impact sound from rain on the roof, ISO 140-18[4], is under development. In the future this will allow comparison of the insulation provided by different roof, window and glazing elements and calculation of the sound pressure level in the space below the roof.

When designing against noise from rain on the roof, consideration should also be given to any glazing (eg roof lights) in the roof. Due to the variety of different roof constructions, advice should be sought from an acoustic consultant who can calculate the sound pressure level in the space due to typical rainfall on the specific roof.

3.2 External walls

For masonry walls, such as a 225 mm solid brick wall, a brick/block cavity wall or a brick-clad timber frame wall, the sound insulation performance will normally be such that the windows, ventilators and, in some cases, the roof will dictate the overall sound insulation of the building envelope.

Timber frame walls with lightweight cladding and other lightweight systems of construction normally provide a lower standard of sound insulation at low frequencies, where road traffic and aircraft often produce high levels of noise. This can result in low airborne sound insulation against these noise sources unless the cladding system has sufficient low frequency sound insulation. The airborne sound insulation can be assessed from laboratory measurements carried out according to BS EN ISO 140-3:1995[5].

3.3 Ventilation

The method of ventilation as well as the type and location of ventilation openings will affect the overall sound insulation of the building envelope. When external noise levels are higher than 60 dB $L_{Aeq,30min}$, simple natural ventilation solutions may not be appropriate as the ventilation openings also let in noise. However, it is possible to use acoustically attenuated natural ventilation rather than full mechanical ventilation when external noise levels are high but do not exceed 70 dB $L_{Aeq,30min}$.

The School Premises Regulations[6] require that:

"All occupied areas in a school building shall have controllable ventilation at a minimum rate of 3 litres of fresh air per second for each of the maximum number of persons the area will accommodate.

All teaching accommodation, medical examination or treatment rooms, sick rooms, sleeping and living accommodation shall also be capable of being ventilated at a minimum rate of 8 litres of fresh air per second for each of the usual number of people in those areas when such areas are occupied."

In densely occupied spaces such as classrooms, 8 litres per second per person is the minimum amount of fresh air that should be provided by a natural or mechanical ventilation system under normal working conditions, in order to maintain good indoor air quality.

In order to satisfy the limits for the indoor ambient noise levels in Table 1.1, it is necessary to consider the sound attenuation of the ventilation openings so that the building envelope can be designed with the appropriate overall sound insulation. In calculations of overall sound insulation the attenuation assumed for the ventilation system should be for normal operating conditions.

The main choices for the natural ventilation of typical classrooms are shown in Figure 3.2. Case Studies 7.8 and 7.9 describe the recent application of two of these design solutions in new secondary school buildings.

Additional ventilation such as openable windows or vents may be required to prevent summertime overheating.

3.3.1 Ventilators

Passive ventilators normally penetrate the walls, but in some cases they penetrate the window frames (eg trickle ventilators) or the windows themselves. Often windows are not used as intended as they cause uncomfortable draughts. For this reason, increased use is being made of purpose designed ventilation systems with or without acoustic attenuation.

Many proprietary products are

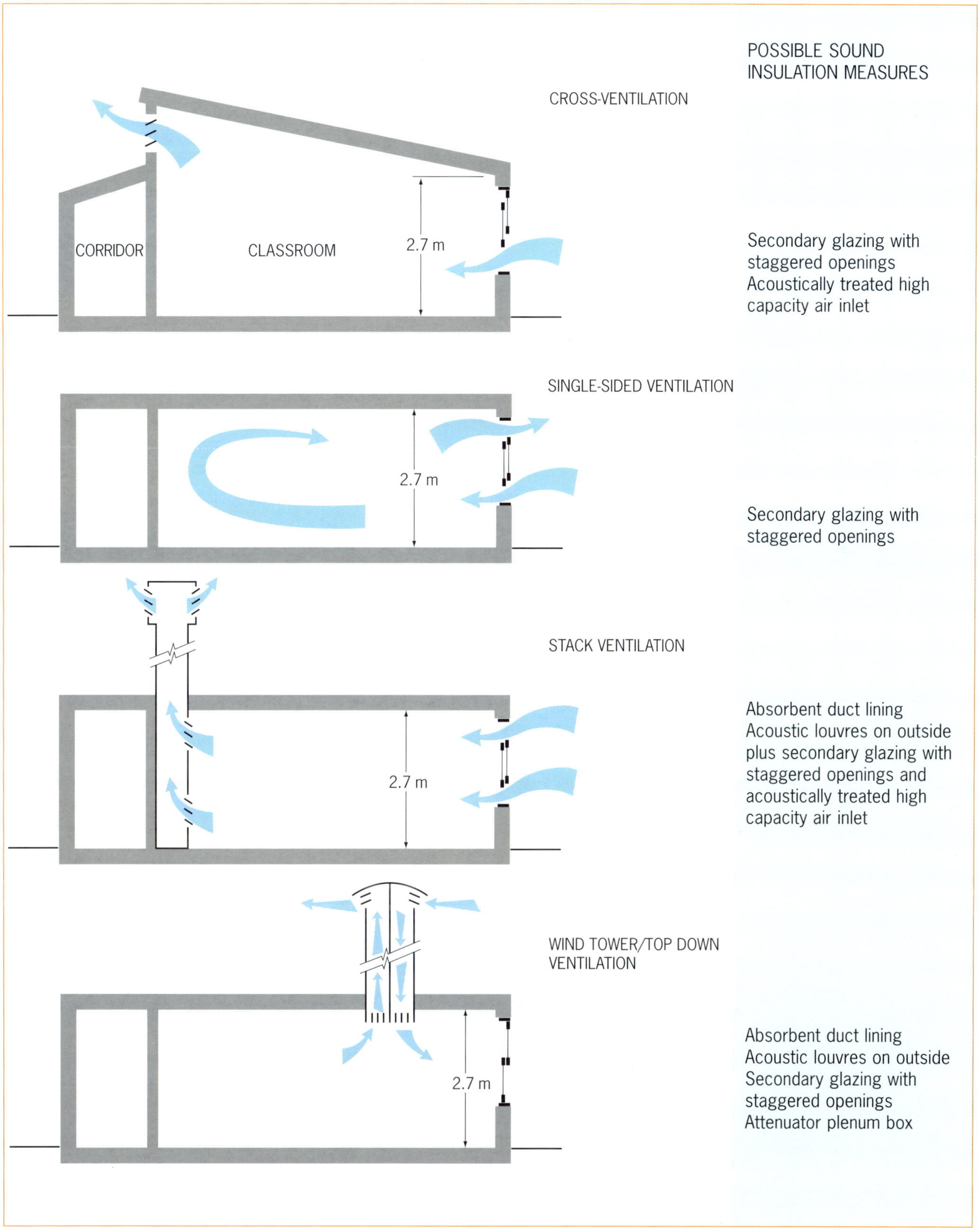

Figure 3.2: Possible types of natural ventilation

designed for the domestic sector and in some cases they do not have large enough openings for classrooms and other large rooms found in schools. The acoustic performance of any ventilator can be assessed with laboratory sound insulation test data measured according to BS EN 20140-10:1992[7]. Because of the complexity of the assessment of the acoustic performance of a ventilator, advice may be needed from a specialist acoustic consultant. To maintain adequate ventilation, it is essential that the effective area of the ventilator is considered as it may be smaller than the free area (see prEN 13141-1[8]).

It is important, particularly in the case of sound-attenuated products, that a good seal is achieved between the penetration through the wall or window and the ventilator unit. Where through-the-wall products are used, the aperture should be cut accurately and the gap around the perimeter of the penetrating duct should be packed with sound insulating material prior to application of a continuous, flexible, airtight seal on both sides.

In some schools bespoke ventilator designs, such as that shown in Figure 3.3, are needed. For more examples of ventilator solutions see Case Studies 7.8 and 7.9.

3.4 External windows

The airborne sound insulation of windows can be assessed from laboratory measurements of the sound reduction index according to BS EN ISO 140-3:1995[5]. When choosing suitable windows using measured data, care must be taken to differentiate between measured data for glazing and measured data for windows. The reason is that the overall sound insulation performance of a window is affected by the window frame and the sealing as well as the glazing.

To achieve the required sound insulation with thin glass it is often necessary to use two panes separated by an air (or other gas) filled cavity. In theory, the wider the gap between the panes, the greater the sound insulation. In practice, the width of the cavity in double glazing makes relatively little difference for cavity widths between 6 mm and 16 mm. Wider cavity widths perform significantly better.

In existing buildings, secondary glazing may be installed as an alternative to replacing existing single glazing with double glazing. The effectiveness of secondary glazing will be determined by the thickness of the glass and the width of the air gap between the panes. Another alternative may be to fit a completely new double-glazed window on the inside of the existing window opening, leaving the original window intact. The use of sound absorbing reveal linings improves the performance of double-glazed windows, but the improvement is mainly in the middle to high frequency region, where it has little effect on road traffic and aircraft noise spectra.

To achieve their optimum performance, it is essential that the glazing in windows makes an airtight seal with its surround, and that opening lights have effective seals around the perimeter of each frame. Neoprene compression seals will provide a more airtight seal than brush seals. The framing of the window should also be assembled to achieve an airtight construction.

It is equally important that an airtight seal is achieved between the perimeter of the window frame and the opening into which it is to be fixed. The opening should be accurately made to receive the window, and the perimeter packed with sound insulating material prior to application of a continuous seal on both sides.

For partially open single-glazed windows or double-glazed windows with opposite opening panes, the laboratory measured airborne sound insulation is approximately 10-15 dB R_w . This increases to 20-25 dB R_w in the open position for a secondary glazing system with partially open ventilation openings, with the openings staggered on plan or elevation, and with absorbent lining of the window reveals (see Figure 3.3). In situ, the degree of attenuation provided by an open window also depends on the spectrum of the noise and the geometry of the situation.

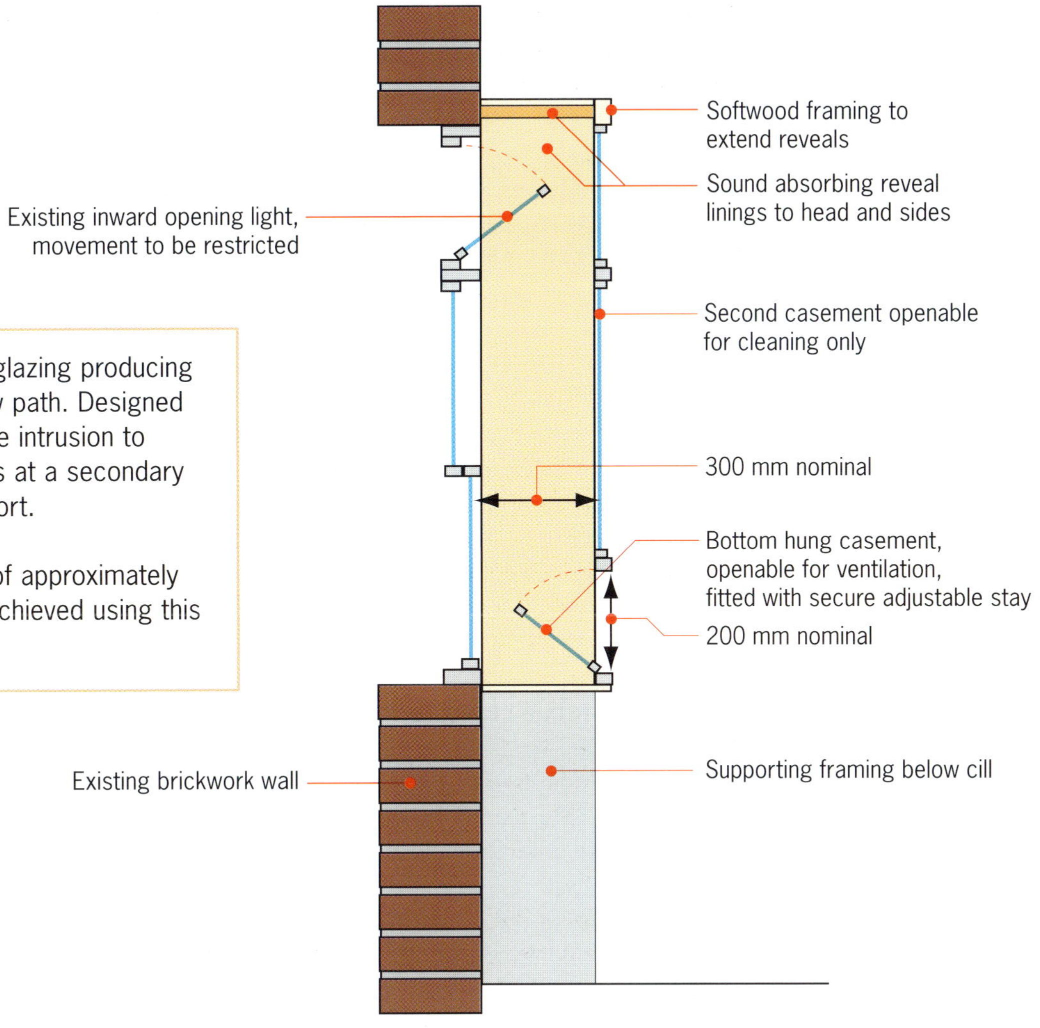

Retrofit secondary glazing producing a staggered air flow path. Designed to limit aircraft noise intrusion to science laboratories at a secondary school near an airport.

A sound reduction of approximately 20-25 dB R_W was achieved using this design.

Figure 3.3: Sound reducing secondary glazing system

The spreadsheet of sound reduction indices on the DfES acoustic website gives values of R_W for various types of window, glazing thickness, and air gap. Indications are also given of the sound reduction indices of open windows.

3.5 External doors

For external doors the airborne sound insulation is determined by the doorset, which is the combination of door and frame. The quality of the seal achieved around the perimeter of the door is crucial in achieving the potential performance of the door itself. Effective seals should be provided at the threshold, jambs and head of the door frame. As with windows, neoprene compression seals are more effective than brush seals, but their effectiveness will be strongly influenced by workmanship on site. Brush seals can however be effective and tend to be more hard wearing than compression seals.

It is also important that an airtight seal is achieved between the perimeter of the door frame and the opening into which it is to be fixed. The opening should be accurately made to receive the door frame and any gaps around the perimeter packed with insulating material prior to application of a continuous, airtight seal on both sides.

A high level of airborne sound insulation is difficult to provide using a single door; however, it can be achieved by using a lobby with two sets of doors, as often provided for energy efficiency, or a specialist acoustic doorset.

SOUND INSULATION OF THE BUILDING ENVELOPE

There are two methods by which it is possible to calculate the indoor ambient noise levels due to external noise.

The first method is to calculate the indoor ambient noise level according to the principles of BS EN 12354-3:2000[9]. An Excel spreadsheet to calculate the sound insulation of building envelopes, based on BS EN 12354-3:2000, is available via the DfES acoustics website. The principles of this calculation spreadsheet are given in Appendix 5.

The second method is to calculate the indoor ambient noise level using the measured façade sound insulation data from an identical construction at another site.

3.6 Subjective characteristics of noise

The indoor ambient noise levels in Table 1.1 provide a reasonable basis for assessment, but some noises have tonal or intermittent characteristics which make them particularly noticeable or disturbing, even below the specified levels. This is most common with industrial noise. At a minority of sites, achieving the levels in Table 1.1 will not prevent disturbance from external industrial sources, and additional noise mitigation may be required. In these cases advice from an acoustic consultant should be sought.

The potentially beneficial masking effect of some types of continuous broadband external noise (such as road traffic noise) must also be borne in mind, see Section 2.12. This noise may partially mask other sounds, such as from neighbouring classrooms, which may be more disturbing than the external noise. There are acoustic benefits, as well as cost benefits, in ensuring that the level of insulation provided is not over-specified but is commensurate with the external noise.

3.7 Variation of noise incident on different facades

It may be convenient to determine the external noise level at the most exposed window (or part of the roof) of a building, and to assume this exposure for other elements too. This may be suitable at the early design stage for large schools. However, where external noise levels vary significantly, this approach can lead to over-specification and unnecessary cost.

3.8 Calculations

A calculation of the internal noise level according to BS EN 12354-3:2000 can be used to estimate whether, for the levels of external noise at any particular site, a proposed construction will achieve the levels in Table 1.1. By estimating the internal levels for various different constructions, designers can determine the most suitable construction in any given situation. BS EN 12354-3:2000 allows the effects of both direct and flanking transmission to be calculated, but in many cases it is appropriate to consider only direct transmission.

3.9 Test method

Field testing of an existing building envelope should be conducted according to BS EN ISO 140-5:1998[10], with reference to the clarifications given in this section.

BS EN ISO 140-5:1998 sets out various test methods. The three 'global' tests using the prevailing external noise source(s) (road traffic, railway traffic, air traffic) are preferable. At most sites road traffic is likely to be the dominant source of noise, and the corresponding standardised level difference is denoted $D_{tr,2m,nT}$. Where aircraft noise is the major concern measurements should be made accordingly, and the standardised level difference denoted $D_{at,2m,nT}$. Similarly the standardised level difference using railway noise as the source is denoted $D_{rt,2m,nT}$.

The global loudspeaker test method (which generates $D_{ls,2m,nT}$ values) may be used only if the prevailing external noise sources are insufficient to generate an adequate internal level.

It is reasonable, under certain conditions as specified below, to use the test results to indicate the likely performance of building envelopes of a similar construction, exposed to similar sources. If the conditions are not met then it is not reasonable to infer the performance from existing sound insulation test results and the calculation procedure should be used.

3.9.1 Conditions for similar constructions

The following features of any untested construction should be similar to those of the tested construction:

- type and number of ventilators
- glazing specification, frame construction and area of windows
- type and number of doors
- external wall construction and area
- roof construction and area.

3.9.2 Conditions for similar sources

Only test results in terms of $D_{tr,2m,nT}$, $D_{at,2m,nT}$, $D_{rt,2m,nT}$ and $D_{ls,2m,nT}$ values are applicable, and these should not be used interchangeably. The following features concerning the prevailing sources of noise should be similar to those of the previously tested construction:

- relative contributions of road traffic, railway and aircraft noise
- orientation of the building relative to the main noise source(s)
- ground height of the building relative to the main noise source(s).

SOUND INSULATION BETWEEN ROOMS

This section describes constructions capable of achieving the different levels of sound insulation specified in Tables 1.2 and 1.4.

Appendix 3 describes how sound insulation between adjacent rooms is measured and calculated.

In addition to the transmission of direct sound through the wall or floor, additional sound is transmitted into the receiving room via indirect, or 'flanking' paths, see Figure 3.4.

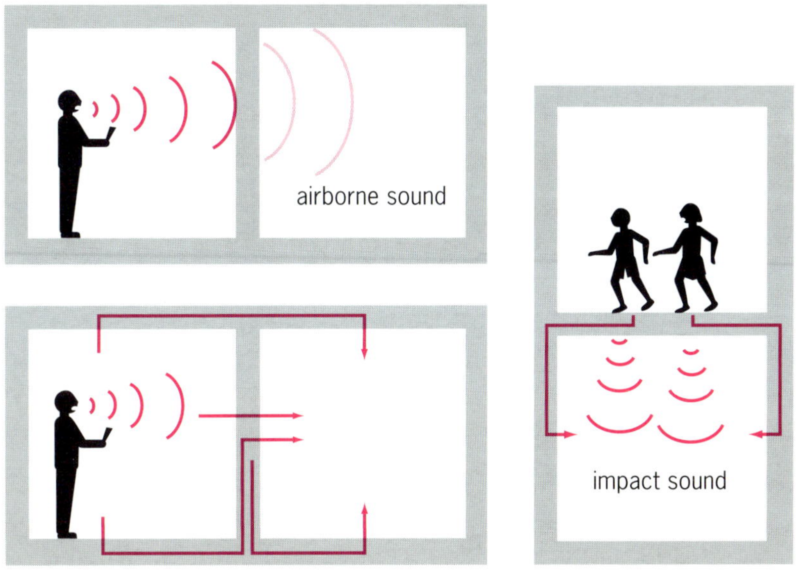

Figure 3.4: Sound transmission paths between adjacent rooms: direct sound paths through the wall and floor and flanking paths through the surrounding ceiling, wall and floor junctions

3.10 Specification of the airborne sound insulation between rooms using R_w

Table 1.2 describes the minimum weighted sound level difference between rooms in terms of $D_{nT(T_{mf,max}),w}$. However, manufacturers provide information for individual building elements based on laboratory airborne sound insulation data measured according to BS EN ISO 140-3:1995[5], in terms of the sound reduction index, R_w. Figure 3.5 shows the values of R_w for some typical building elements.

This section provides some basic guidance for the designer on how to use laboratory R_w values to choose a suitable separating wall or floor for the initial design. However, specialist advice should always be sought from an acoustic consultant early on in the design stage to assess whether the combination of the separating and flanking walls is likely to achieve the performance standard in Table 1.2. An acoustic consultant can use advanced methods of calculation to predict the sound insulation (eg Statistical Energy Analysis or BS EN 12354-1:2000[11]). The correct specification of flanking walls and floors is of high importance because incorrect specification of flanking details can lead to reductions in the expected performance of up to 30 dB.

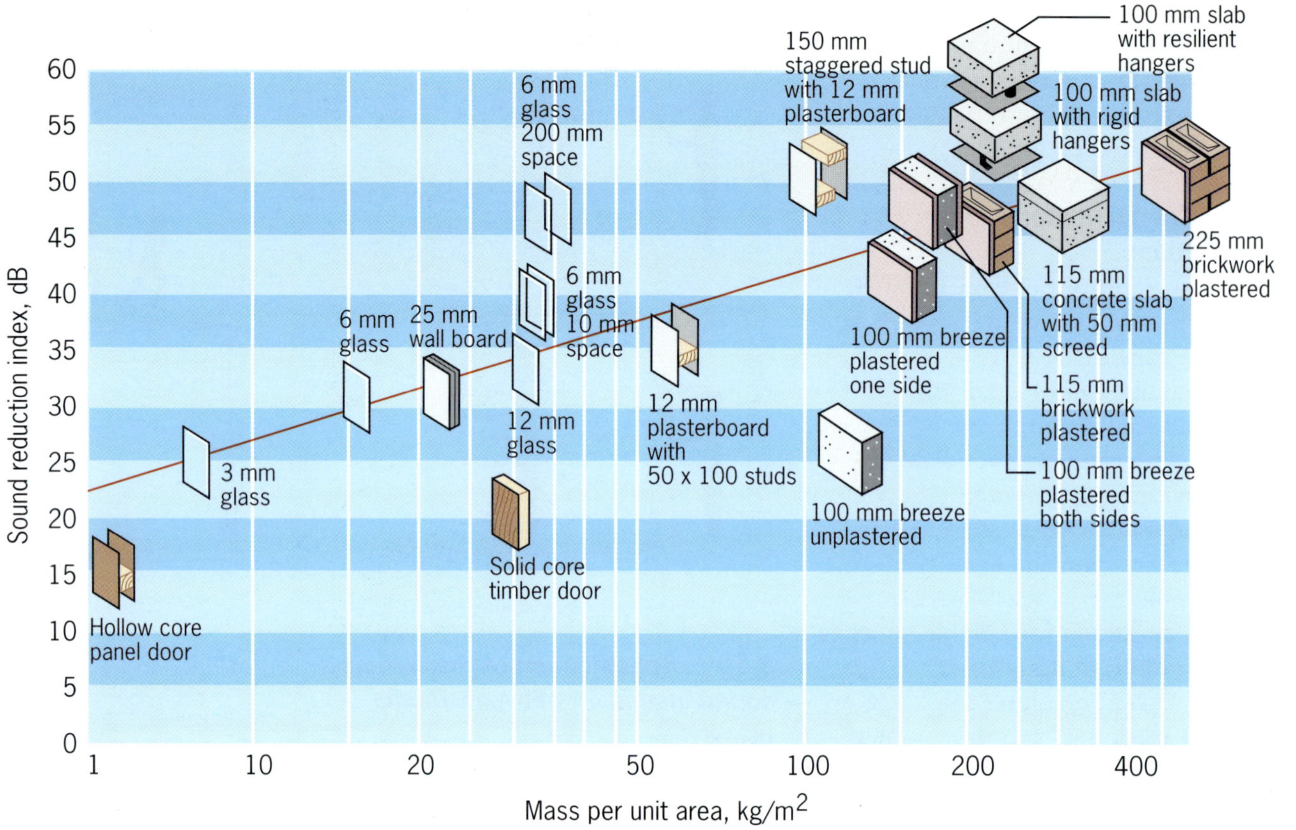

Figure 3.5: Typical sound reduction indices for construction elements

The following procedure can be used to choose an appropriate type of separating wall or floor before seeking specialist advice on appropriate flanking details.

1. From Table 1.2 determine the required minimum weighted BB93 standardized sound level difference between rooms, $D_{nT(T_{mf,max}),w}$.

2. Estimate the required weighted sound reduction index for the separating wall or floor, as follows:

a. Use the following formula to provide an initial estimate of the measured sound reduction index ($R_{w,est}$) that should be achieved by the separating wall or floor in the laboratory.

$$R_{w,est} = D_{nT(T_{mf,max}),w} + 10 \lg \left(\frac{S T_{mf,max}}{V} \right) + 8 \text{ dB}$$

where $D_{nT(T_{mf,max}),w}$ is the minimum weighted BB93 standardized level difference between rooms from Table 1.2
S is the surface area of the separating element (m^2)
$T_{mf,max}$ is the maximum value of the reverberation time T_{mf} for the receiving room from Table 1.5 (s)
V is the volume of the receiving room (m^3).

b. Estimate the likely reduction, X dB, in the airborne sound insulation that would occur in the field, to account for less favourable mounting conditions and workmanship than in the laboratory test. X can be estimated to be 5 dB assuming that flanking walls and floors are specified with the correct junction details. However, if flanking walls and floors are not carefully designed then poor detailing can cause the airborne sound insulation to be reduced by up to 30 dB. To allow the designer to choose a suitable separating wall for the initial design it is recommended that X is assumed to be 5 dB and an acoustic consultant is used to check the choice of separating element and ensure that the correct flanking details are specified.

c. Calculate the final estimate for the

weighted sound reduction index R_w that should be used to select the separating wall or floor from laboratory test data:

$$R_w = R_{w,est} + X \text{ dB}$$

3.10.1 Flanking details

A simplified diagram indicating the main flanking transmission paths is shown in Figure 3.6. General guidance on flanking details for both masonry and framed constructions can be found in Approved Document E[1]. Specific guidance on flanking details for products can also sometimes be found from manufacturers' data sheets, or by contacting manufacturers' technical advisers.

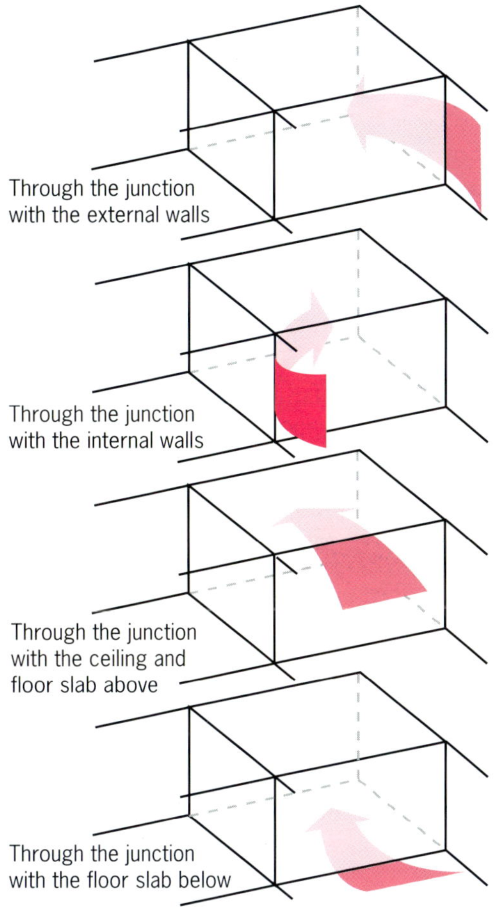

Figure 3.6: The main flanking transmission paths

3.10.2 Examples of problematic flanking details

In some buildings it is considered desirable to lay a floating screed (eg a sand-cement screed laid upon a resilient material) across an entire concrete floor and build lightweight partitions off the screed to form the rooms, see Figure 3.7(a). This allows the flexibility to change the room spaces. However, a continuous floating screed can transmit a significant quantity of structure-borne flanking sound from one room to another.

For example, if a lightweight partition with 54 dB R_w was built off a continuous floating screed the actual sound insulation could be as low as 40 dB $D_{nT(T_{mf,max}),w}$. In fact, even if a more expensive partition with a higher performance of 64 dB R_w was built, the actual sound insulation would still be 40 dB $D_{nT(T_{mf,max}),w}$, because the majority of sound is being transmitted via the screed, which is the dominant flanking path. This demonstrates the importance of detailing the junction between the screed and the lightweight partition. To reduce the flanking transmission, the floating screed should stop at the lightweight partition, see Figure 3.7(b).

Another flanking detail that can cause problems is where a lightweight profiled metal roof deck runs across the top of a separating partition wall. With profiles such as trapezoidal sections, it is very difficult for builders to ensure that they do not leave air paths between the top of the partition wall and the roof.

3.10.3 Junctions between ceilings and internal walls

Ceilings should be designed in relation to internal walls to achieve the required combined performance in respect of sound insulation, fire compartmentation and support.

In the case of suspended ceiling systems the preferred construction is one in which

Figure 3.7: Flanking transmission via floating screed
(a) Incorrect detail
(b) Correct detail

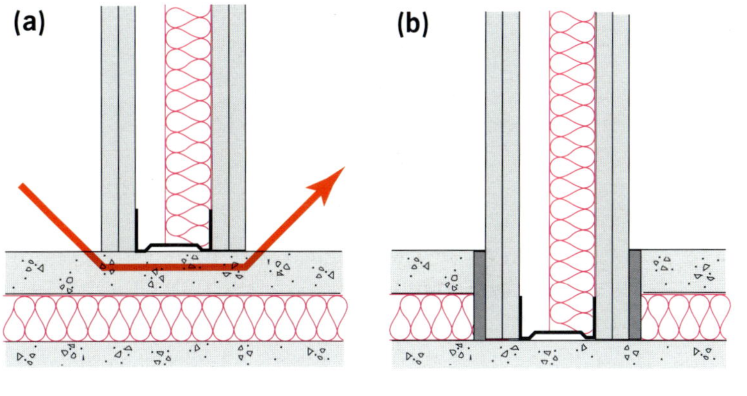

partitions or walls pass through the suspended ceiling membrane, do not require support from the ceiling system, and combine with the structural soffit above to provide fire resisting compartmentation and sound insulation.

The alternative construction in which partitions or walls terminate at, or just above the soffit of a suspended ceiling, is not recommended as it demands a ceiling performance in respect of fire resistance and sound insulation which is difficult to achieve and maintain in practice in school buildings. This is because the number of fittings required at ceiling level is incompatible with testing of fire resistance to BS 476[12], which is based on a test specimen area of ceilings without fittings. Furthermore, the scale and frequency of access to engineering services in the ceiling void through the membrane (in respect of fire) and through insulation backing the membrane (in respect of sound) is incompatible with maintenance of these aspects of performance.

3.10.4 Flanking transmission through windows

Flanking transmission can occur between adjacent rooms via open windows in the external walls. Side opening casement windows near the separating wall should have their hinges on the separating wall side to minimise airborne sound transmitted from one room to another. Where possible, windows in external walls should be located away from the junction between the external walls and the separating wall or floor. In particular, windows in the external walls of noise-sensitive rooms and in the external walls of rooms adjacent to them should be as far as possible from the separating wall or floor.

3.11 Specification of the impact sound insulation between rooms using $L_{n,w}$

Table 1.4 describes the minimum impact sound insulation between rooms in terms of $L'_{nT(T_{mf,max}),w}$. However, manufacturers usually provide information for floors based on laboratory impact sound insulation data measured according to BS EN ISO 140-6:1998[13], in terms of $L_{n,w}$.

This section provides some basic guidance for the designer on how to use laboratory $L_{n,w}$ values to design a suitable separating floor. However, specialist advice should always be sought from an acoustic consultant early on in the design process to assess whether the combination of the separating floor and flanking walls is likely to achieve the performance standard in Table 1.4. An acoustic consultant can use advanced methods of calculation to predict the sound insulation (eg Statistical Energy Analysis or BS EN 12354-2:2000[14]).

The following procedure can be used to choose an appropriate type of separating floor before seeking specialist advice on flanking details from an acoustic consultant.

1. Determine the maximum weighted BB93 standardized impact sound pressure level, $L'_{nT(T_{mf,max}),w}$ from Table 1.4.

2. Estimate the required weighted normalized impact sound pressure level for the separating floor, as follows:

a. Use the following formula to provide an initial estimate of the weighted normalized impact sound pressure level ($L_{n,w,est}$) that should be achieved by the separating floor in the laboratory:

$$L_{n,w,est} = L'_{nT(T_{mf,max}),w} + 10\lg\frac{V}{T_{mf,max}} - 18\text{ dB}$$

where $L'_{nT(T_{mf,max}),w}$ is the maximum weighted BB93 standardized impact sound pressure level from Table 1.4

V is the volume of the receiving room (m^3)

$T_{mf,max}$ is the maximum value of the reverberation time T_{mf} for the receiving room from Table 1.5 (s).

b. Estimate the likely increase, X dB, in the impact sound pressure level that would occur in the field to account for less favourable mounting conditions and good workmanship than in the laboratory test.

X can be 5 dB assuming that flanking walls are specified with the correct junction details. However, if flanking walls are not carefully designed the impact sound pressure level can increase by up to 10 dB. To allow the designer to choose a suitable separating floor for the initial design it is suggested that X is assumed to be 5 dB and an acoustic consultant is used to check the choice of separating floor and ensure that the correct flanking details are specified.

c. Calculate the final estimate for the weighted normalised impact sound pressure level $L_{n,w}$ that should be used to select the separating wall or floor from laboratory test data.

$$L_{n,w} = L_{n,w,est} - X \text{ dB}$$

3.12 Internal walls and partitions

3.12.1 General principles

Figure 3.5 shows typical values of the sound reduction index (R_w) for different wall constructions. For comparison the performance of other constructions including doors, glazing and floors is included.

The solid line shows the theoretical value based purely on the mass law. For single leaf elements (eg walls, single glazing, doors, etc) the mass law states that doubling the mass of the element will give an increase of 5 to 6 dB in R_w. When constructions provide less sound insulation than predicted by the mass law it is usually because they are not airtight.

In general, lightweight double-leaf constructions such as double glazing, cavity masonry or double-leaf plasterboard partitions provide better sound insulation than the mass law would indicate. At medium and high frequencies, double-leaf constructions benefit from the separation of the two leaves, with performance increasing with the width of the air gap between the leaves and the physical separation of the leaves. (Note that for double-leaf plasterboard constructions, timber studwork is rarely used to achieve high standards of sound insulation because lightweight metal studs provide better mechanical isolation between the leaves.)

Figure 3.8: Chart to estimate R_w for a composite wall consisting of two elements with different transmission losses

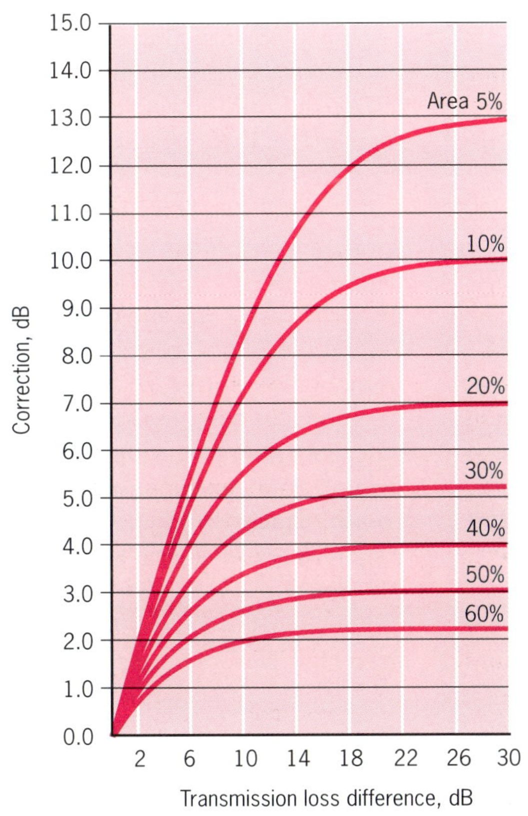

The percentage of the total area of the wall occupied by the element with the lower transmission loss, eg a door, and the difference between the higher R_w and the lower R_w, are used to calculate the correction in dB which is added to the lower R_w to give the R_w of the whole wall.

For example: Assume a classroom to corridor wall has an R_w of 45 dB and a door in the wall has an R_w of 30 dB. If the area of the door is 0.85 m x 2.1 m = 1.785 m^2 and the area of the wall is 7 m x 2.7 m = 18.9 m^2, then the percentage of the wall occupied by the door is 1.785/18.9 x 100 = 9.4%

The difference in R_w = 15 dB.

Therefore reading from the chart gives a correction of about 9 dB to be added to the lower R_w, giving a composite R_w of 39 dB.

If a higher performance door of say 35 dB had been used, the composite R_w would be 35 + 7 = 42 dB.

At low frequencies the performance of plasterboard partitions is limited by the mass and stiffness of the partition. Masonry walls can provide better low frequency sound insulation simply because of their mass. This is not obvious from the R_w figures, as the R_w rating system lends more importance to insulation at medium and high frequencies rather than low frequencies. This is not normally a problem in general classroom applications where sound insulation is mainly required at speech frequencies. However, it can be important in music rooms and in other cases where low frequency sound insulation is important.

A combination of masonry and dry-lining can be very effective in providing reasonable low frequency performance with good sound insulation at higher frequencies. This combination is often useful when increasing the sound insulation of existing masonry walls.

While partition walls may be provided as a means of achieving sound reduction, it should be remembered that sound insulation is no better than that provided by the weakest element.

Figure 3.8 can be used to assess the overall effect of a composite construction such as a partition with a window, door, hole or gap in it. The sound insulation of the composite structure is obtained by relating the areas and sound insulation values of the component parts using the graph.

Partitions should be well sealed, as small gaps, holes, etc. significantly reduce sound insulation. (Note that this applies to porous materials, eg porous blockwork, which can transmit a significant amount of sound energy through the pores.)

3.12.2 Sound insulation of common constructions

Figure 3.9 shows the approximate weighted sound reduction index R_w for masonry and plasterboard constructions.

Using the procedures given in Section 3.10, it is possible to determine which constructions are capable of meeting the requirements between different types of rooms.

The values in Figure 3.9 are necessarily approximate and will depend on the precise constructions and materials used. Many blockwork and plasterboard manufacturers provide data for specific constructions.

More sound reduction indices, both single value and octave band data, and further references to specific manufacturers' data are in the sound reduction indices spreadsheet included on the DfES acoustics website.

3.12.3 Flanking transmission

In general, a weighted sound level difference of up to 50 dB $D_{nT(T_{mf,max}),w}$ can be achieved between adjacent rooms by a single partition wall using one of the constructions described above, provided that there are no doors, windows or other weaknesses in that partition wall, and that flanking walls/floors with their junction details are carefully designed. Flanking transmission is critical in determining the actual performance and specialist advice should be sought from an acoustic consultant.

3.12.4 High performance constructions – flanking transmission

High-performance plasterboard partitions or masonry walls with independent linings can provide airborne sound insulation as high as 70 dB R_w in the laboratory. However, to achieve high performance in practice (ie above 50 dB $D_{nT(T_{mf,max}),w}$), flanking walls/floors with their junction details must be carefully designed. Airborne sound insulation as high as 65 dB $D_{nT(T_{mf,max}),w}$ can be achieved on site using high performance plasterboard partitions, or masonry walls with independent linings with lightweight isolated floors and independent ceilings to control flanking transmission. This will require specialist advice from an acoustic consultant.

For rooms which would otherwise need high-performance partitions it may be possible to use circulation spaces, stores and other less noise-sensitive rooms to act as buffer zones between rooms such that partitions with lower levels of sound insulation can be used. Case Study

Figure 3.9: Walls – airborne sound insulation for some typical wall constructions

Performance R_w (dB)		**Walls - typical forms of construction**
35–40	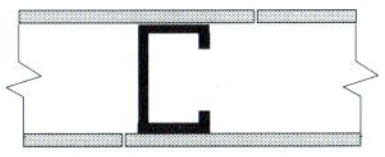	1x12.5 mm plasterboard each side of a metal stud (total width 75 mm)
		100 mm block (low density 52 kg/m^2) plastered/rendered 12 mm one side
40–45	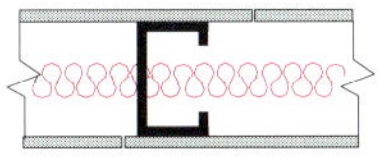	1x12.5 mm plasterboard each side of a 48 mm metal stud with glass fibre/mineral wool in cavity (total width 75 mm)
		100 mm block (medium density 140 kg/m^2) plastered/rendered 12 mm one side
45–50	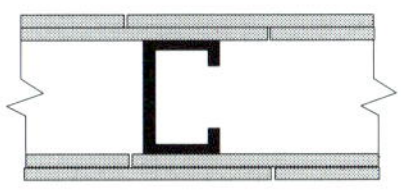	2x12.5 mm plasterboard each side of a 70 mm metal stud (total width 122 mm)
	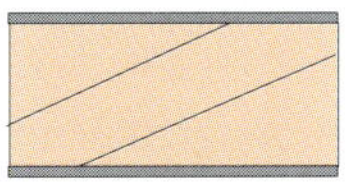	115 mm brickwork plastered/rendered 12 mm both sides
		100 mm block (medium density 140 kg/m^2) plastered/rendered 12 mm both sides
50–55	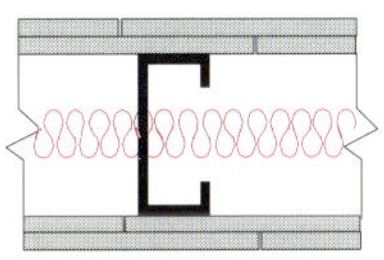	2x12.5 mm plasterboard each side of a 150 mm metal stud with glass fibre/mineral wool in cavity (total width 198 mm)
	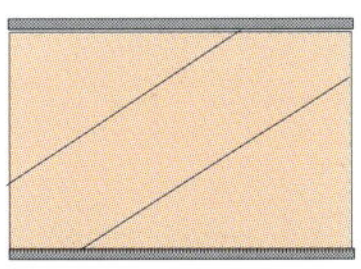	225 mm brickwork plastered/rendered 12 mm both sides
	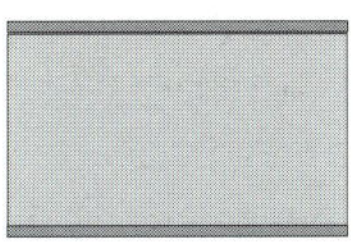	215 mm block (high density 430 kg/m^2) plastered/rendered 12 mm both sides
55–60	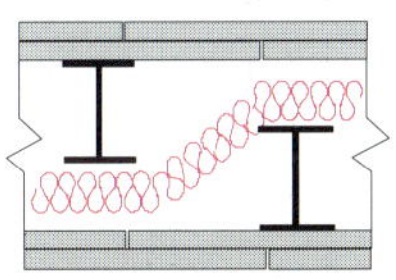	2x12.5 mm plasterboard each side of a staggered 60 mm metal stud with glass fibre/mineral wool in cavity (total width 178 mm)
	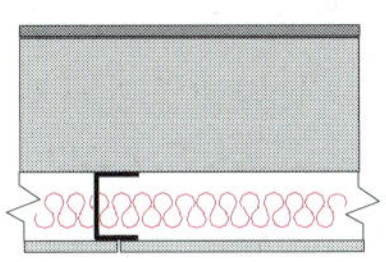	100 mm block (high density 200 kg/m^2) with 12 mm plaster on one side and 1x12.5 mm plasterboard on metal frame with a 50 mm cavity filled with glass fibre/mineral wool on other side

Figure 3.10: Glazing – airborne sound insulation for some typical glazing

Performance R_w (dB)		Glazing - typical forms of construction
25		4 mm single float (sealed)
28		6 mm single float (sealed)
		4/12/4: 4 mm glass/12 mm air gap/4 mm glass
30		6/12/6: 6 mm glass/12 mm air gap/6 mm glass
		10 mm single float (sealed)
33		12 mm single float (sealed)
		16/12/8: 16 mm glass/12 mm air gap/8 mm glass
35		10 mm laminated single float (sealed)
		4/12/10: 4 mm glass/12 mm air gap/10 mm glass
38		6/12/10: 6 mm glass/12 mm air gap/10 mm glass
		12 mm laminated single float (sealed)
40		10/12/6 lam: 10 mm glass/12 mm air gap/6 mm laminated glass
		19 mm laminated single float (sealed)
		10/50/6: 10 mm glass/50 mm air gap/6 mm glass
43		10/100/6: 10 mm glass/100 mm air gap/6 mm glass
		12 lam/12/10: 12 mm laminated glass/12 mm air gap/10 mm glass
45		6 lam/200/10: 6 mm laminated glass/200 mm air gap/10 mm + absorptive reveals
		17 lam/12/10: 17 mm laminated glass/12 mm air gap/10 mm glass

7.5 (see also Figure 2.4) describes a purpose built music suite which uses buffer zones effectively. In some cases, such as the refurbishment of music facilities in existing buildings, room layout may not allow this, and in these cases high levels of sound insulation between adjacent rooms will be required.

3.12.5 Corridor walls and doors

The R_w values in Table 1.3 should be used to specify wall (including any glazing) and door constructions between corridors or stairwells and other spaces. To ensure that the door achieves its potential in terms of its airborne sound insulation, it must have good perimeter sealing, including the threshold where practical.

Note that a lightweight fire door will usually give lower sound insulation than a heavier, sealed acoustic door.

Greatly improved sound insulation will be obtained by having a lobby door arrangement between corridors or stairwells and other spaces. However, this is not often practicable between classrooms and corridors. Some noise transmission from corridors into classrooms is inevitable, but this may not be important if all lesson changes occur simultaneously.

For some types of room, such as music rooms, studios and halls for music and drama performance, lobby doors should generally be used.

3.13 Internal doors, glazing, windows and folding partitions

Internal doors, glazing and windows are normally the weakest part of any separating wall. Figures 3.10 and 3.11 show the performance of a number of different types of window and door. In general, rooms which require at least 35 dB $D_{nT(T_{mf,max}),w}$ should not have doors or single glazing in the separating wall or partition.

3.13.1 Doors

The choice of appropriate doors with good door seals is critical to maintaining effective sound reduction, and controlling the transfer of sound between spaces.

Internal doors are often of lightweight hollow core construction, providing only around 15 dB R_w which is about 30 dB less than for a typical masonry wall (see Figure 3.5). The sound insulation of an existing door can be improved by increasing its mass (eg by adding two layers of 9 mm plywood or steel facings) as long as the frame and hinges can support the additional weight. However, it is often simpler to fit a new door.

The mass of a door is not the only variable that ensures good sound insulation. Good sealing around the frame is crucial. Air gaps should be minimised by providing continuous grounds to the frame which are fully sealed to the masonry opening. There should be a generous frame rebate and a proper edge seal all around the door leaf. Acoustic seals can eliminate gaps between the door and the door frame to ensure that the door achieves its potential in terms of its airborne sound insulation.

As a rule of thumb, even a good quality acoustically sealed door in a 55 dB R_w wall between two classrooms will reduce the R_w of the wall so that the $D_{nT(T_{mf,max}),w}$ is only 30-35 dB. Two such doors, separated by a door lobby, are necessary to maintain the sound insulation of the wall. Figure 3.12 shows the effect of different doors on the overall sound insulation of different types of wall. In a conventional layout with access to classrooms from a corridor, the corridor acts as a lobby between the two classroom doors.

3.13.2 Lobbies

The greater the distance between the lobby doors, the better the sound insulation, particularly at low frequencies. Maximum benefit from a lobby is associated with offset door openings as shown in Figure 3.13(a) and acoustically absorbent wall and/or ceiling finishes.

A lobby is useful between a performance space and a busy entrance hall. Where limitations of space preclude a lobby, a double door in a single wall will be more effective than a single door; this configuration is illustrated in Figure 3.13(b).

Inter-connecting doors between two music spaces should be avoided and a

Acoustic performance	Typical construction
30 dB R_w 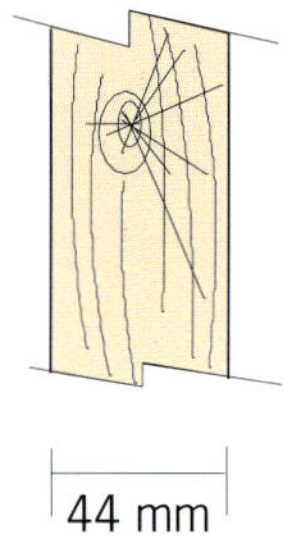 44 mm thick timber door, half hour fire rated	This acoustic performance can be achieved by a well fitted solid core doorset where the door is sealed effectively around its perimeter in a substantial frame with an effective stop. A 30 minute fire doorset (FD30) can be suitable. Timber FD30 doors often have particle cores or laminated softwood cores with a mass per unit area ≈ 27 kg/m^2 and a thickness of ≈ 44 mm. Frames for FD30 doors often have a 90 mm x 40 mm section with a stop of at least 15 mm. Compression or wipe seals should be used around the door's perimeter along with a threshold seal beneath. A drop-down or wipe type threshold seal is suitable. Doors incorporating 900 mm x 175 mm vision panels comprising 7 mm fire resistant glass can meet this acoustic performance.
35 dB R_w 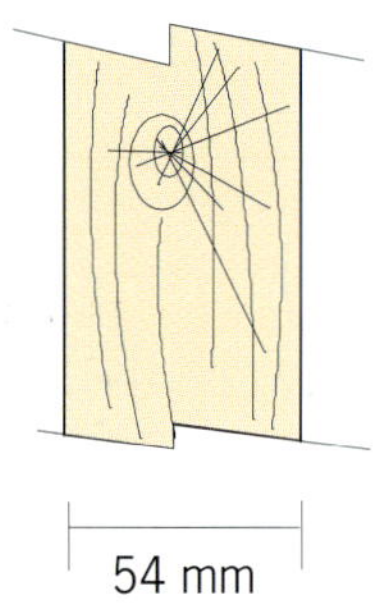 54 mm thick timber door, one hour fire rated	This acoustic performance can be achieved by specialist doorsets although it can also be achieved by a well fitted FD60 fire doorset where the door is sealed effectively around its perimeter in a substantial frame with an effective stop. Timber FD60 doors often have particle core or laminated softwood cores with a mass per unit area ≈ 29 kg/m^2 and a thickness of ≈ 54 mm. Using a core material with greater density than particle or laminated softwood can result in a door thickness of ≈ 44 mm. Frames for FD60 doors can have a 90 mm x 40 mm section with stops of at least 15 mm. Compression or wipe seals should be used around the door's perimeter along with a threshold seal beneath. A drop-down or wipe type threshold seal is suitable. Doors incorporating 900 mm x 175 mm vision panels comprising 7 mm fire resistant glass can meet this performance.

NOTES ON FIGURE 3.11

1 Care should be taken to ensure that the force required to open doors used in schools is not excessive for children. To minimise opening forces, doors should be fitted correctly and good quality hinges and latches used. Door closers should be selected with care.

2 The opening force at the handles of doors used by children aged 5–12 should not exceed 45 N.

3 Manufacturers should be asked to provide test data to enable the specification of doorsets.

4 Gaps between door frames and the walls in which they are fixed should be ≤ 10 mm.

5 Gaps between door frames and the walls in which they are fixed should be filled to the full depth of the wall with ram-packed mineral wool and sealed on both sides of the wall with a non-hardening sealant.

6 Seals on doors should be regularly inspected and replaced when worn.

Figure 3.11: Doors – airborne sound insulation for some typical door constructions

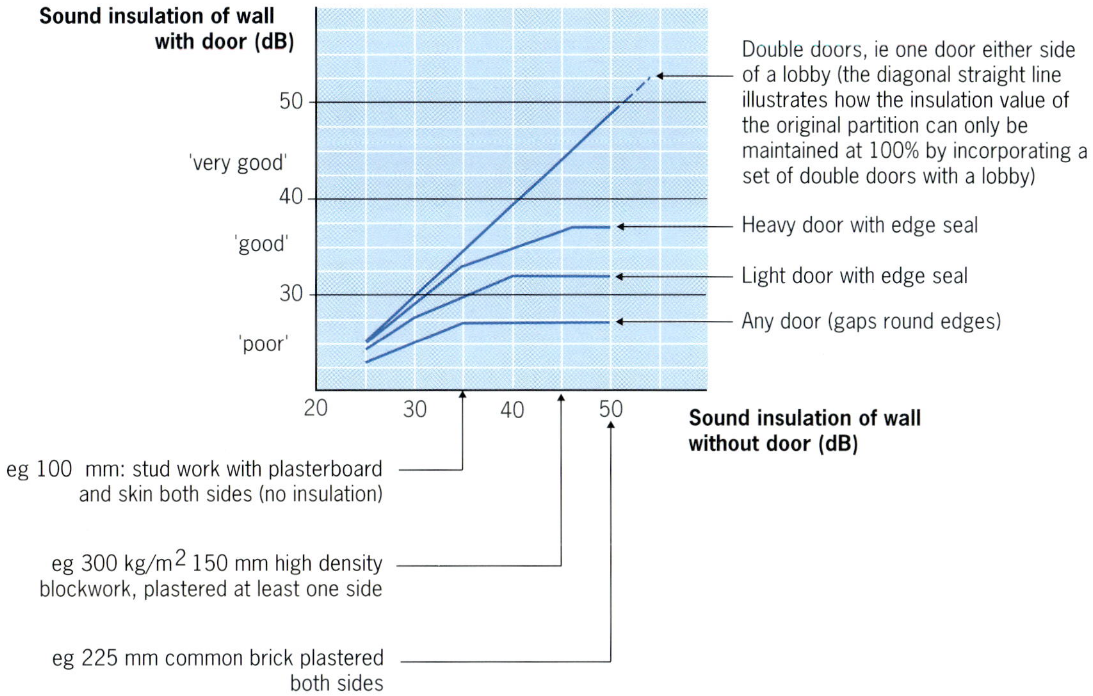

Figure 3.12: Reduction of sound insulation of a wall incorporating different types of door

lobby used to provide the necessary airborne sound insulation.

3.13.3 Folding walls and operable partitions

Folding walls and operable partitions are sometimes used to provide flexibility in teaching spaces or to divide open-plan areas. A standard folding partition with no acoustic seals or detailing may provide a value as low as 25 dB R_w . However, folding partitions are available that can provide up to 55 dB R_w . The sound insulation depends on effective acoustic sealing and deteriorates if seals or tracks are worn or damaged.

It is important that the specification of folding partitions takes into account their weight, ease of opening and maintenance. Regular inspection and servicing will extend the life of a partition and ensure that it achieves the required sound insulation.

Folding partitions are useful in many applications but they should only be used when necessary and not as a response to a non-specific desire for flexibility in layout of teaching areas.

3.13.4 Roller shutters

Roller shutters are sometimes used to separate kitchens from multi-purpose spaces used for dining. Because roller shutters typically only provide sound insulation of around 20 dB R_w it is common for noise from the kitchen to disturb teaching activities. One solution

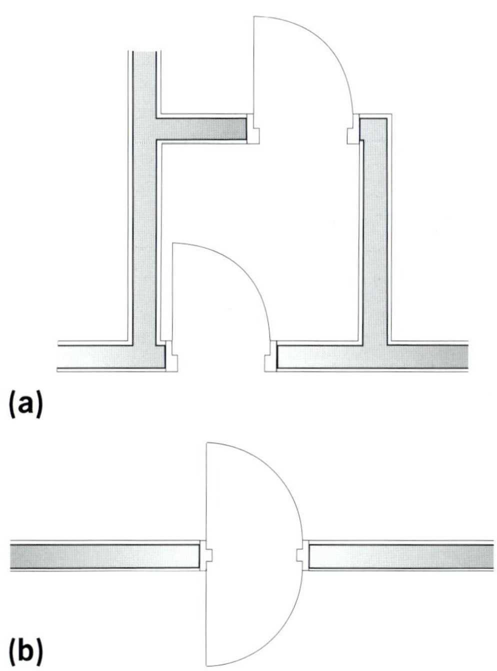

Figure 3.13: Use of lobbies and double doors
(a) Lobbied doorway
(b) Double door

Figure 3.14: Existing timber floors – airborne and impact sound insulation for some typical floor/ceiling

Option	**Construction – timber floors**		R_w **(dB)**	$L_{n,w}$ **(dB)**	Depth (mm)
1	Basic timber floor consisting of 15 mm floorboards on 150-200 mm wooden joists, plaster or plasterboard ceiling fixed to joists	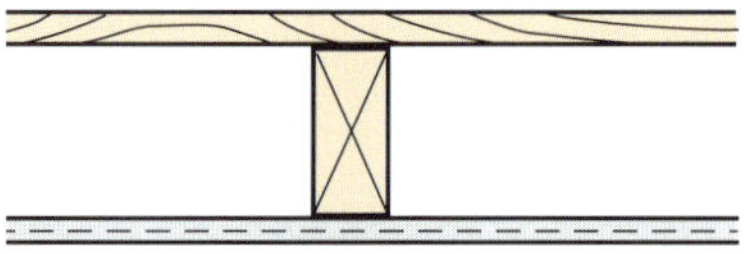	35–40	80–85	180–230
2	As 1, ceiling consisting of one layer of 15 mm plasterboard and one layer of 12.5 mm dense plasterboard fixed to proprietary resilient bars on underside of joists	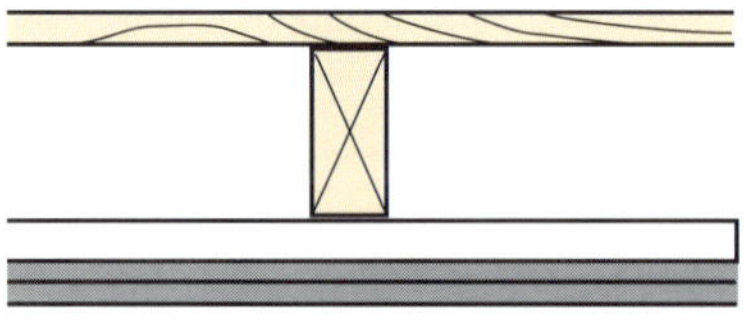	50–55	65–70	220–270
3	As 1, ceiling retained, with suspended ceiling consisting of 2 layers of 15 mm wallboard or 2 layers of 12.5 mm dense plasterboard, suspended on a proprietary metal ceiling system to give 240 mm cavity containing 80-100 mm mineral wool (>10 kg/m^3)	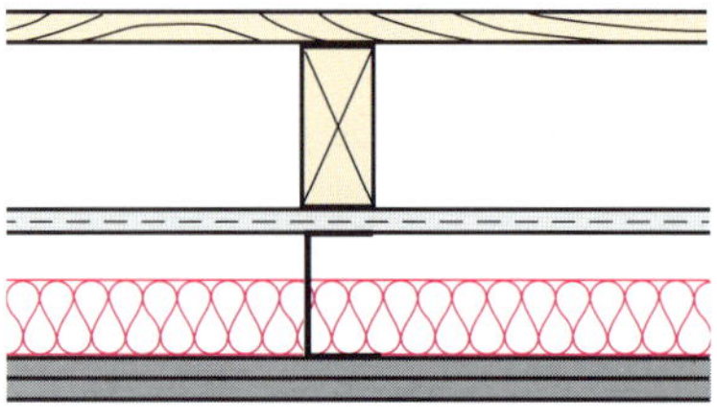	55–60	60–65	450–500
4	As 1, ceiling removed, with suspended ceiling consisting of 2 layers of 15 mm wallboard or 2 layers of 12.5 mm dense plasterboard, suspended on a proprietary metal ceiling system to give 275 mm cavity containing 80-100 mm mineral wool (>10 kg/m^3)	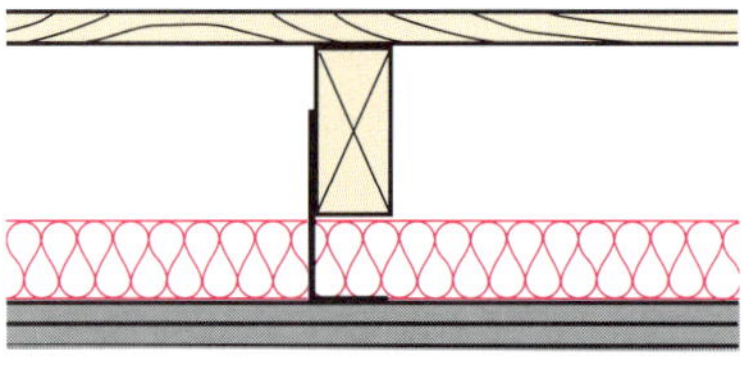	55–60	60–65	450–500
5	As 1, ceiling removed, with suspended ceiling consisting of 2 layers of 15 mm wallboard or 2 layers of 12.5 mm dense plasterboard, suspended special resilient hangers to give 275 mm cavity containing 80-100 mm mineral wool (>10 kg/m^3)	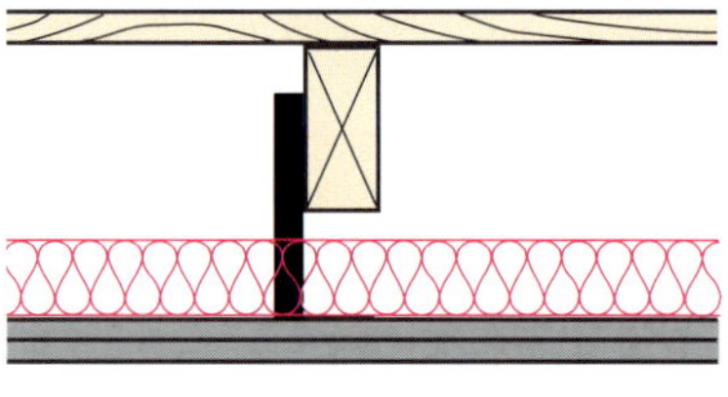	60–65	55–60	450–500
6	As 1 with proprietary lightweight floating floor using resilient pads or strips (eg 15 mm tongue-and-groove floorboards on a 15 mm plywood, chipboard or fibre-bond board supported on 45 mm softwood battens laid on 25 mm thick open-cell foam pads). 80-100 mm mineral wool (>10 kg/m^3) laid on top of existing floorboards	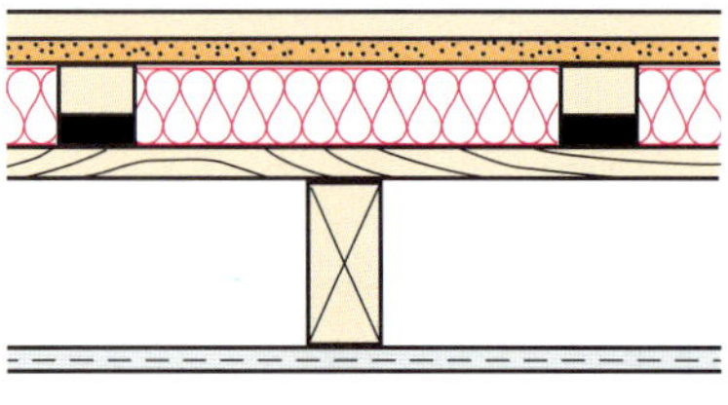	50–55	60–65	270–320
7	As 1, floorboards removed and replaced with 15 mm tongue-and-groove floorboards on a 15 mm plywood, chipboard or fibre-bond board supported on 12 mm softwood battens laid on 25 mm thick open-cell foam pads bonded to the joists, 80-100 mm mineral wool (>10 kg/m^3) laid on top of existing ceiling	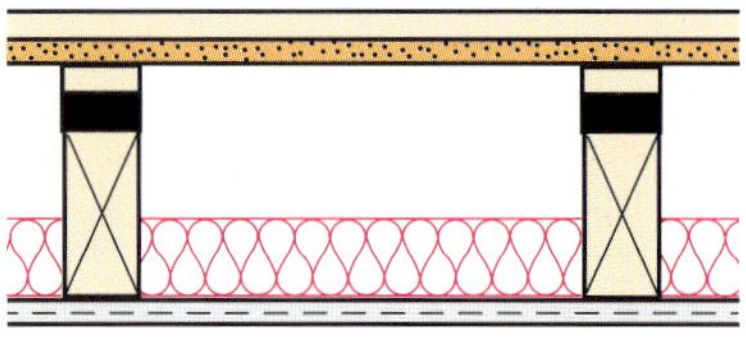	55–60	55–60	240–290

Option	Construction – timber floors		R_w (dB)	$L_{n,w}$ (dB)	Depth (mm)
8	As 7 but mineral wool replaced by 100 mm pugging (80 kg/m^2) on lining laid on top of ceiling	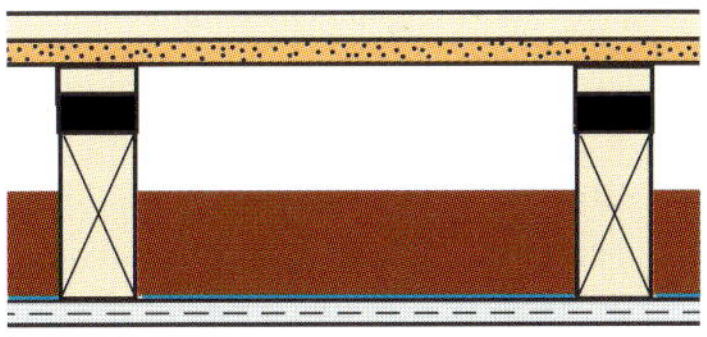	55–60	50–55	240–290
9	As 8 but with 75 mm pugging laid on top of board fixed to sides of joists	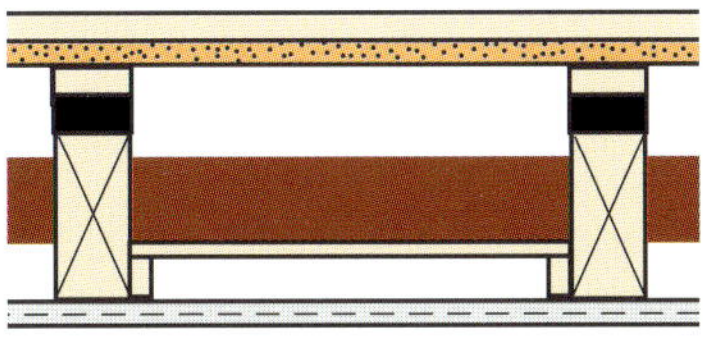	50–55	55–60	240–290
10	As 1 with proprietary lightweight floating floor using a continuous layer (eg 15 mm tongue-and-groove floorboards on a 15 mm plywood, chipboard or fibre-bond board on 6-12 mm thick continuous open-cell foam mat)	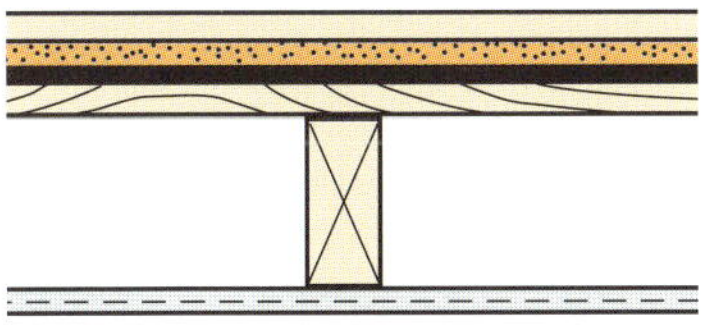	50–55	55–60	220–270
11	As 10, ceiling removed and replaced with suspended ceiling consisting of 2 layers of 15 mm wallboard or 2 layers of 12.5 mm dense plasterboard, suspended on a proprietary metal ceiling system to give 275 mm cavity containing 80-100 mm mineral wool (>10 kg/m^3)	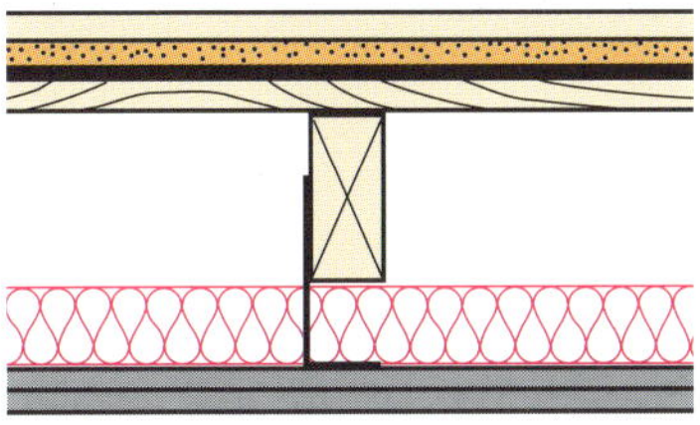	60–65	50–55	360–410

NOTES ON FIGURE 3.14

1 Where resilient floor materials are used, the material must be selected to provide the necessary sound insulation under the full range of loadings likely to be encountered in that room and must not become over-compressed, break down or suffer from long-term 'creep' under the higher loads likely to be encountered. Where large ranges of loading are encountered, or where there are high point loads such as pianos, heavy furniture or operable partitions, the pad stiffness may have to be varied across the floor to take account of these.

2 All figures are approximate guidelines and will vary between different products and constructions. Manufacturers' data should be obtained for all proprietary systems and constructions. These must be installed in accordance with good practice and manufacturers' recommendations and all gaps sealed.

Figure 3.14 Continued

is to provide doors in front of the shutters to improve the sound insulation.

3.14 Floors and ceilings

Both airborne and impact noise can be transmitted between vertically adjacent rooms through the separating floor and its associated flanking constructions.

Vertical noise transmission between classrooms can be a problem in older multi-storey buildings with wooden floors, such as traditional Victorian school buildings. Both airborne noise and impact noise can be problematic with wooden floors, and both problems need to be considered when dealing with vertically adjacent spaces. Adding carpets or other soft coverings to wooden floors reduces impact noise but has very little effect on airborne noise transmission.

Impact noise can also be a problem with concrete floors (although airborne noise may not be a problem); this can sometimes be solved by adding a carpet.

Where the use of carpet is proposed

Option	Construction – lightweight concrete floors		R_w (dB)	$L_{n,w}$ (dB)	Depth (mm)
1	Lightweight floor consisting of concrete planks (solid or hollow) or beam and blocks, with 30-50 mm screed, overall weight approximately 100 kg/m^2, no ceiling or floor covering	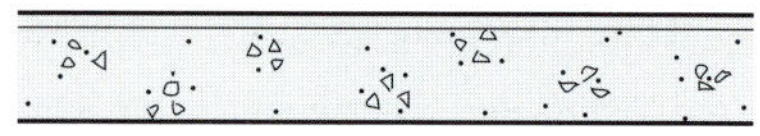	35–40	90–95	100–150
2	As 1 with soft floor covering >5 mm thick	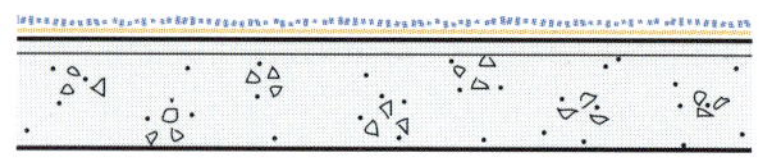	35–40	75–85	105–155
3	As 1 with suspended ceiling consisting of 2 layers of 15 mm wallboard or 2 layers of 12.5 mm dense plasterboard, suspended on a proprietary metal ceiling system to give 240 mm cavity containing 80-100 mm lightweight mineral wool (>10 kg/m^3)	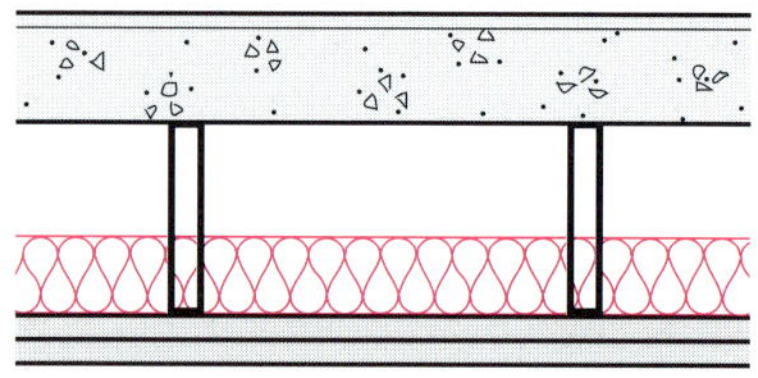	60–65	55–60	370–420
4	As 3 with soft floor covering >5 mm thick	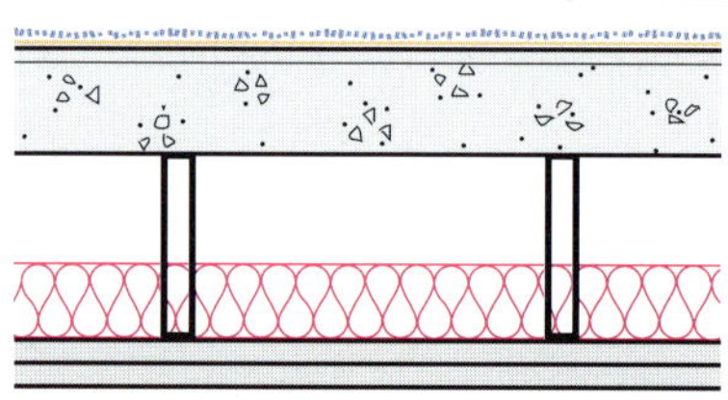	60–65	50–55	375–425
5	As 1 with proprietary lightweight floating floor using resilient pads or strips (eg 15 mm tongue-and-groove floorboards on a 15 mm plywood, chipboard or fibre-bond board on 25 mm thick open-cell foam pads)	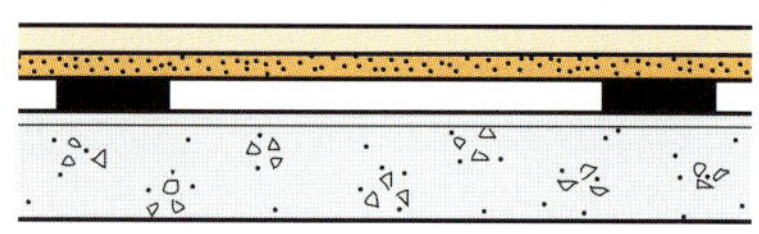	50–60	50–60	155–205
6	As 1 with proprietary lightweight floating floor using a continuous layer (eg 15 mm tongue-and-groove floorboards on a 15 mm plywood, chipboard or fibre-bond board on 6-12 mm thick continuous open-cell foam mat)	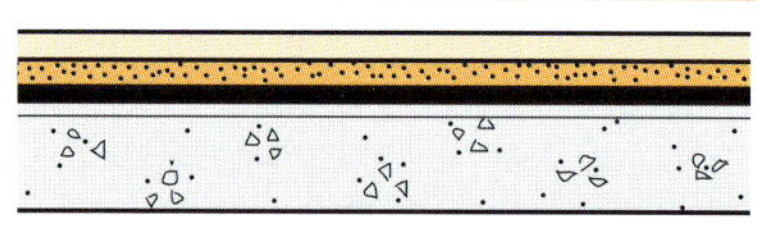	50–55	55–60	150–200
7	As 1 with heavyweight proprietary suspended sound insulating ceiling tile system	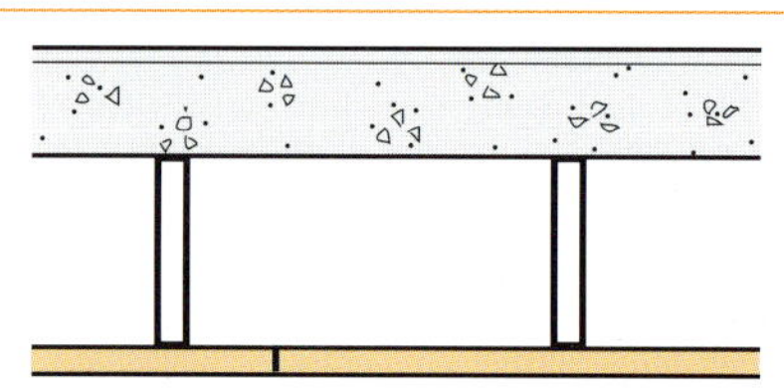	45–55	60–70	250–500

Figure 3.15: Lightweight concrete floors – airborne and impact sound insulation of some typical constructions

issues of cleaning, maintenance and effects on air quality may need to be considered.

3.14.1 Impact sound insulation

Impact noise on floors may arise from:

- foot traffic, particularly in corridors at break times/lesson changeover
- percussion rooms
- areas for dance or movement
- loading/unloading areas (eg in kitchens and workshops)
- machinery.

Where possible, impact noise should be reduced at source through use of soft floor coverings or floating floors. Carpets are not an option in practical spaces but other soft floor coverings, such as acoustic

Option	Construction – heavyweight concrete floors		R_w (dB)	$L_{n,w}$ (dB)	Depth (mm)
1	Solid concrete floor consisting of reinforced concrete with or without shuttering, concrete beams with infill blocks and screed, hollow or solid concrete planks with screed, of thickness and density to give a total mass of at least 365 kg/m^2, with soft floor covering >5 mm thick	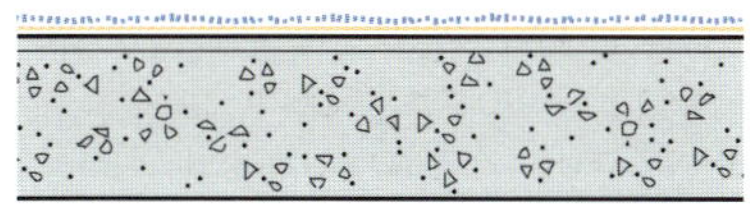	50–55	60–65	150–200
2	As 1 with proprietary lightweight floating floor using resilient pads or strips (eg 15 mm tongue-and-groove floorboards on a 15 mm plywood, chipboard or fibre-bond board on 25 mm thick open-cell foam pads)	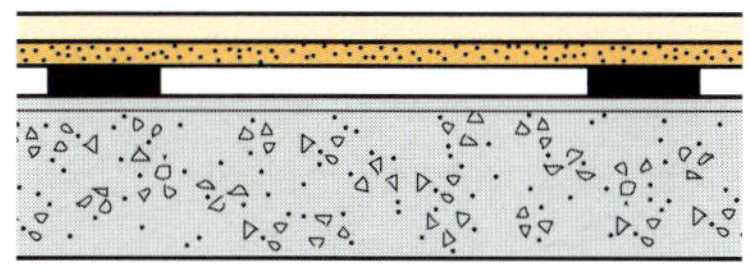	55–60	50–55	200–250
3	As 1 with proprietary lightweight floating floor using a continuous layer (eg 15 mm tongue-and-groove floorboards on a 15 mm plywood, chipboard or fibre-bond board on 6-12 mm thick continuous open-cell foam mat)	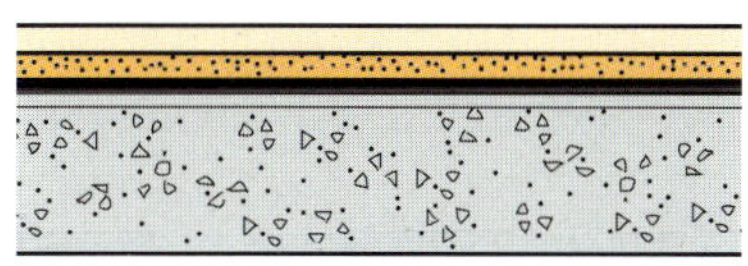	55–60	50–60	175–230
4	As 1 with suspended ceiling consisting of 2 layers of 15 mm wallboard or 2 layers of 12.5 mm dense plasterboard, suspended on a proprietary metal ceiling system to give 240 mm cavity containing 80-100 mm mineral wool (>10 kg/m^3)	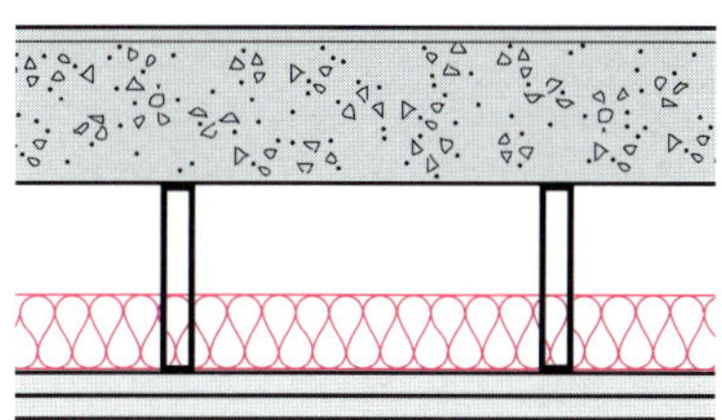	60–70	55–60	420–470
5	As 4 with soft floor covering >5 mm thick	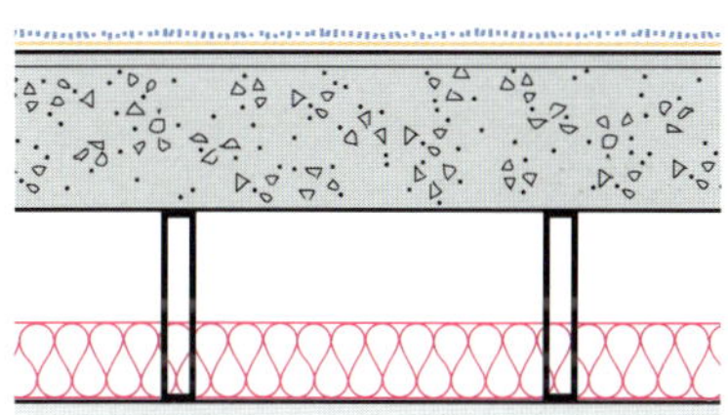	60–70	50–55	425–475

Figure 3.16: Heavyweight concrete floors – airborne and impact sound insulation of some typical constructions

NOTES ON FIGURES 3.15 AND 3.16

1 Where soft floor covering is referred to this should be a resilient material or a material with a resilient base, with an overall uncompressed thickness of at least 4.5 mm; or any floor covering with a weighted reduction in impact sound pressure level of not less than 17 dB when measured in accordance with BS EN ISO 140-8:1998[15] and calculated in accordance with BS EN ISO 717-2:1997[16].

2 Where resilient floor materials are used, the material must be selected to provide the necessary sound insulation under the full range of loadings likely to be encountered in that room and must not become over-compressed, break down or suffer from long-term 'creep' under the higher loads likely to be encountered. Where large ranges of loading are encountered, or where there are high point loads such as pianos, heavy furniture or operable partitions, the pad stiffness may have to be varied across the floor to take account of these.

3 All figures are approximate guidelines and will vary between different products and constructions. Manufacturers' data should be obtained for all proprietary systems and constructions. These must be installed in accordance with good practice and manufacturers' recommendations and all gaps sealed.

vinyl floor or vinyl flooring laid on an acoustic mat, may be suitable.

Planning and room layout can be used to avoid impact noise sources on floors above noise-sensitive rooms. Soft floor coverings and floating floor constructions and independent ceilings are the most effective means of isolation, and resilient floor finishes are also appropriate for some sources.

Typical airborne and impact noise performance are listed for a number of constructions in Figures 3.14, 3.15 and 3.16. Note that, unlike airborne sound insulation, impact sound insulation is measured in terms of an absolute sound level, so that a lower figure indicates a better standard of insulation.

3.14.2 Voids above suspended ceilings

Where partitions run up to the underside of lightweight suspended ceilings, the airborne sound insulation will be limited by flanking transmission across the ceiling void, which will often prevent the minimum values for airborne sound insulation in Table 1.2 being achieved. Therefore, partitions should either be continued through the ceiling up to the soffit, or a plenum barrier should be used.

3.14.3 Upgrading existing wooden floors using suspended plasterboard ceilings

Figure 3.14 shows the airborne and impact noise performance of a standard wooden floor with various forms of suspended plasterboard ceiling.

Option 2 is possibly the most widely used system of increasing both impact and airborne sound insulation, with or without the original plaster ceiling. In small rooms good results can be achieved using timber studs fixed only to the walls, but large timber sections are needed to span wider rooms.

In wider span rooms it is generally more convenient to suspend the plasterboard from the floor joists above, fixing through the existing ceiling if this is retained, using a proprietary suspension and grid system (option 4). The grid can be hung from simple metal strips or, for higher performance, special flexible ceiling hangers.

The major manufacturers of dry-lining systems all provide their own systems for these options, and provide sound insulation data and specifications for a variety of configurations. The performance for both airborne and impact sound improves with the depth of the ceiling void, with the mass of the ceiling and with the deflection of the ceiling hangers under the mass of the ceiling. Adding a layer of lightweight acoustically absorbent glass wool or mineral wool in the ceiling void increases the sound insulation, typically by 2-3 dB, but there is no point in adding more than specified.

Performance on site is strongly dependent on good workmanship to avoid air gaps, so careful attention should be given to ensuring that joints are close-butted, taped and filled and that all gaps are properly sealed. At the perimeter a small gap should be left between the plasterboard and the walls, and this should be sealed using non-setting mastic to allow a small amount of movement without cracking.

Penetrations through the ceiling need to be properly detailed to maintain an airtight seal while allowing movement, and services should not be allowed to provide a rigid link between the ceiling and the floor above. This can be a particular problem with sprinkler pipes. A problem with these constructions is that recessed light fittings, grilles and diffusers significantly reduce the sound insulation so any services should be surface-mounted.

A plasterboard finish is acoustically reflective whereas in some rooms an acoustically absorbent ceiling is required, to meet the specifications for room acoustics and reverberation times. One solution to this, if there is sufficient height, is to suspend a separate lightweight sound absorbing ceiling under the sound insulating plasterboard ceiling. This can be a standard lightweight composite or perforated metal tile system. These lightweight, acoustically absorbent, ceilings add very little to the sound insulation but do provide acoustic absorption. Lights and services can be recessed in the absorbent ceiling.

The term 'acoustic ceiling' generally refers to lightweight acoustically absorbent ceiling tile systems, designed to provide acoustic absorption. Note that these systems do not always increase the sound insulation as well.

There are, however, some systems which use relatively heavy ceiling tiles which are designed to fit into ceiling grids to provide a reasonably airtight fit. These may consist of dense plasterboard or mineral fibre products, or perforated metal tiles with metal or plasterboard backing plates. If properly installed and maintained these can provide a useful increase in sound insulation as well as acoustic absorption. Manufacturers of these systems can provide both airborne and impact sound insulation figures, as well as acoustic absorption coefficients. If no measured sound insulation data are provided, it is better to err on the side of caution and assume that the tile will not provide a significant increase in sound insulation.

The sound insulation performance figures quoted in Figure 3.14 all assume that the floorboards are in good condition and reasonably airtight, with thin carpet laid on top. If retaining the original floorboards it is good practice to fill in any gaps with glued wooden strips, caulking or mastic, or to lay hardboard on top, to provide an airtight seal. If not retaining the original boards, 18 mm tongue-and-grooved chipboard can be used to achieve the same effect, with all joints and gaps properly sealed, especially at the perimeters.

3.14.4 Upgrading existing wooden floors using platform and ribbed floors

The systems discussed in Section 3.14.3 all maintain the original wooden floor mounted directly on joists. This has the advantage of maintaining the original floor level at the expense of loss of ceiling height below. An alternative approach is to provide a floating floor system either on top of the existing floorboards (a platform floor) or to remove the existing floorboards and build a new floor on resilient material placed on top of the floor joists (a ribbed floor). In both cases the increase in both airborne and impact sound insulation relies on the mechanical isolation of the floor from the joists using resilient material.

Figure 3.14 shows a number of typical lightweight floating floor constructions and indicative sound insulation figures. There are many proprietary systems using a wide range of isolating materials and manufacturers should supply test data in accordance with ISO 140 measurements.

The isolating layer will typically consist of rubber, neoprene, open-cell or closed-cell foams, mineral fibre or composite materials. The isolating layer can be in the form of individual pads, strips or a continuous layer of material.

The sound insulation increases with the deflection of the resilient layer (up to the limit of elasticity for the material), with the mass of the floating layer and with the depth of the cavity. Adding a layer of lightweight acoustically absorbent glass wool or mineral wool in the ceiling void increases the sound insulation, typically by 2-3 dB, but there is no point in adding more than specified. In each case the deflection of the material under the permanent 'dead' load of the floating layer and the varying 'live' loads of occupants and furniture must be considered. If the material is too resilient and the floating layer is insufficiently heavy or rigid, the floor will deflect under the varying loads as people move about the room. For this reason it is advantageous for the floating layer to be as heavy and as stiff as practicable, in some cases using ply or fibre-bond board (for mass) laid on top of the resilient layer, with tongue-and-grooved chipboard on top of this.

If there are likely to be very heavy local loads in the room (eg pianos) it may be necessary to increase the stiffness of the resilient material, or, in the case of pads, to space the pads more closely together to support these loads.

Junctions with walls and at doors need to be designed to maintain an effectively airtight seal while allowing movement of the floating layer. Manufacturers generally provide their own proprietary solutions for this, with or without skirtings.

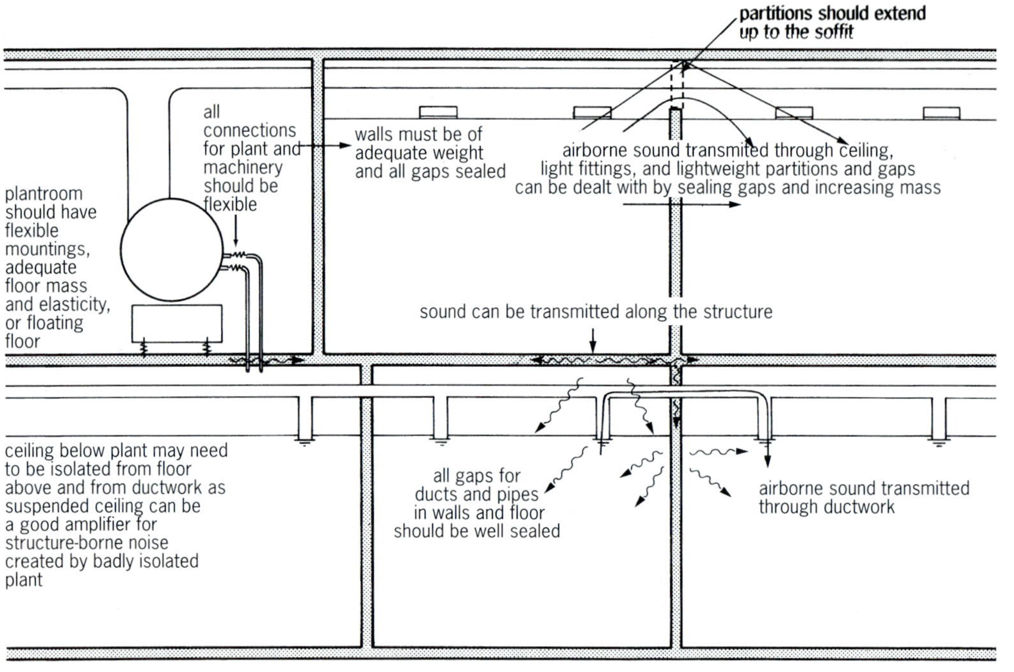

Figure 3.17: Possible sound transmission paths and their prevention

Lightweight floating floors are quite specialist constructions, and achieving the correct deflection under varying live loads without overloading the resilient material can be difficult. Most materials suffer from long-term loss of elasticity or 'creep' under permanent loads and this should be taken into account in the design and selection of materials. The system manufacturer should normally be provided with all of the relevant information and required to specify a system to meet all of the acoustic and structural requirements over the expected lifetime of the floor. In difficult cases the advice of an acoustics consultant and/or structural engineer should be sought.

3.14.5 Concrete floors

In general, concrete floors provide much greater low frequency airborne sound insulation than wooden floors by virtue of their greater mass. There are, however, considerable variations in performance between dense poured concrete floors and comparatively lightweight precast concrete plank floors. Impact sound transmission can be a problem even in heavy concrete floors because of the lack of damping in concrete, and a soft or resilient floor covering is generally required. This may simply be carpet on suitable underlay.

Figures 3.15 and 3.16 show airborne sound insulation and impact sound transmission data for a number of typical concrete floor constructions, with and without suspended ceilings and floating floors.

3.15 Design and detailing of building elements

Important points to remember when designing constructions to achieve adequate sound insulation are:

- Weak elements (eg doors and glazing, service penetrations, etc) will reduce the effectiveness of the walls in which they are located.
- Impact sound will travel with little reduction through a continuous member such as a steel beam or servicing pipe.
- Partitions between sensitive spaces should normally continue beyond the ceiling up to the structural soffit or roof layer, to prevent noise passing over the top of the partition above the ceiling or through a loft space.
- Openings in walls caused by essential services passing through should be acoustically sealed. Pipework passing between noise sensitive spaces should be appropriately boxed-in (see Approved Document E[1]).

Figure 3.17 shows how possible transmission paths through the structure of a building can be prevented.

References

[1] Approved Document E - Resistance to the passage of sound. The Stationery Office, 2003, ISBN 01 753 642 3 www.safety.odpm.gov.uk

[2] Sound Control for Homes (BRE report 238, CIRIA report 127), 1993. Available from CRC Ltd. 1993, BRE ISBN 0 85125 559 0, CIRIA ISBN 0 86017 362 3, CIRIA ISBN 0305 408 X.

[3] J McLoughlin, D J Saunders and R D Ford. Noise generated by simulated rainfall on profiled steel roof structures. Applied Acoustics 42 239-255, 1994

[4] ISO 140-18 Acoustics - Measurement of sound insulation in buildings and of building elements - Part 18: Laboratory measurement of sound generated by rainfall on building elements (in preparation).

[5] BS EN ISO 140-3: 1995 Measurement of sound insulation in buildings and of building elements. Part 3. Laboratory measurement of airborne sound insulation of building elements.

[6] The Education (School Premises) Regulations 1999. (Statutory Instrument 1999 No 2, Education, England & Wales). The Stationery Offiice, 1999. ISBN 0 11 080331 0 www.hmso.gov.uk

[7] BS EN 20140-10: 1992 Acoustics - Measurement of sound insulation in buildings and of building elements. Part 10. Laboratory measurement of airborne sound insulation of small building elements.

[8] BS 98/704582 DC. Ventilation for buildings. Performance testing of components/products for residential ventilation. Part 1. Externally and internally mounted air transfer devices. Draft for public comment (prEN 13141-1 Current Euronorm under approval).

[9] BS EN 12354-3:2000 Building Acoustics - Estimation of acoustic performance in buildings from the performance of elements. Part 3. Airborne sound insulation against outdoor sound.

[10] BS EN ISO 140-5: 1998 Measurement of sound insulation in buildings and of building elements. Part 5. Field measurements of airborne sound insulation of façade elements and facades.

[11] BS EN 12354–1: 2000 Building Acoustics. Estimation of acoustic performance in building from the performance of elements. Part 1. Airborne sound insulation between rooms.

[12] BS 476 Fire tests on building materials and structures.

[13] BS EN ISO 140-6: 1998, Acoustics - Measurement of sound insulation in buildings and of building elements. Part 6. Laboratory measurement of impact sound insulation of floors.

[14] BS EN 12354-2: 2000 Building Acoustics. Estimation of acoustic performance in building from the performance of elements. Part 2. Impact sound insulation between rooms.

[15] BS EN ISO 140-8: 1998 Acoustics. Measurements of sound insulation in buildings and of building elements. Part 8. Laboratory measurements of the reduction of transmitted impact noise by floor coverings on a heavyweight standard floor.

[16] BS EN ISO 717-2: 1997 Acoustics - Rating of sound insulation in buildings and of building elements. Part 2. Impact sound insulation.

The design of rooms for speech

SECTION 4

The design of rooms for speech is a critical aspect of the acoustic design of a school. Rooms must be designed to facilitate clear communication of speech between teachers and students, and between students.

4.1 Approach to acoustic design

The vast majority of rooms in schools are designed for speech. A structured approach to the acoustic design of these rooms would consider the following subjects in the order given:

1 Indoor ambient noise levels (Table 1.1)
2 Room size - floor area, shape and volume and hence, required reverberation time (Table 1.5)
3 Amount of acoustic absorption required for reverberation time
4 Type, location, and distribution of that acoustic absorption
5 Special considerations for non-standard rooms (eg reflectors and diffusers)
6 Use of electronic sound reinforcement systems.

Figure 4.1: Sound pressure levels of speech at 1 m

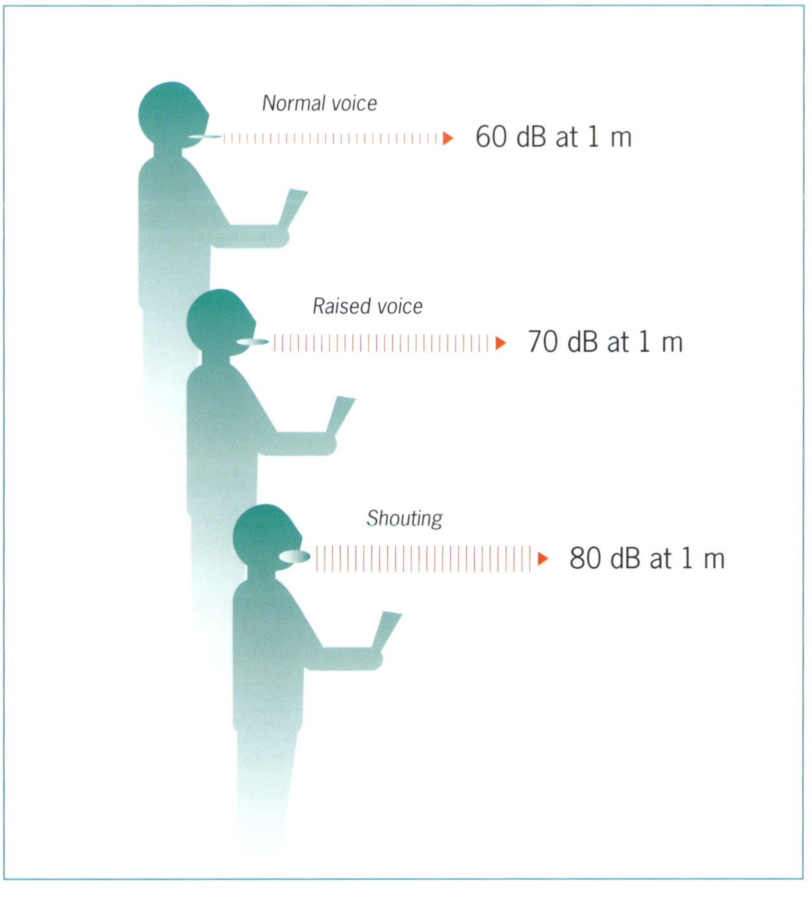

4.2 Internal ambient noise levels and speech clarity

The internal ambient noise level is very important in teaching spaces as the teacher's voice needs to be clearly heard above the background noise. The sound power output of conversational speech is typically 10 microwatts which results in a sound pressure level of about 60 dB at 1 m in front of the speaker. This output power can be raised to around 100 microwatts when the speaker talks as loudly as possible without straining the voice, which increases the sound pressure level at 1m to about 70 dB. By shouting, the output power can be further raised to around 1000 microwatts with a consequent further increase in sound pressure level to about 80 dB. In subjective terms, this means that a speaker can approximately double the loudness of the voice by speaking very loudly, and then double it again by shouting, see Figure 4.1.

It is also desirable to preserve the character, or nuances and intonations, of speech. This is particularly relevant to language teaching and to the performance of drama. The frequencies of sound in speech range from bass to treble, that is from below 125 Hz to above 8 kHz. Vowels have a sustained, tonal sound which contains characteristics of the speaker's voice. Men's voices have the lowest characteristic pitch (120 Hz), women an intermediate pitch (225 Hz), and children the highest pitch (265 Hz). Vowels contain most of the sound energy in speech but recognition of the consonants, whose energy is generally concentrated towards the higher frequency end of the speech spectrum, is the key factor for high intelligibility.

The intelligibility of speech depends upon its audibility as well as its clarity. Audibility is affected by the loudness of the speech relative to the background noise level. An increase in the background noise will cause greater masking of speech and hence will decrease intelligibility. It is possible to speak louder but this effect is limited and can also lead to voice strain. The indoor ambient noise levels for different rooms specified in Table 1.1 have been chosen to ensure that an adequate signal to noise ratio can be achieved without undue strain on the teacher's voice, and to minimise the effects of distraction from other sources.

4.3 Reverberation times

A classroom with a long reverberation time of several seconds will cause syllables to be prolonged so that they overlap and hence degrade speech intelligibility. Long reverberation times occur in large rooms with hard wall and ceiling surfaces. Adding acoustic absorption and reducing the ceiling height will reduce the reverberation time and will improve speech intelligibility. Table 1.5 specifies the reverberation times required for various teaching spaces ranging from teaching classrooms to assembly halls.

Appendix 2 describes the theory and terminology of reverberation time and acoustic absorption. The method for calculation of reverberation time in rooms other than corridors, entrance halls and stairwells is described in Appendix 6, together with a worked example. There is a link to a façade sound insulation and reverberation time computer spreadsheet for schools from the DfES acoustics website.

Long reverberation times also increase reverberant noise levels within a room, which further decrease speech intelligibility. To compensate for this, in reverberant rooms people tend to increase their voice levels to make themselves heard over the reverberant noise, which further exacerbates the situation. This is a common feature of many school dining rooms and sports halls.

4.4 Amount of acoustic absorption required

The method described in Appendix 6 allows the amount and type of acoustic absorption to be calculated.

In general, in rooms for music performance, the reverberation time calculations will show that relatively little absorption is required in addition to that provided by the audience.

In classrooms and other rooms for speech, however, larger amounts of fixed acoustic absorption are often required, particularly where rooms have high ceilings as often occurs in older buildings.

When calculating reverberation times to comply with the specified values in Table 1.5 in rooms for speech, the absorption due to furnishings such as chairs, school desks and benches, may be ignored. Accurate absorption data for such items can be difficult to identify and if the furnishings have any effect it is likely to result in shorter, rather than longer, reverberation times.

4.5 Distribution of absorbent materials

The location of acoustic absorption within a room is important. The traditional calculation of reverberation time assumes that the absorbent surfaces in a room are reasonably evenly distributed. If this is not so, the reverberation time equation is not valid and undesirable local variations in the acoustics can occur, particularly in large rooms or halls. Large areas of acoustically reflective material can also lead to echoes, focusing and standing waves.

4.6 Room geometry

To achieve adequate loudness for all listeners in a room, it is necessary that the direct sound from speaker to listener has a clear unobstructed path. The loudness of the direct sound can be enhanced by strong, short delay reflections from room surfaces. These short delay reflections should arrive at the listener within one twentieth of a second (50 milliseconds) of the direct sound, which is approximately the time required for the ear to integrate such reflections with the direct sound. Strong reflections after 50 milliseconds tend to be detrimental to speech intelligibility, and ultimately, if the delay is long enough, they will be perceived as distinct echoes.

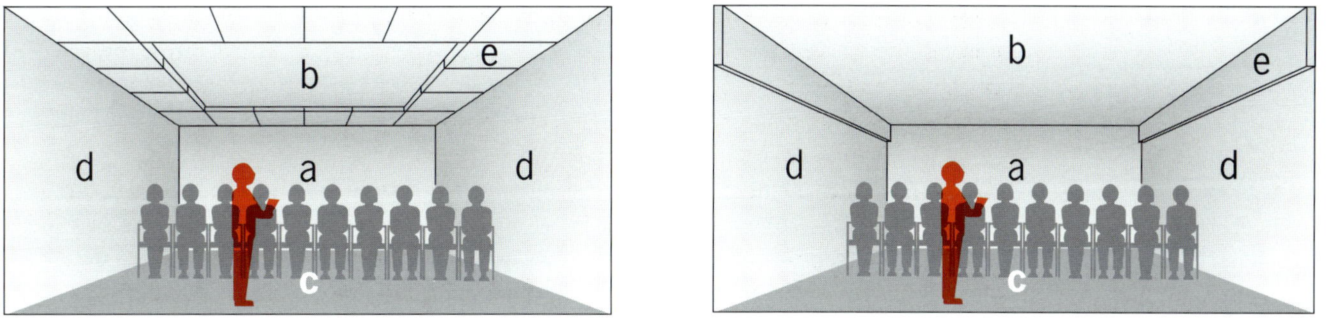

(a) Surface finishes in classroom or lecture theatre:
a. Rear wall - sound absorbing or diffusing
b. Ceiling - sound reflective (eg plasterboard)
c. Floor - sound absorbing (eg carpet)
d. Walls - sound reflective
e. Ceiling - sound absorbing

(b) Surface finishes in classroom or lecture theatre:
a. Rear wall - sound absorbing or diffusing
b. Ceiling - sound reflective (eg plasterboard)
c. Floor - sound absorbing (eg carpet)
d. Walls - sound reflective
e. Top of walls - sound absorbing or diffusing

Figure 4.2: Surface finishes in classroom or lecture theatre

4.7 Classrooms

For classrooms and other rooms for speech, there are two approaches to locating the acoustic absorption:

1. To make the ceiling predominantly absorbent. In most cases a standard acoustically absorbent suspended ceiling will provide all of the necessary absorption. In the case of rooms with exposed concrete soffits (providing thermal mass to limit overheating in summertime) acoustically absorbent suspended baffles may be used. The ideal case is often to have the central part of the ceiling reflective with absorption at the edges, see Figure 4.2(a).

2. To leave the ceiling acoustically reflective (plaster, plasterboard, concrete, etc) and to add acoustic absorption to the walls. In these cases it is advisable to locate most of the absorption at high level and some on the back wall facing the teacher to prevent 'slap echo' off the back wall. This is particularly important if the rear wall is concave or the distance from the speaker to the rear wall is greater than 8.5 m, see Figure 4.2(b).

In large rooms, reflections from the rear wall can be disturbing for a speaker if they arrive later than 50 milliseconds after the speech has been voiced. This can occur if the speaker to rear wall distance is greater than 8.5 m. To avoid this problem, the rear wall should be made acoustically absorbent, or acoustically diffusing.

There are instances where provision of sound field amplification can improve speech intelligibility, see Section 6.

4.8 Assembly halls, auditoria and lecture theatres

Most school halls are used primarily for speech functions such as assemblies, meetings and drama, and use for music is less frequent. The most common problem in school halls is excessive reverberation resulting in high noise levels and poor speech intelligibility.

4.8.1 Room geometry

The direct sound from speaker to listener must be as strong as possible at all positions. Because this sound weakens rapidly with distance according to the inverse square law (the intensity is reduced by a factor of four and the sound level falls by 6 dB when the speaker to receiver distance is doubled), the average distance between speaker and listener should be kept as small as possible. Furthermore, there should be no obstructions along the direct sound path.

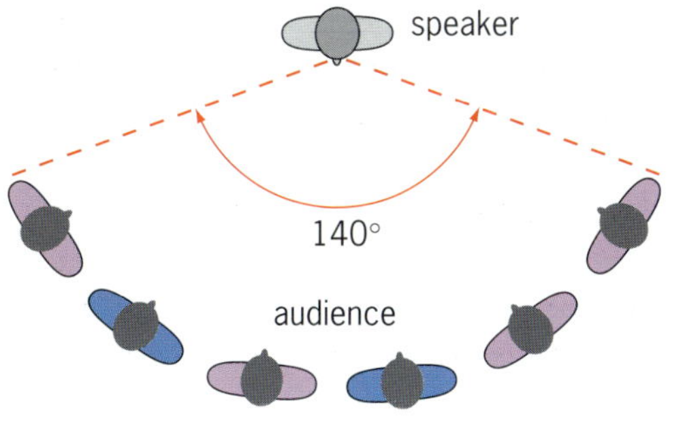

Figure 4.3: Ideal seating plan

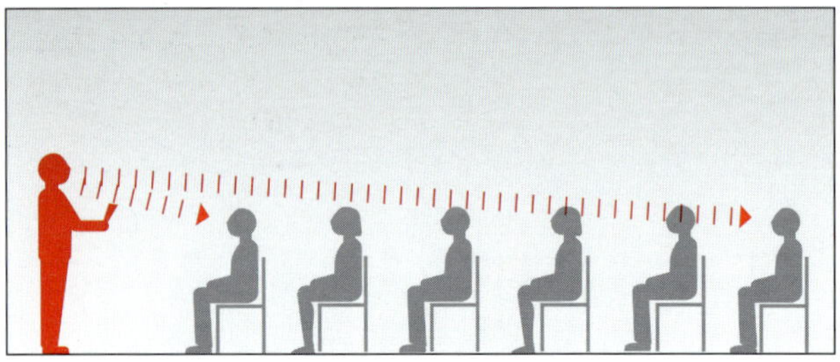

(a) Adequate loudness is essential, direct sound must have a clear unobstructed path.

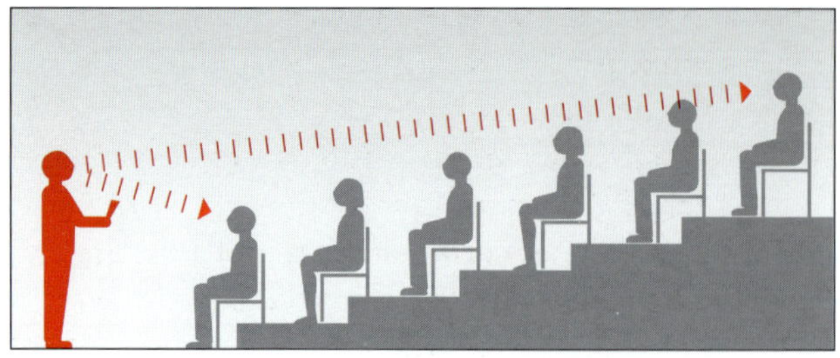

(b) Loudness of direct sound towards rear is increased with raked seating.

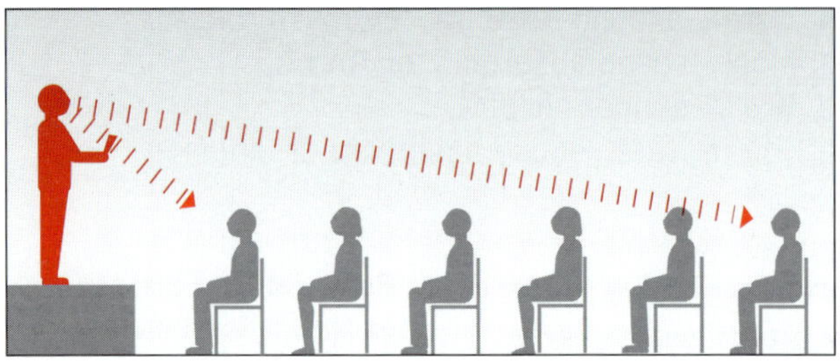

(c) Loudness of direct sound can be increased by putting the speaker on a platform.

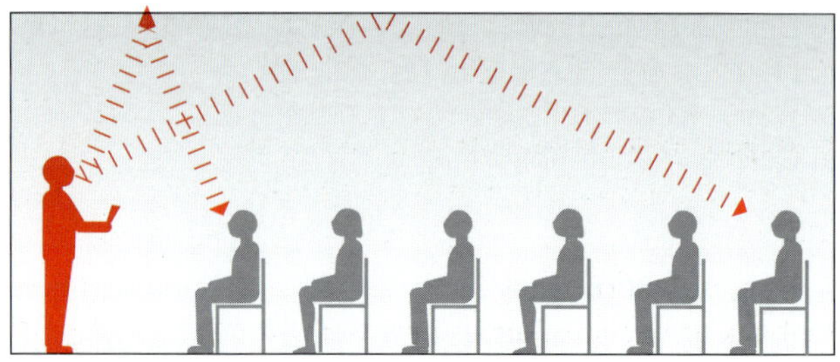

(d) Reflected sound enhances direct sound if time delay is less than 50 milliseconds.

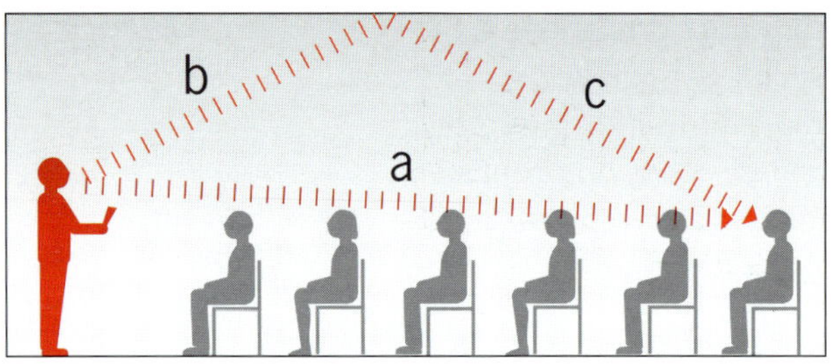

(e) For useful sound reflections, additional path travelled by reflected sound must be less than 17 m: b+c – a<17 m.

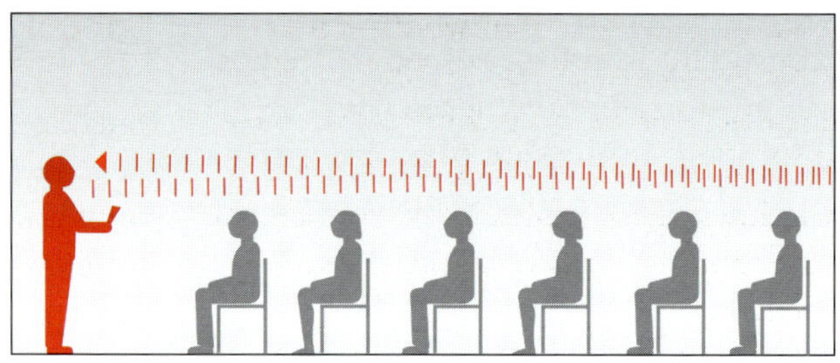

(f) Rear wall can cause a disturbing echo for speakers if over 8.5 m away. Rear wall should be absorbing or diffusing.

Figure 4.4: Effects of room geometry on speech.

For large rooms such as school halls, additional factors need to be considered in relation to the direct sound. First, the seating plan should be arranged to fall within an angle of about 140° subtended at the position of the speaker, see Figure 4.3. This is because speech is directional, and the power of the higher frequencies on which intelligibility largely depends falls off fairly rapidly outside this angle. Secondly, sound is weakened as it passes over seated people at grazing incidence. Therefore, if possible, listeners should be seated on a rake where a clearance of around 100 mm is provided between the sightline from one row and the sightline from the next, see Figures 4.4(a) and 4.4(b). It is known that if people can not see the speaker well, they will not hear well either. It is frequently necessary in schools to have a flat floor in a school hall. In these cases, speakers should be raised on a platform which is sufficiently high to ensure that minimum clearance is obtained at the rear rows of the hall, see Figure 4.4(c).

The direct sound from speaker to listener can be enhanced by strong early reflections that arrive within 50 milliseconds, see Figure 4.4(d). These early reflections increase the loudness of the direct sound and therefore increase speech intelligibility. They are particularly useful at the furthest seats where the loudness of the direct sound has been reduced by distance. To provide reflections within 50 milliseconds of the direct sound, hard surfaces must be located within a certain distance of the speaker and listener. In most rooms, the centre part of the ceiling is the most important reflecting surface and should be of hard, sound-reflecting material. Other useful surfaces providing early reflections are side walls near the speaker, overhead reflecting panels and angled ceiling panels.

The additional path travelled by the reflected sound should be no greater than 17 m more than the direct sound path between speaker and the seating area where the reflection arrives, see Figure 4.4(e).

Any reflection that arrives at a listener, or back at the speaker, more than 50 milliseconds after the direct sound is likely to be disturbing, see Figure 4.4(f). These are most probable in school halls where late reflections can occur from the rear wall or a control room window at the rear. Rear walls can be rendered sound absorbing or sound diffusing to avoid this. In the case of control room windows, these can be tilted to direct the reflection away from speakers and listeners.

Focusing of sound by domes or barrel vaults illustrated in Figure 4.5(a), can be a serious fault which can cause strong late reflections or echoes. If the dome or barrel vault is above a flat, hard floor as in a school hall, flutter echoes can occur which can be disturbing for speaker and listener alike. This effect can also occur with shallow pitched reflective roofs above a flat floor, see Figure 4.5(b) and the assembly hall in Case Study 7.1. The same effect can also occur on plan where a room has a curved or segmented rear wall opposite a flat front wall, see Figure 4.5(c).

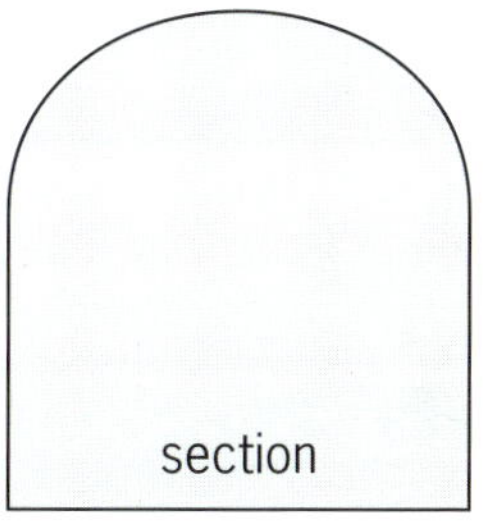

(a) Barrel vault can cause focusing and flutter echoes.

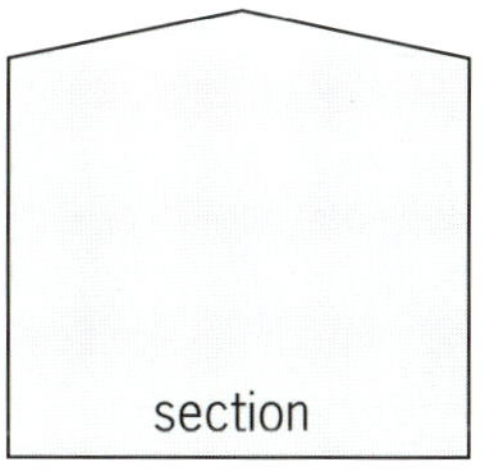

(b) Shallow hipped roof can cause focusing and flutter echoes.

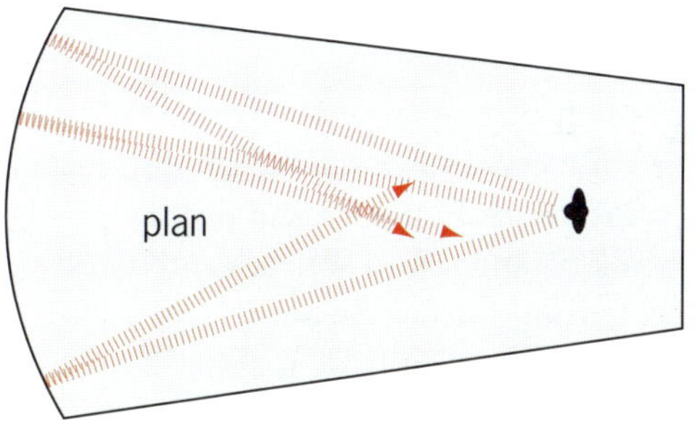

(c) Curved rear wall can cause focusing.

Figure 4.5: Room shapes which can cause focusing and echoes

4.8.2 Sound reinforcement

With an acoustically well designed room it is possible for strong speakers to achieve good speech intelligibility for large audiences. Quieter and untrained speakers, however, will not be able to do this and a speech reinforcement system is likely to be required for some functions.

The key aim of such a system is to increase the loudness of the direct sound, particularly for more distant listeners, whilst keeping the sound as natural as possible.

The distribution of loudspeakers and their directional characteristics is a key issue in achieving high speech intelligibility. For large teaching rooms and lecture theatres, loudspeakers can be distributed in the ceiling or on the walls at high level. In school halls, column loudspeakers can be located on sidewalls, or in a central cluster as shown in Figure 4.6.

The design of sound reinforcement systems is a specialist field and specialist advice should be sought.

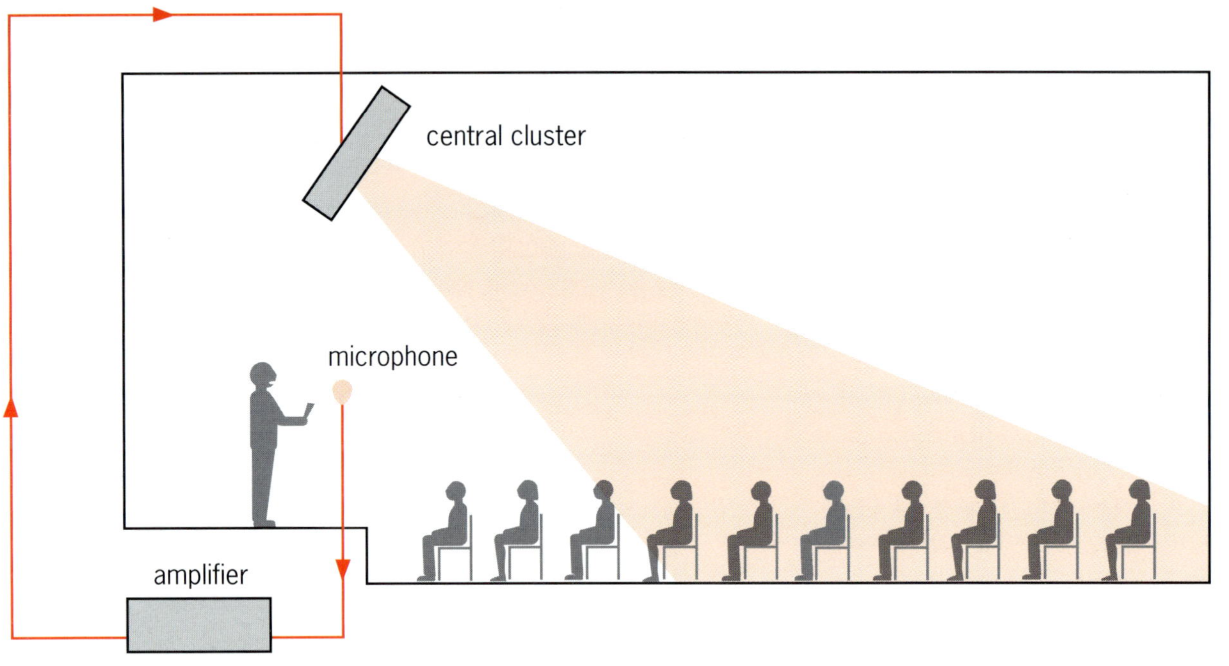

Figure 4.6: An arrangement of loudspeakers in a school hall

4.9 Open-plan teaching and learning areas

In open-plan areas it is essential to provide good speech intelligibility and to secure freedom from aural distraction by more distant sound sources and by background noise. Section 1 contains performance standards for speech intelligibility in open-plan spaces. Some degree of acoustic privacy is also desirable. This can be difficult to achieve in practice and there have been many instances of distraction and disturbance between class groups in open-plan areas. Case Studies 7.2, 7.3 and 7.10 describe surveys of the acoustics of open-plan teaching areas in primary and secondary schools.

In open-plan areas, a carpeted floor is recommended together with a sound absorbing ceiling. In addition, sound absorbing screens should be interposed between class groups. Screens should be at least 1.7 m high and ideally should reach to within 0.5 m of the ceiling, see Figure 4.7.

A major improvement in the acoustic privacy between spaces in open-plan areas can be realised by installing full height moveable walls which, if fitted with seals, can provide a moderate degree of sound insulation between the divided spaces. In general however it is found that such screens are rarely used because of the time and effort required to open and close them. While in theory it is possible to achieve adequate sound insulation between classrooms using high-performance moveable walls, there are issues of cost, weight, complexity of installation and maintenance to consider. Specialist advice from an independent consultant should always be sought when using such partitions to comply with the sound insulation requirements set out in Section 1.

Research has shown that in many large open-plan 'flexible' areas certain activities are severely restricted or have to be dropped because of noise interference. Indeed, it must be recognised that there are but a small number of activities that can share a degree of acoustic linkage and even then the timetable has to be designed to allow this.

Those plans which provide a generous range of spaces in a variety of sizes can be seen to give far more opportunities in teaching than those with large open spaces and moveable screens, because in the former it is possible to achieve good sound insulation standards between spaces.

When designing open-plan areas it is important to provide plenty of acoustically absorbent surfaces and to use screens to block direct sound paths.

4.10 Practical spaces

Spaces for teaching practical subjects have particular requirements which need careful design in order to comply with the acoustic requirements for teaching and learning. This section addresses the needs of Design and Technology spaces and Art rooms. Music rooms are considered separately in Section 5. Although Science involves a significant amount of practical activity, the general approach described for classrooms (Section 4.7) can be applied to spaces for the teaching of Science. For further information on Design and Technology spaces see Building Bulletin 81[1] and the DfES acoustics website.

4.10.1 Design and Technology spaces

Design and Technology departments in secondary schools contain timetabled spaces for a variety of practical activities, eg graphics, resistant materials (wood, metal and plastics), electronics/control, food and textiles. They also include non-timetabled learning spaces, typically shared design/ ICT resource areas.

The equipment and the activities in these spaces can vary widely depending on the type and size of department. Activity noise and noise tolerance classifications for different spaces are given in Table 1.1. It is important to establish what activities will take place in any one space and what equipment will be used before calculating required levels of sound insulation to minimise the background noise in nearby spaces.

Resistant materials areas containing wood or metal working machinery can produce high noise levels. Machines extracting dust particles, CADCAM and other noisy equipment will increase the activity noise level of a space. It is important to consider the effects of such equipment on teaching activities both within the space containing the equipment and in adjoining rooms. Where possible, it is advisable to locate noisy equipment in spaces away from rooms housing quieter activities. CADCAM equipment is sometimes housed in a separate room or within purpose designed enclosures which can reduce the noise level.

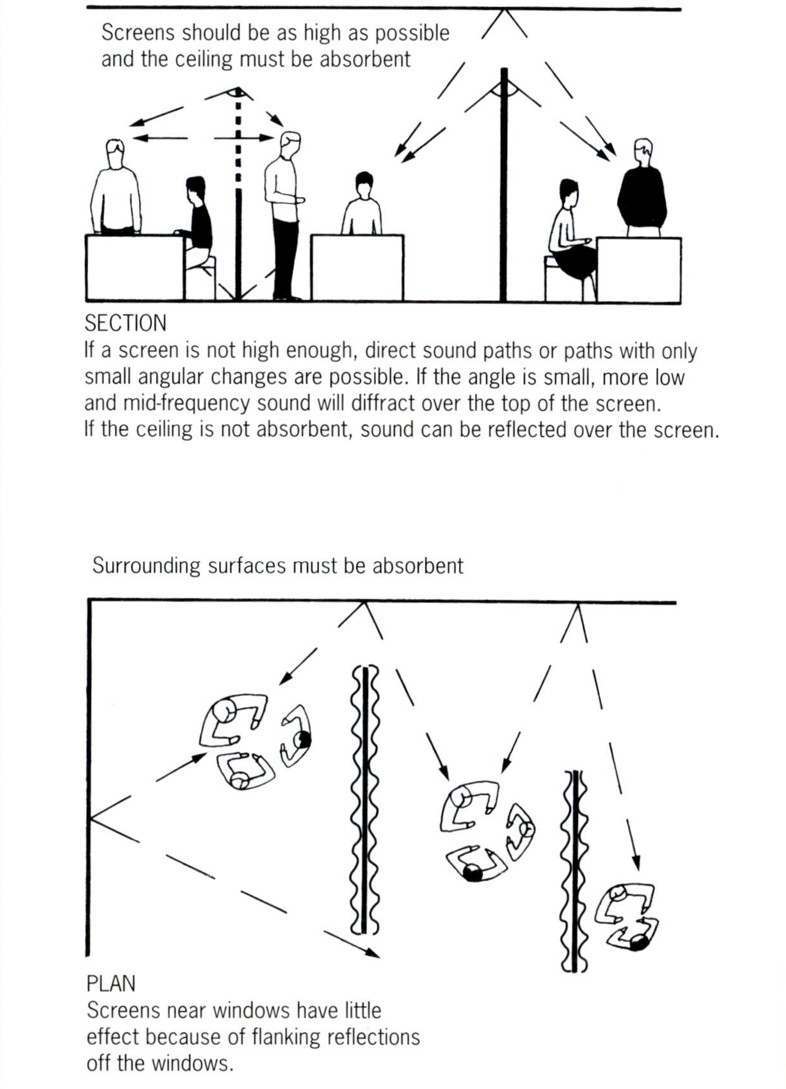

Figure 4.7: Positioning of indoor screens

The effect of noise from machines within the space is not required to be included in the indoor ambient noise level calculations submitted for approval by Building Control Bodies, except in the case of open-plan arrangements. Often machines can be switched off when quieter learning activities such as group presentations are taking place, but this may not always be possible. The location and use of noisy equipment needs to be discussed and agreed with the user.

Partially glazed partitions have commonly been used between design and technology spaces, particularly between timetabled and non-timetabled spaces. This is both for ease of supervision and to emphasise the link between related design activities. However, these considerations must be balanced against the acoustic requirements. Large areas of glazing will both increase the reverberation time within a space and reduce the sound insulation of a partition. Both of these factors will have a detrimental effect upon the speech intelligibility within the space and other nearby spaces.

Similarly, if interconnecting doors are used between neighbouring rooms the doorsets must be chosen to provide adequate sound insulation.

Central resource areas are often located adjoining the circulation spaces of design and technology departments. A common arrangement uses the central resource area predominantly for individual and small group work but such areas are not generally suitable for whole class teaching. Usually, there are areas of glazing and doors between the central resource area and adjacent practical spaces. The central area should be suitable for most design/resource activities as long as the circulation is restricted to the department and does not include access to other parts of the school.

Where spaces are open-plan or divided by moveable or extensively glazed partitions, it may be appropriate to adopt alternative acoustic performance standards in accordance with Section 1.2.1. This will need to be based on an activity plan for the area which has been agreed with the user.

The speech intelligibility in open-plan spaces will need to be assessed using computer prediction models, as described in Section 1.1.7. This may apply to a shared design/ICT resource area where group presentation could take place at the same time as other activities.

4.10.2 Art rooms

Art classes in secondary schools involve independent and group activities which are in general quieter than those in other practical areas. Noise levels in secondary school art spaces are likely to be similar to those in a general classroom. However, art departments tend to have a more informal environment reflecting the nature of the activity, and are often of open-plan design. There may be more glazing in partitions in art departments than in other parts of the school, to emphasise the importance of the visual environment.

4.10.3 Floor finishes in practical spaces

Carpets are not appropriate for most practical areas and so cannot be used as a way of increasing sound absorption or reducing the impact sound transmission through floors in science, art and design and technology spaces. They may however be suitable in some design/resource areas.

Acoustic vinyl flooring or a vinyl floor laid on top of an acoustic mat may be suitable for practical spaces where improved impact sound insulation is required. Resistance to indentation will need to be considered and a change of flooring may be necessary underneath fixed heavy machinery such as floor mounted machine tools.

4.11 Drama rooms

There are three types of drama room in common use:

1 Rooms for small scale drama teaching and practical work
2 Drama studios – for rehearsal, teaching and small-scale performance
3 Theatres and flexible spaces primarily for performance.

Rooms for small scale drama teaching and practical work are usually little more than classrooms, which may be fitted with curtains both for blackout and to reduce reverberation time. They may also be provided with a basic set of lights and dimmable main lighting.

Drama studios tend to be larger spaces dedicated to drama, with special equipment such as moveable staging, seating rostra, lighting and sound systems. They do not normally have fixed stages or platforms and the acoustics will tend to change with the layout, seating and audience. They may be fitted with heavy curtains on some or all walls, to allow some control of reverberation time, for blackout, and to allow some flexibility in the room's appearance. In this case the wall finishes will generally be hard (masonry or plasterboard). Studios usually have wooden floors and acoustically absorbent ceilings, although large amounts of permanent lighting and rigging also provide useful diffusion.

Theatres and spaces primarily for performance vary considerably in form and size from the conventional assembly hall to adaptable theatres. They can be traditional theatres with fixed proscenium and stage, open stages, thrust stages or in the round, see Figure 4.8. Adaptable theatres can be converted from one arrangement to another depending on the type of performance.

Each type has different acoustic characteristics. The basic acoustic requirements for auditoria are discussed in Section 4.8, however spaces designed specifically for public performance are specialised rooms and the advice both of an acoustician and a theatre consultant should normally be sought.

For successful drama it is necessary for the audience to see and hear considerably better than in most school halls, because of the close relationship between actors and their audience. In principle, to achieve close communication between actor and the audience it is necessary to restrict the size of the auditorium so that the maximum distance from any member of the audience to the stage does not exceed 20 m. In small theatres this is not generally a problem, but for larger audiences it may require the use of balconies and galleries, giving rise to the traditional fan-shaped theatre (which is, however, very bad acoustically for music). Deep balconies are to be avoided as the space under these can be acoustically 'dead' and considerable care is required to ensure that reflections from the ceilings and walls compensate for the lack of direct sound in such areas.

It is common for theatres in schools to be used not only for drama, but also for lectures, films, meetings and music, which all have different acoustic requirements. The acoustics of multi-purpose halls are discussed in the following section.

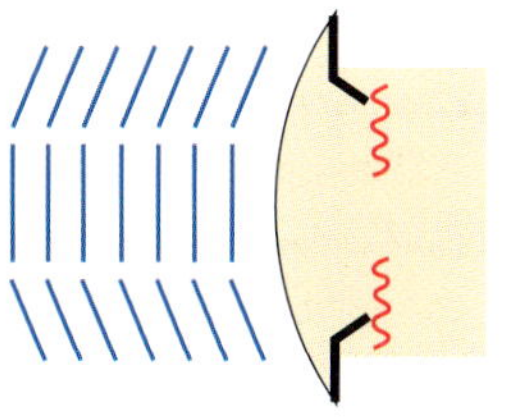

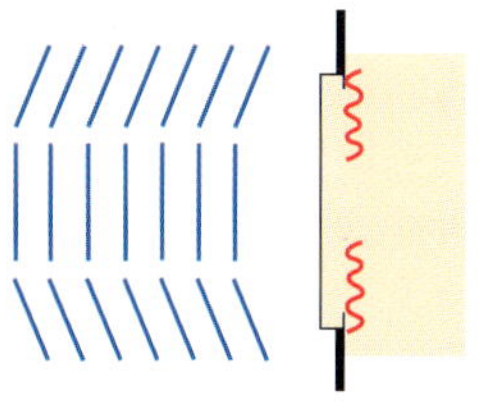

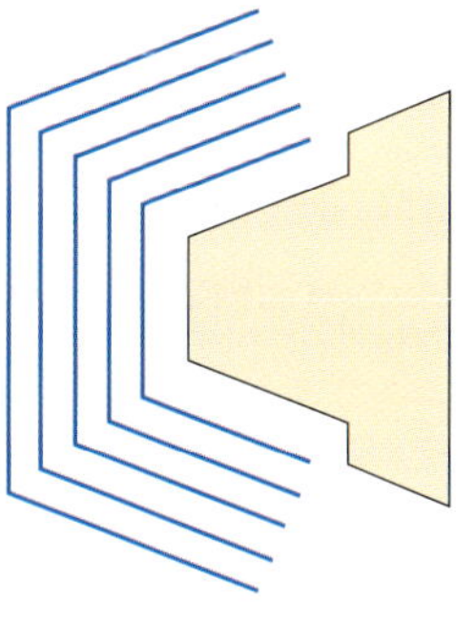

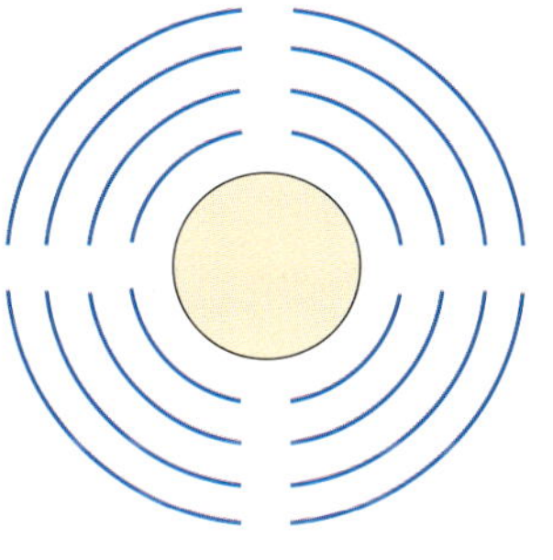

Figure 4.8: Typical performance spaces for drama

4.12 Multi-purpose halls

In large schools the multi-purpose space, intended to act as assembly hall, theatre, concert hall and gymnasium, is passing out of favour as it is difficult for a single hall to fulfil all of these functions well. None the less, in some cases a single flexible hall is required for a variety of uses and this gives rise to specific acoustic problems.

The different uses of multi-purpose halls often have conflicting acoustic requirements, making it difficult to provide a space with optimum acoustics for all uses. The main conflict is that between speech and unamplified music.

Speech	Music
'Dry' acoustic	'Live' or 'warm' acoustic
Short reverberation time	Long reverberation time
Good clarity, loudness and intelligibility of speech	Even decay of sound
Sound must appear to come from stage with some contribution from room reflections but no perceptible reverberation	Good 'envelopment' – audience should feel surrounded by the sound, and musicians should be able to hear themselves and each other easily
Small volume	Large volume

Table 4.1: General acoustic requirements for speech and music

Table 4.1 shows the general acoustic requirements for speech and music. (See also Section 5.7.)

Where regular performances of music are expected, reverberation time is sometimes changed using moveable areas of absorption (typically curtains) without changing the volume of the space. Although this can successfully change the reverberation time at medium and high frequencies, it often has little effect at low frequencies, resulting in an acoustic which is less than ideal for either speech or music.

(Note that the 'dry' acoustic required for speech is also generally suitable for amplified music.)

Further information regarding the design of multi-purpose auditoria is given in Section 5.

4.13 Other large spaces

Sports halls, gymnasia and especially swimming pools have long reverberation times through the nature of their construction and surfaces necessary to their function. This results in high noise levels and poor speech intelligibility.

A variety of relatively rigid, robust and hygienic, acoustically absorbent materials are available and can be used. In general, these materials are installed on ceilings and at high level on walls or as hanging baffles. If there are large areas of acoustically hard parallel surfaces, flutter echoes can occur, significantly increasing the reverberation time and reducing speech intelligibility. A reasonable distribution of acoustic absorption or diffusion (such as provided by wallbars against gymnasium walls) will eliminate this effect.

4.14 Dining areas

Dining areas suffer from excessive activity noise. The high activity noise interferes with conversation leading to increasing noise levels. Therefore, sound absorption is required in these areas to reduce the reverberant noise level. The most practical place to position sound absorption is on the ceiling and the walls. Shapes in section or on plan that produce focusing, such as barrel vaulted roofs and circular walls, should be avoided unless treated with sound absorbent material.

References

[1] Building Bulletin 81, Design and Technology Accommodation in Secondary schools, to be published January 2004 (replacing 1986 edition).

The design of rooms for music

SECTION 5

Music rooms require special attention in the acoustic design of a school. It is important to establish the user's expectations of the acoustic performance of the spaces. Musical activities range from playing, listening and composing in group rooms to orchestral performances in school halls, and a music room can be anything from a small practice room to a large room for rehearsing and performing music.

5.1 Aspects of acoustic design

Building Bulletin 86 Music Accommodation in Secondary Schools[1] gives detailed design advice on the range of types of music spaces found in schools. The performance standards of the most common music room types are listed in the tables in Section 1.

Some non-specialist classrooms may be used for teaching music theory to large groups, with only occasional live or recorded music. In these rooms the majority of activity depends on good speech intelligibility rather than an enhanced acoustic for music and in these cases classrooms with the same acoustic criteria as normal classrooms may be used.

A brief, outlining the client's acoustic requirements, should be obtained before starting the design of any specialist music facility. The main problems are noise transfer between spaces, unsuitable reverberation times, flutter echoes, standing waves, and high noise levels.

5.2 Ambient noise

The requirements for indoor ambient noise levels in music rooms are set out in Table 1.1. To control noise from mechanical ventilation, it is important to select quiet fans or air handling units which are connected to appropriately sized silencers (attenuators). Typical primary attenuator lengths will be in the range 2.4 to 3.0 m. Air velocities in the duct system should be kept low and should not generally exceed 5 m/s in main ducts, 4.5 m/s in branch ducts and 2.5 m/s at runouts. Terminal units (grilles etc) should be selected for low noise output.

Noise from hot water radiator systems should be minimised by good design. Equipment, particularly the valves and pumps, should be designed and selected for quiet operation, with vibration isolation where appropriate.

In noise-sensitive spaces, such as music performance spaces and recording spaces, hot water pipes should not come into rigid contact with the building construction. Resilient pipe brackets and flexible penetration details should be adopted to prevent clicking noises resulting from expansion and contraction.

Lighting can cause disturbing buzzing and occasionally sharp cracks from expansion or contraction of metal fittings. In music rooms, 50 Hz fluorescent lights should not be used because they are inherently prone to buzzing and mains hum which is audible to some people. These effects do not occur with high frequency (HF) fittings, which should in general be specified on energy efficiency and cost saving grounds. HF fittings are acceptable for most general music spaces. Where the quietest conditions are required, lighting should be restricted to tungsten or similar lamps. In certain spaces such as a recording/control room, the sound caused by transformers used with low voltage spotlights can be distracting.

5.3 Sound insulation

Standards for sound insulation between different types of room are given in Table 1.2. To avoid excessive noise transfer between music rooms Table 1.2 specifies a minimum of 55 dB $D_{nT(T_{mf,max}),w}$ between most music rooms. These are minimum requirements and will not always prevent interference between adjacent rooms. It is beneficial to increase these figures, especially when the indoor ambient noise level is significantly below the level in Table 1.1. This can occur in naturally ventilated rooms on quiet sites where the indoor ambient noise level is too low to provide useful masking of distracting noise from adjacent rooms.

The level of sound and possible disturbance between music spaces will vary depending on the instruments being played. Clearly, as the loudness of the instruments varies from group to group, so the room-to-room sound insulation requirement will also vary. An important question is that of cost versus flexibility. High flexibility is desirable so that any instrument can occupy any room. However it is expensive to provide sound insulation to satisfy the most stringent requirement at all locations throughout the building. Alternatively, designating groups of rooms to groups of instruments severely limits flexibility but concentrates investment in sound insulation where it is most required.

Rooms for percussion and brass will generate high noise levels and great care is needed in choosing their location. Rooms for percussion should, if possible, be located at ground level to minimise the transmission of impact vibration into the building structure. Otherwise floating floor constructions may be required.

Figures 2.4 and 7.5.1 illustrate the principles of good planning, using corridors and storage areas as 'buffer zones' between music rooms where possible. This allows the sound insulation requirements to be met without resorting to very high performance constructions. However, in some cases such as refurbishments of existing buildings the provision of special sound insulating constructions, as discussed in Section 3, is the only option.

Background noise must be controlled in circulation areas. However, limited break-out of musical sounds into circulation routes is acceptable since it allows teachers to monitor, from a distance, unsupervised small group musical activities.

5.3.1 Sound insulation between music rooms

The sound insulation required between the different types of music room can be determined from Tables 1.1 and 1.2.

Other criteria such as those of Miller[2], which take account of both sound insulation and indoor ambient noise level, are sometimes used in the specification of sound insulation between music rooms; however, the normal way of satisfying Requirement E4 of The Building Regulations is to meet the performance standards in Table 1.2 for airborne sound insulation between rooms.

Case Study 7.5 gives an example of the acoustic design of a purpose built music suite in a secondary school.

5.4 Room acoustics

5.4.1 Reverberation time, loudness and room volume

In general, rooms for the performance of non-amplified music require longer reverberation times than rooms for speech. Figure 5.1 shows optimum mid frequency reverberation times for speech and music as a function of room volume.

The volume of a room has a direct effect on the reverberation time (RT) and early decay time; in general, the larger the volume, the longer the RT. The reverberation times should be in the ranges given in Table 1.5 and should be constant over the mid to high frequency range. An increase of up to 50% is permissible, and indeed is preferred, at low frequencies as indicated in Figure 5.2.

To achieve this it is generally necessary for the volume of music rooms to be greater than for normal classrooms and this generally requires higher ceilings. These also help with the distribution of room modes as described in the section on room geometry below.

If the volume of a room is too small, even with the correct reverberation time the sound will be very loud. This is a common problem in small practice rooms with insufficient acoustic absorption, and can give rise to sound levels which could, in the long term, lead to hearing damage. Many professional orchestral musicians have noise-induced hearing loss due to extended exposure to high noise levels both from their own instruments and, to a lesser extent, from other instruments nearby. Under the Noise at Work Regulations 1989 (see Appendix 9) there is a general requirement to minimise noise exposure of employees in the school context, who for this purpose include full-time, part-time and freelance peripatetic music teachers. It is therefore important to ensure that practice, rehearsal and teaching rooms are neither excessively reverberant nor excessively small for a given occupancy.

Setting the floor area and ceiling height is normally the first step in designing a music room. The floor area is usually determined by the number of occupants and guidelines are given in Building Bulletin 86[1], as are methods of curriculum analysis to determine the needs of a secondary school music department. A typical suite of music rooms in a secondary school might consist of:

Large performance/teaching room	85 m^2
Second teaching room	65 m^2
Ensemble room	20 m^2
Practice/group rooms	8 m^2
Control room for recording	10 m^2

Ceiling heights and consequently volumes for halls and recital rooms are generally equivalent to two storeys, around 6 m. For group rooms and practice rooms, a full storey height (at least 3 m) is normally required.

5.4.2 Distribution of acoustic absorption

The acoustically absorbent material required to achieve the correct RT should be distributed reasonably evenly about the room. Where absorption occurs only on the floor and ceiling – for example in a simple solution employing acoustic ceiling tiles and carpeted floor – users may experience an over-emphasis on sound reflections in a horizontal plane. This often leads to 'flutter echoes' between walls, which result in the actual RT being considerably longer than the calculated RT. A better solution, especially in large rooms, is to distribute some of the absorptive material about the walls.

Although the RT requirements in Table 1.5 are for unoccupied rooms, it is important to remember that the occupants will present a significant amount of absorption which will be in the lower half of the room. To give a reasonably even distribution of absorptive material therefore, acoustic absorption is

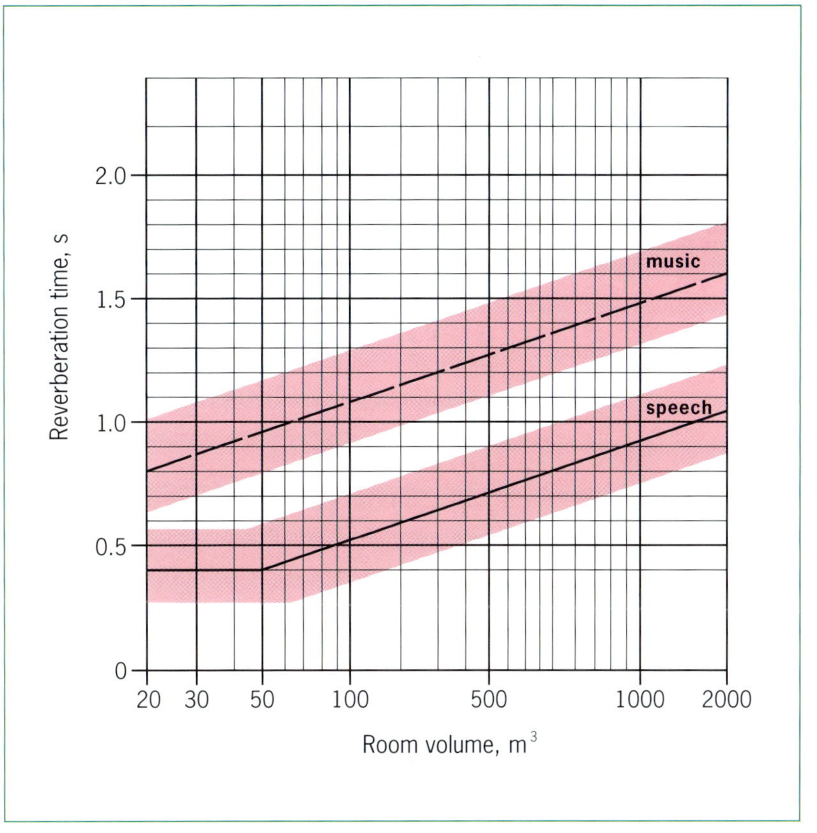

Figure 5.1: Optimum mid-frequency reverberation times for speech and music, for unoccupied spaces

Figure 5.2: Recommended percentage increase in reverberation times at lower frequencies for rooms specifically for music

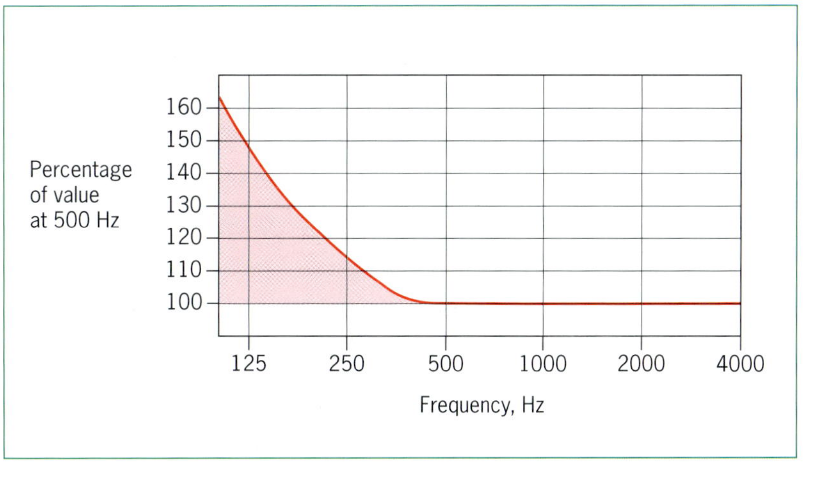

often located at high level on the walls.

Because of the absorption of the audience, there can be large variations in RT depending on the presence or absence of an audience. To reduce this effect, acoustically absorbent seats with upholstered backs can be used and in large halls the acoustic absorption of the seats has to be determined and specified quite carefully. An acceptable alternative in smaller halls can be the use of retractable curtains to reduce the RT during rehearsals when no audience is present.

In auditoria and music rooms, surfaces around and above the stage or performing area are normally reflective to provide feedback to the performers.

Floors on stage should be reflective although carpet in an auditorium may be permissible.

5.4.3 Room geometry

It is important to consider both room shape and proportion. In large rooms such as halls and recital rooms, the geometry of the room surfaces will determine the sequence of sound reflections arriving at the listener from a given sound source. Early reflections, that is those arriving within approximately 80 milliseconds of the direct sound, will be integrated by the listener's hearing system and will generally enhance the original sound for music (50 milliseconds is the corresponding figure for speech, see Section 4).

Prominent reflections with a longer delay (late reflections) may be perceived as disturbing echoes. This is often encountered where the rear wall in a hall has a large flat area of glass or masonry. Strong individual reflections can also lead to 'image shifting' where early reflections can be so strong that the ear perceives the sound as coming from the reflecting surface and not the sound source.

This problem can be exacerbated if late reflections are particularly strong. This can occur when sound is focused from large concave surfaces such as curved rear walls, barrel vaults, domes, etc. Furthermore, focusing results in an uneven distribution of sound throughout the room. Consequently, large concave surfaces are not generally recommended in music spaces.

In small rooms, such as group rooms and music practice rooms, geometry affects the distribution of standing waves or room modes throughout the sound spectrum, particularly at low frequencies. Where the distance between two parallel walls coincides with or is a multiple of a particular wavelength of sound, a standing wave can be set up and the balance of sound will be affected, see Figure 5.3. Certain notes will be amplified more than the rest leading to an unbalanced tonal sound, sometimes called colouration. Bathrooms with tiled walls are a good example of rooms with prominent room modes. Although such rooms can enhance certain notes of a singer's voice, they will not produce a balanced sound and will tend to sound boomy. The effect is exaggerated if distances are the same in more than one dimension. Thus rooms which are square, hexagonal or octagonal in plan should be avoided. The same effect occurs if the room width is the same as the room height, or is a simple multiple of it.

Ideally, the distribution and strength of room modes should be reasonably uniform. Perhaps the best way to control

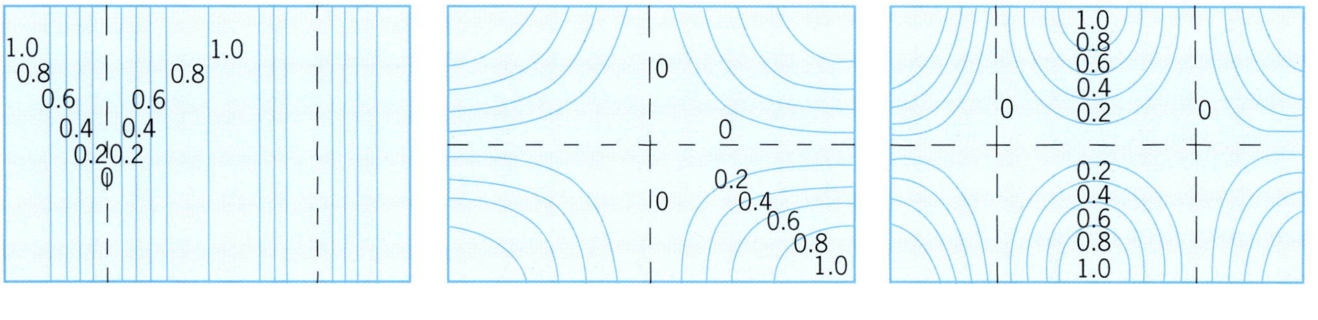

Figure 5.3: Standing waves in different modes
0 – No sound pressure
1.0 – Maximum sound pressure

these low frequency modes is to select room dimensions that are not in simple ratios. It should not be possible to express any of the room dimensional ratios as whole numbers, for example, a proposed space 7 m wide, 10.5 m long and 3.5 m high (2:3:1) would not be considered an advisable shape from an acoustic point of view. Mathematically, an ideal ratio is 1.25 : 1 : 1.6; (sometimes referred to as the 'golden ratio') but many other ratios work equally well.

Both flutter echoes and room modes can also be controlled by using non-parallel facing walls, but this is often impractical for architectural reasons; the use of absorption or diffusion is equally effective.

5.4.4 Diffusion

In addition to the correct RT, the room should be free from echoes, flutter echoes, and standing waves and the sound should be uniformly distributed throughout the room, both in the performance and listening areas. To achieve this without introducing too much absorption, it may be necessary to introduce diffusing hard surfaces to diffuse, or scatter, the sound. These are normally angled or convex curved surfaces but bookshelves, balcony fronts or other shapes can also provide diffusion, see Figure 5.4. Acoustic diffusion is a complex subject, and if calculation of diffusion is likely to be required a specialist should be consulted.

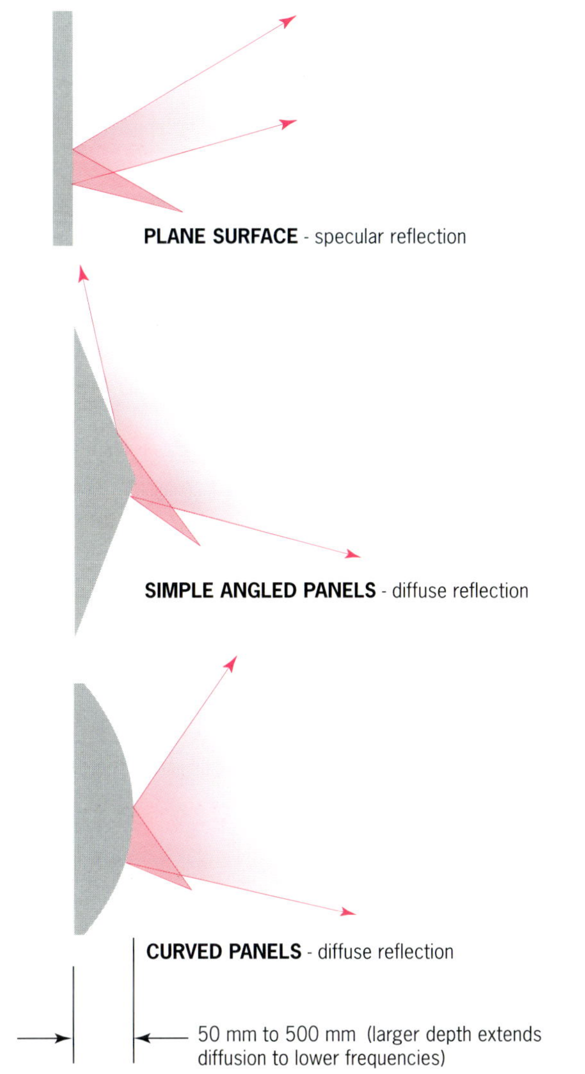

Figure 5.4: Surfaces which provide specular and diffuse reflections

5.5 Types of room

5.5.1 Music classrooms

Figure 5.5 shows a 65 m^2 music classroom for a range of class-based activities involving a number of different instruments. The room proportion avoids an exact square. The height is assumed to be between 2.7 m and 3.5 m, creating a reasonable volume for the activities (see Section 5.4.3). The main points to note about the acoustic treatment of the space are described below.

- To minimise the possibility of flutter echoes or standing waves occurring between opposing parallel walls, surfaces are modelled to promote sound diffusion. On the side wall this takes the form of shelving to store percussion instruments, etc. On the back wall, framed pinboards (with non-absorptive covering) are set at an angle, breaking up an otherwise plain surface.
- Full length heavy drapes along the back wall can be drawn across to vary the acoustics of the space.
- The observation window into the adjacent control room is detailed to ensure a high level of sound insulation between the two spaces (see Figure 5.6 and the discussion of control rooms below).
- The door into the room is of solid core construction with a small vision panel. The door and frame details, Figures 5.7

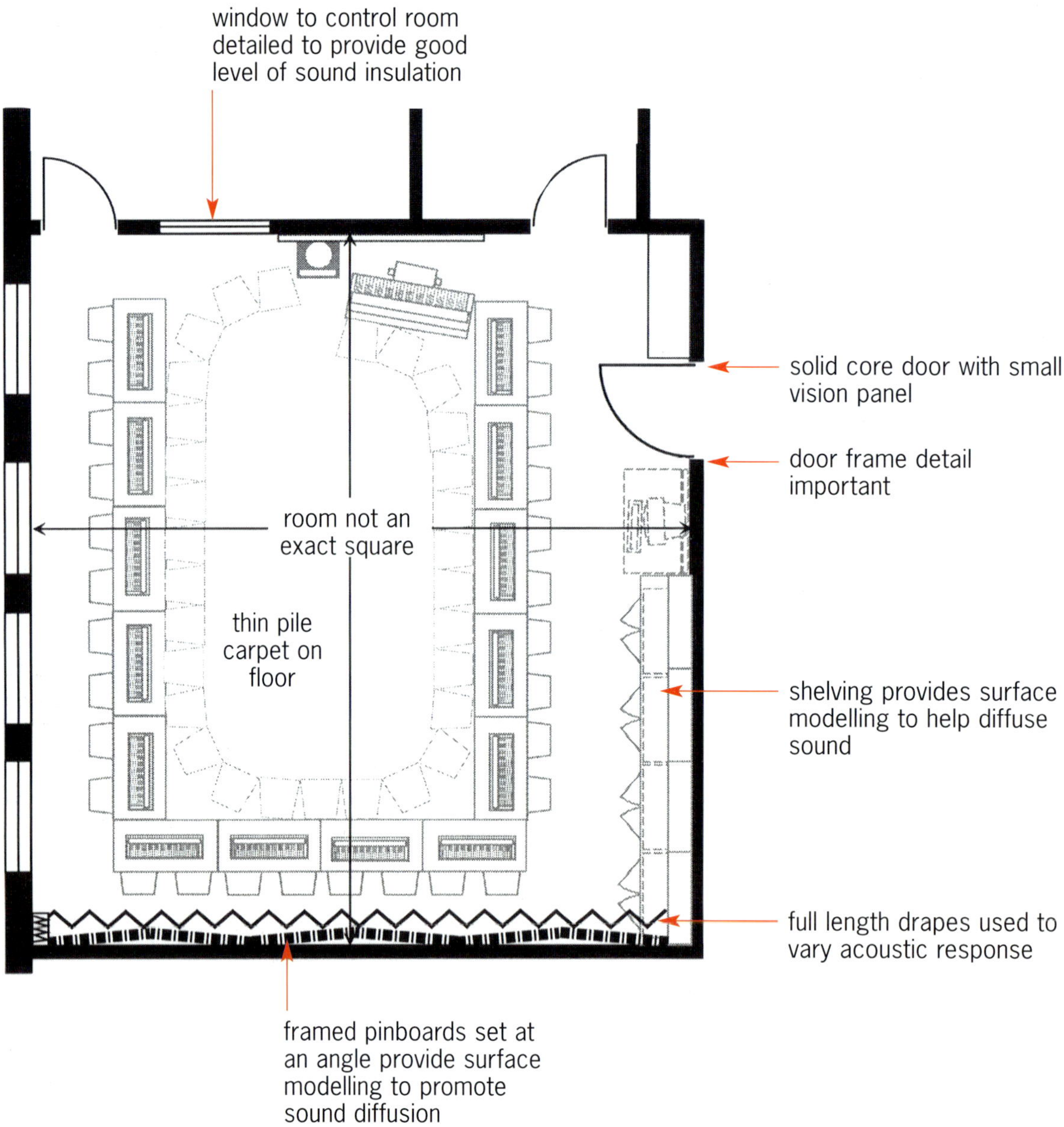

Figure 5.5: Acoustic treatment to 65 m^2 music classroom

and 5.8, are designed to maximise the sound insulation properties of the wall as a whole.

- The floor is fitted with a thin pile carpet providing an absorbent surface while the ceiling has a hard reflective surface. The type of carpet can have a significant effect on the overall RT in a room. It is worthwhile checking the precise absorption coefficient of any surface finish. (A spreadsheet of indicative absorption coefficients for common materials is on the DfES acoustics website.)

5.5.2 Music classroom/recital room

Figure 5.9 shows a larger, 85 m^2, classroom. The proportions of the room are in a ratio of fractional numbers (2.6 : 3.8 : 1) with the height between 2.7 m and 3.5 m as for the 65 m^2 music classroom. The acoustic treatment is similar to that for the 65 m^2 room but as this space is larger, and bigger groups are likely to rehearse and perform here, drapes are provided on two adjacent walls.

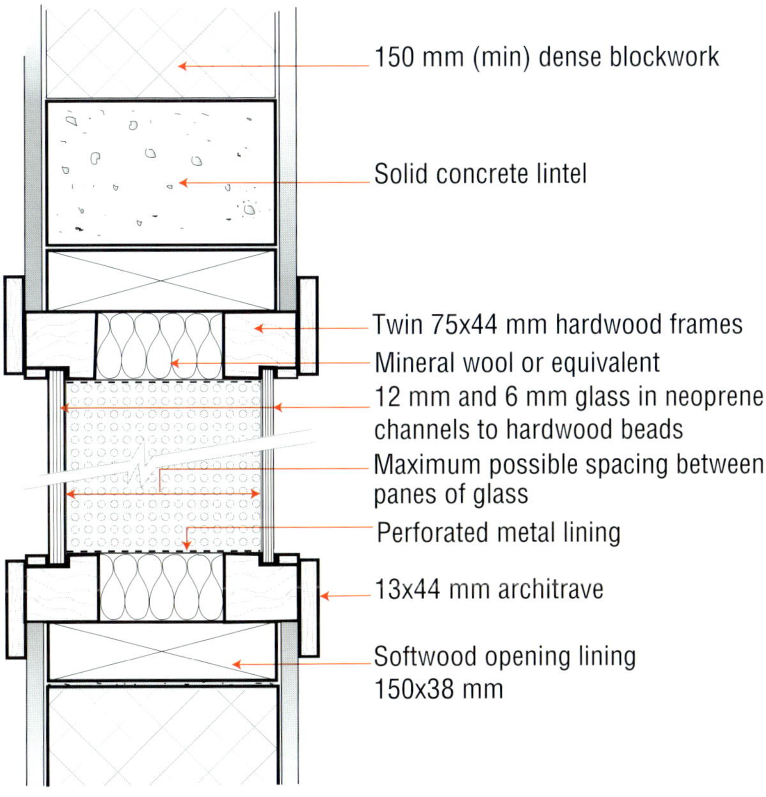

Figure 5.6: Section through control room window

Figure 5.7: Desirable features of an acoustic door installation

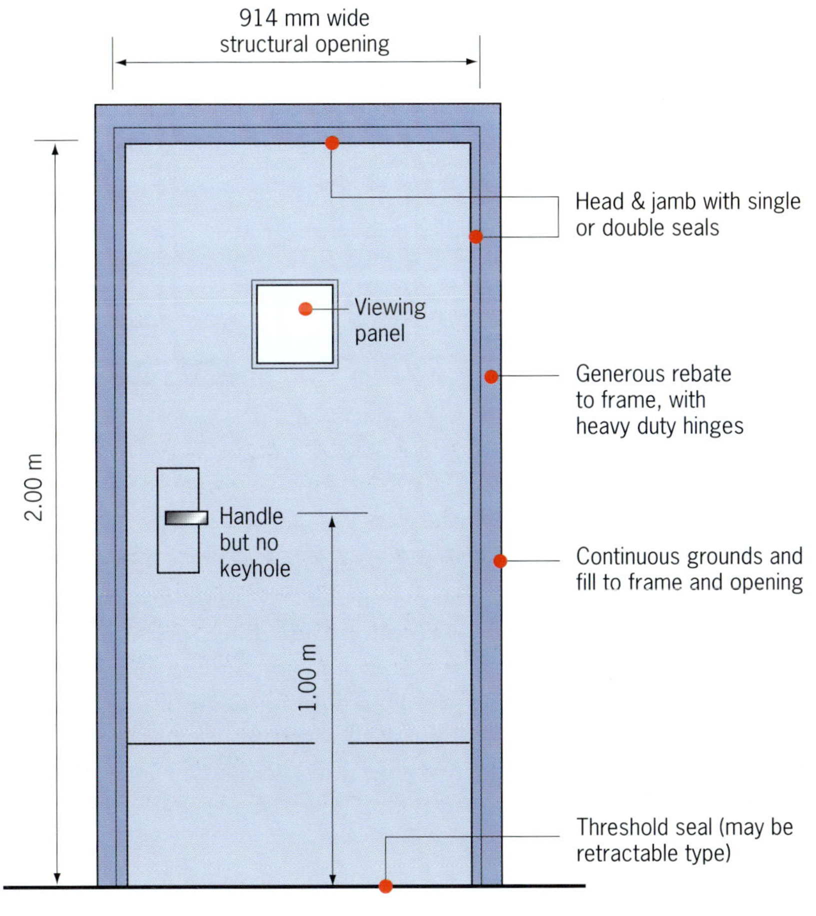

Solid core timber door assembly
Elevation showing how the performance of a door assembly depends on a number of features of the construction, not just on the mass of the door leaf

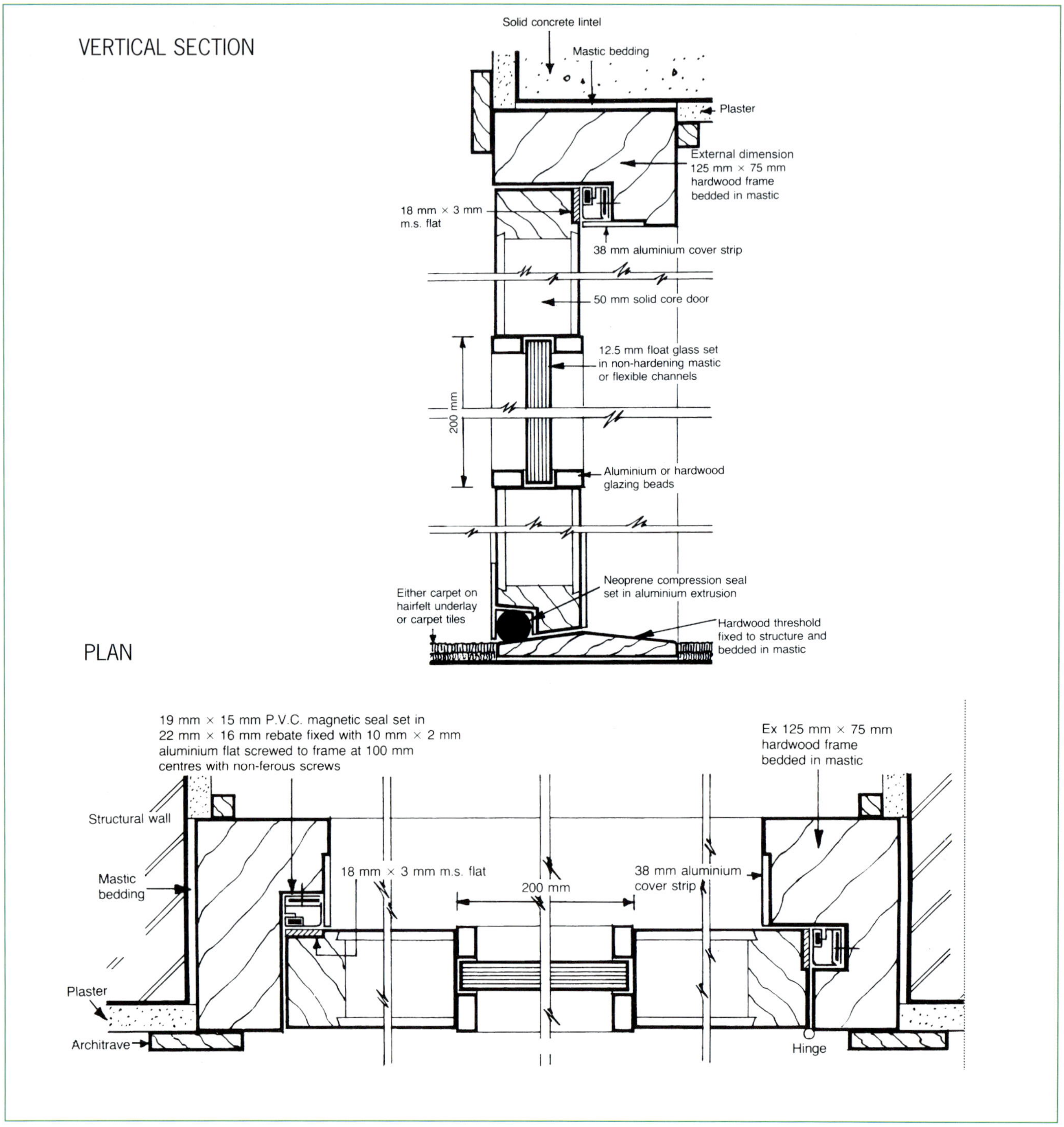

Figure 5.8: Vertical and horizontal sections through a door installation. Taken from BBC Engineering Guide to Acoustic Practice, 2nd Edition 1990. *These drawings are reproduced here with the kind permission and co-operation of the BBC*

5.5.3 Practice rooms / group rooms

Figure 5.10 shows a typical 8 m^2 group room which will accommodate both instrumental lessons and composition groups and which can be used for individual practice. Points to note are as follows.

- One wall is at an angle of 7° to avoid flutter echoes (a particular issue in small rooms) and prominent standing waves. Window and door reveals provide useful diffusion to other walls.
- A full length drape can be pulled across the window to increase surface absorption and reduce loudness.
- The window is fairly small and positioned in the centre of the wall to control the amount of external noise reaching the space and avoid sound travelling between adjacent group rooms.
- Floor and ceiling finishes are as for the larger rooms.

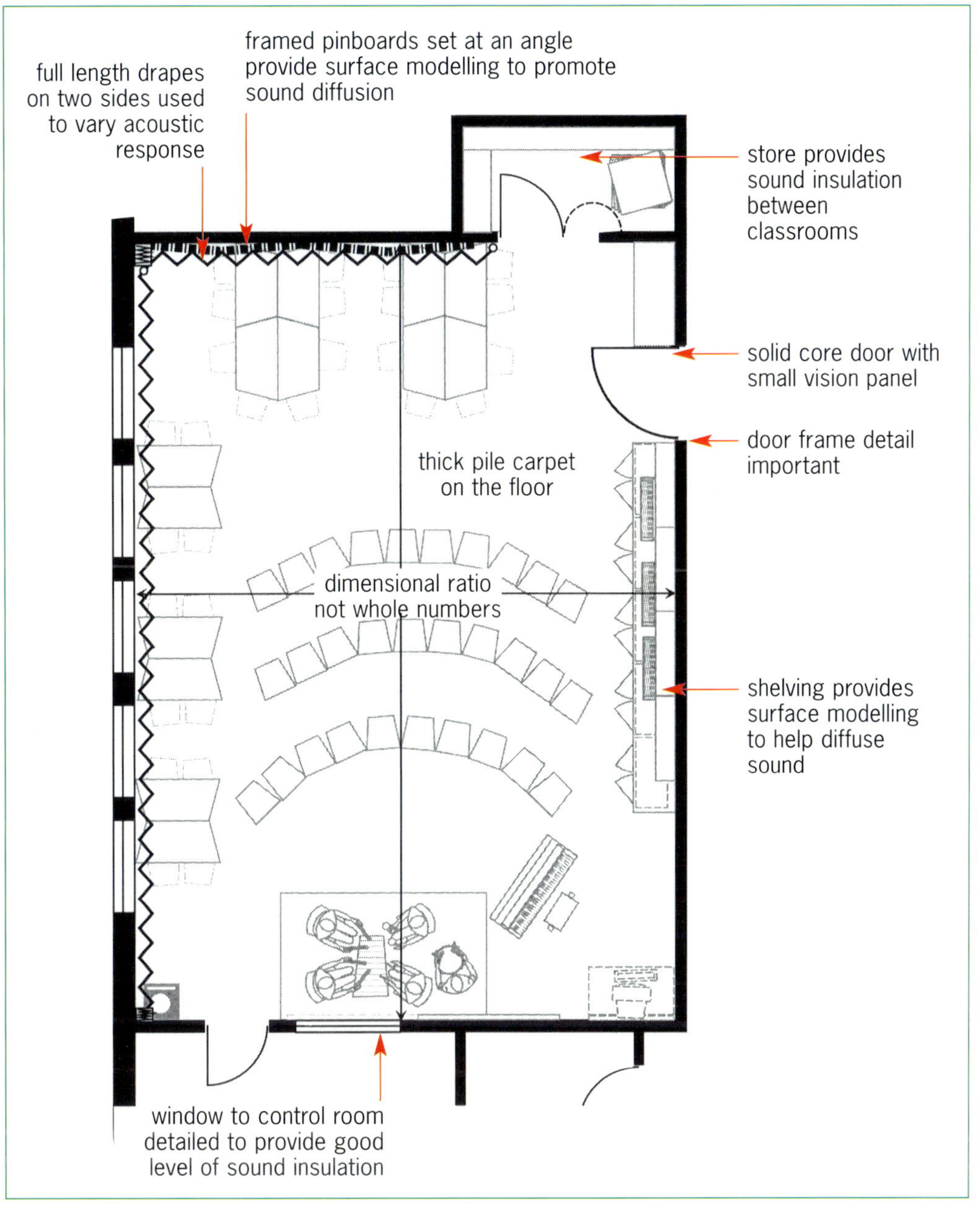

Figure 5.9: Acoustic treatment to 85 m^2 music classroom/recital room

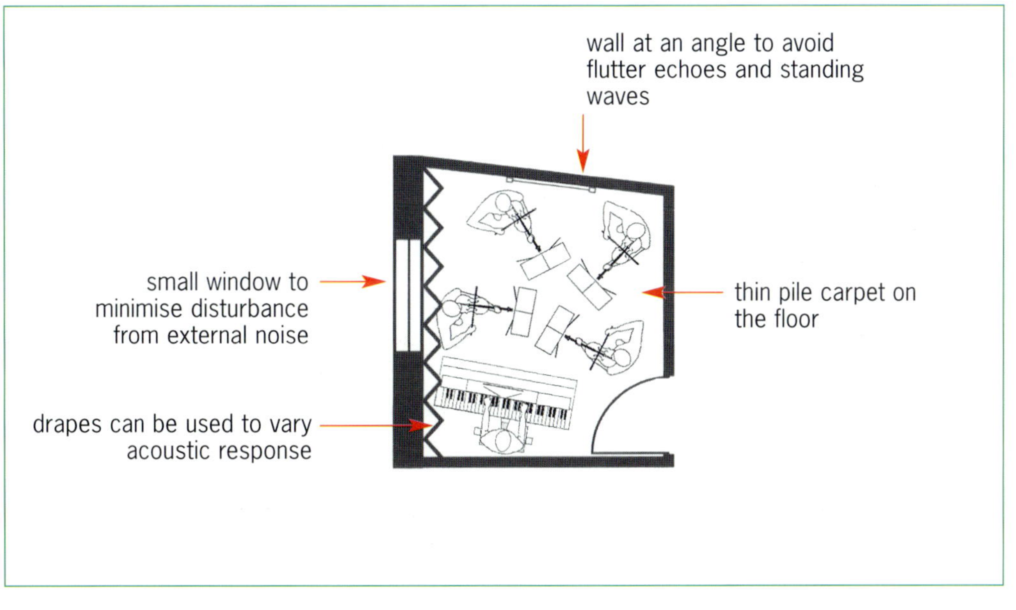

Figure 5.10: Acoustic treatment to 8 m^2 group room

5.5.4 Ensemble rooms

Figure 5.11 shows a plan of a 25 m^2 ensemble room. In terms of shape, the same rules apply as for larger music spaces. Ceilings should be high, of the order of 3 m or more. Surface finishes may comprise carpet on the floor, a suspended plasterboard ceiling to provide the necessary bass absorption, and a mixture of hard and soft wall finishes to provide the required RT. An acoustic drape along one wall can provide a degree of acoustic variability.

5.5.5 Control rooms for recording

Control rooms for recording have assumed a much greater significance due to the need to prepare tapes of compositions for GCSE assessment. Figure 5.12 shows an 11 m^2 control room for recording. A teacher or pupil can record a music performance taking place in an adjacent space after which the recording may be heard on headphones or loudspeakers. The RT specified in Table 1.5 is < 0.5 s.

Notable aspects of the acoustic treatment are as follows:

- Sound absorbing panels on the walls behind the monitor loudspeakers are used to control strong early sound reflections which could distort loudspeaker sound.
- Shelving units on the window wall provide surface diffusion.
- Drapes are fitted on all three observation windows. If a curtain is pulled across one window, problems of flutter echoes and prominent resonances associated with two facing hard parallel surfaces are reduced. Ideally, the effect can be avoided by installing glazing in one of each pair of windows at 5° off parallel. Drapes also provide additional privacy.
- The external window is small to minimise disturbance from external noise. A venetian blind can be used to control sunlight, or a blackout blind may be provided if required.
- The floor is carpeted.
- Figure 5.6 shows a detail of a typical

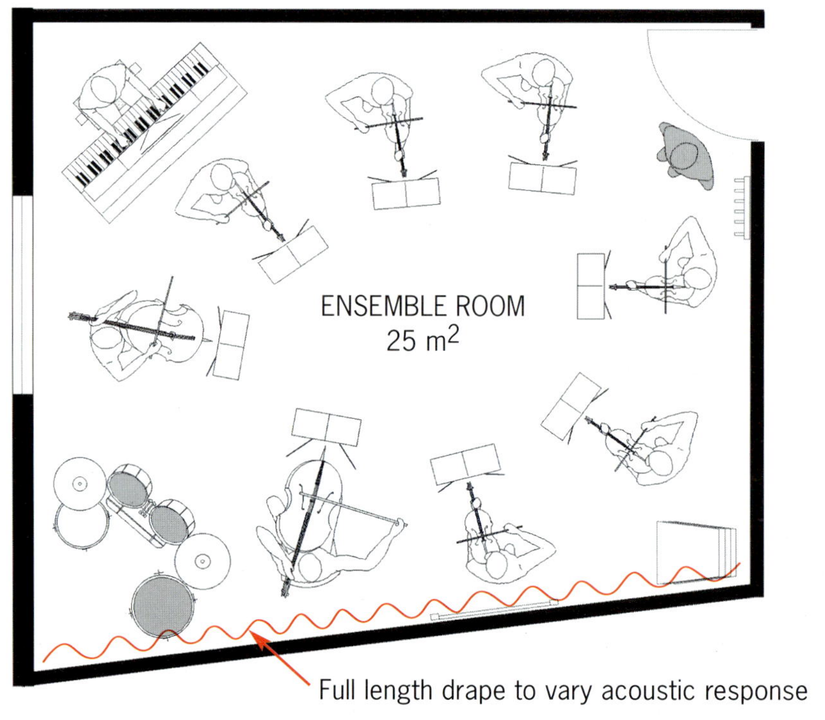

Figure 5.11: Acoustic treatment to 25 m^2 ensemble room

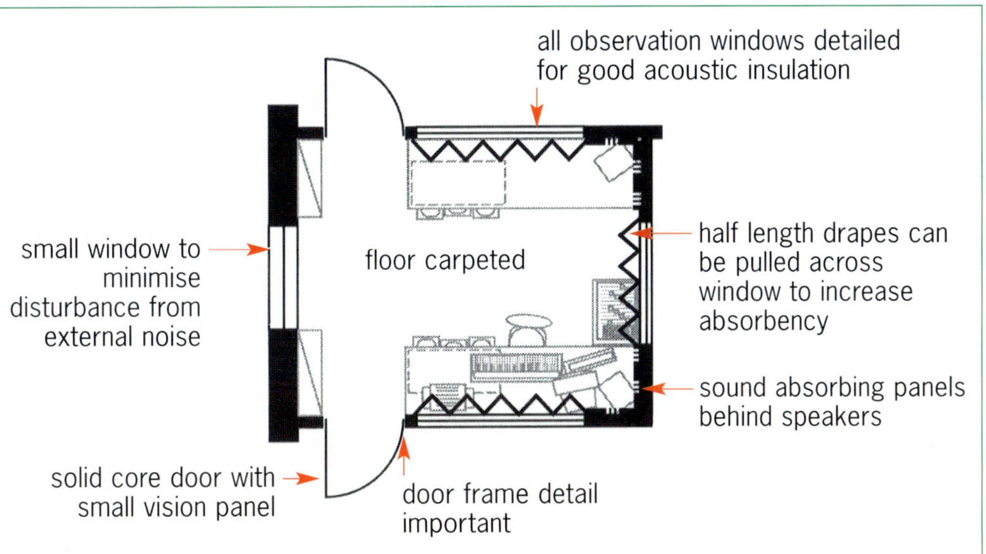

Figure 5.12: Acoustic treatment to recording/control room

control room window. Two panes of heavy plate glass (of different thicknesses to avoid the same resonances) are separated by an air gap of about 100-200 mm. Such a large gap may not always be possible but 50 mm should be considered a minimum. Each pane of glass is mounted into a separate frame to avoid a direct sound path. The glass is mounted in a neoprene gasket to isolate it from the wooden frame. Acoustically absorbent material, such as mineral wool or melamine foam, is incorporated into the reveal to absorb any energy that enters the air gap.

5.5.6 Recording studios

A recording studio as such rarely exists in a school. The control room for recording may have an observation window onto an ordinary ensemble room or professional/recital room. A professional type recording studio would require a lower indoor ambient noise level than that given in Table 1.1, and specialist advice should be sought.

5.5.7 Audio equipment

The design and selection of recording equipment and audio systems is a fast-evolving subject and guidance on specific technologies would be rapidly out of date. Although members of staff within a school will have their own preferences for specific items of equipment, these may be based on experience of only a few systems and alternatives should at least be considered. Advice from an independent designer or consultant familiar with the full range of available equipment should be sought.

5.6 Acoustic design of large halls for music performance

Large halls designed primarily for music are rare in schools, where the main use of any large hall is likely to be for assemblies and other speech-related uses. Assembly halls, theatres and multi-purpose halls are discussed in Section 4. If a purpose-built concert hall is required a specialist acoustics designer should always be consulted early in the project, but this section sets out some general principles which can be considered at the concept stage.

5.6.1 Shape and size

Key acoustic requirements are sufficient volume to provide adequate reverberation and a shape that will provide a uniform sound field with strong reflections off the side walls. A rule of thumb is that the volume of a concert hall should be at least 8 m^3 per member of audience, which is typically twice that for a theatre or cinema. In most cases this will lead to a rectangular floor plan with a relatively high ceiling. Other shapes, such as the elongated hexagon or asymmetrical

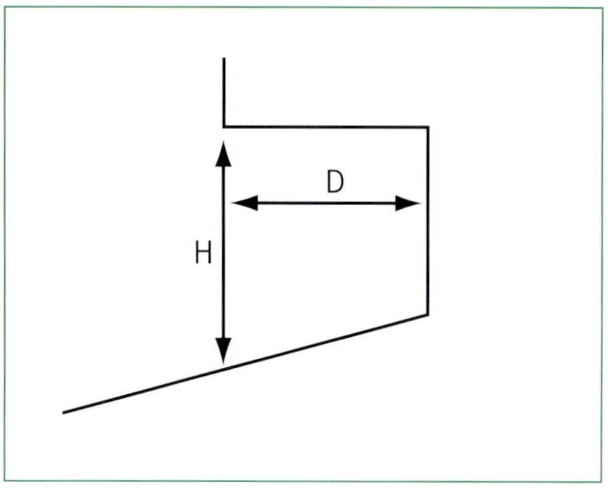

Figure 5.13: Recommended balcony overhang proportions where the depth D should not exeed the height H.

shapes, can work well but require very advanced acoustic design. Fan-shaped halls generally do not provide the lateral reflections beneficial to listening to music.

Balconies and side-wall boxes or galleries may be used although they tend to reduce the volume of the hall for a given audience size. Any overhangs must be kept small to allow reasonable sound to seats under the balcony. Figure 5.13 indicates the recommended proportions of an overhang so that good acoustic conditions are maintained beneath the overhang. Balcony, gallery and box fronts can be used to break up large areas of flat wall and provide essential diffusion, especially on parallel side walls where flutter echoes may otherwise occur.

Ceilings can be flat with some surface modelling, or can be more complex shapes to direct sound towards the audience. A steeply pitched ceiling (around 45° assuming the ridge runs along the length of the auditorium) can also be good. Shallow pitches can cause 'flutter' echoes between a flat floor and the ceiling, see Case Study 7.1.

Shapes with concave surfaces, such as domes and barrel vaults, cause focusing of sound which can result in problematic acoustics and these are best avoided. Where concave surfaces are unavoidable and cause a focus near the audience they should be treated with absorbent or diffusing finishes.

If seating is on a rake, this should not be too steep as musicians find it difficult performing into a highly absorbent audience block - in effect, they receive very little feedback. Generally, rakes which provide adequate sightlines will give satisfactory acoustic conditions. This rake will generally be less than in a theatre or cinema.

The size and shape of the concert platform is of great importance. A full 90-piece symphony orchestra requires a stage at least 12 x 10 m, with allowance for choir risers behind. The front of the platform will not generally be as high as a theatre stage and may be only 400 mm above stalls floor level, but orchestral players will require risers or rostra so that players at the rear of the platform can see the conductor. Surfaces around the stage should be acoustically reflective and should be designed to provide some reflected sound back to the players, so that they can hear themselves and each other, as well as directing some sound towards the audience. This design requires computer or physical scale modelling by a specialist acoustician.

5.6.2 Surface Finishes

Unlike in theatres and assembly halls, the surface finishes in a concert hall with the correct volume will generally be acoustically reflective, for example plastered or fair-faced brick or blockwork. Large areas of flat lightweight panelling, such as wood or plasterboard, tend to be absorbent at low frequencies, which results in inadequate reverberation at these frequencies. The result tends to be a lack of 'warmth' or 'bass response' and is a common problem in many halls. Wood panelling, if used, must be very heavy or stiff. Curved wooden panels are often used as acoustic reflectors because their curvature gives added stiffness, reduces their inherent panel absorption and provides acoustic diffusion.

In most performance venues the seating and the audience provide the majority of the absorption and, therefore, constitute a controlling factor in the room acoustic conditions. The selection of seats and, particularly, the relative absorption of occupied and unoccupied seats is of great importance. In general, it is helpful if the room acoustics are relatively unaffected by the number of occupants. This, however, tends to mean that seating must be very absorptive and probably not a preferred type for school use. A seat which is moderately upholstered on the seat and back is likely to be a good compromise. Where tip-up seats are provided they should be upholstered underneath as well as on the seat; otherwise acoustic conditions will be very different during rehearsal and performance. Most auditorium seating manufacturers supply acoustic test data. Where there is no fixed seating, large areas of acoustic drapes or other operable acoustic absorption can be used to reduce reverberation in rehearsal conditions when the seats are removed.

5.7 Design of large auditoria for music and speech

Table 5.1 lists the general acoustic characteristics that are required for a multi-purpose auditorium.

There are four commonly considered approaches to designing these spaces:

1. To design a concert hall with a large volume ($\approx$10 m^3 per seat), and to reduce the volume of the auditorium when needed for speech. This approach is recommended when the overwhelming requirement is for a good musical acoustic, with a relatively small proportion of theatre or other speech use. Unless the volume can be reduced substantially, this approach requires large amounts of absorbent material to be deployed, which in turn can reduce loudness to the extent at which a speech reinforcement system is needed. Nearly all auditoria adopting this approach depend on high-quality speech reinforcement systems, which are difficult to design in a reverberant hall.

2. To design a small volume (not more than 6 m^3 per seat) with acoustics suitable for a theatre, with additional reverberant volumes accessed by openable flaps or moveable ceilings. As the volume needs to be increased by up to 80%, with reasonably even distribution of absorption, this is often impracticable. In the few cases where this approach has been tried, the results have been poor because it is difficult to provide openings large enough to be transparent to the long wavelengths of low frequency sound.

3. To design to a compromise volume and RT, often with curtains or other moveable acoustic material to provide some variation in RT. The result tends to be an auditorium which is acceptable for a range of uses, but not particularly good for any of them - especially music. Very large areas

Table 5.1: Acoustic characteristics for a multi-purpose auditorium

Low ambient noise levels	Low noise levels from plant, ventilation, lighting and stage machinery are required. Noise from outside the auditorium should ideally be imperceptible.
Even distribution of sound	The acoustic should not change significantly from one seat to another.
Lack of acoustic defects	There should be no echoes or focusing effects.
Loudness or acoustic efficiency	The sound level reaching the listener should be as high as possible without compromising other requirements.
Good direct sound	The sightlines to the source should not be impeded and distances should be as short as possible.
Good early reflections	Reflecting surfaces around and close to the stage, and reflections off the side walls and off the ceiling are required.
Feedback to performers	Some sound from the stage should be reflected back to the source. This gives confidence to the performers and helps with musical ensemble.

of curtain are required to have any significant effect in a large hall, and these will have relatively little effect at low frequencies, resulting in a room that is either 'boomy' for speech or 'dead' for music.

4. To design a small volume (≈6 m^3 per seat) with acoustics suitable for a theatre, with an electro-acoustic enhancement system to introduce more reflected sound. These systems were originally designed to enhance the acoustics of naturally poor auditoria, but their success has recently led to their being built in to new auditoria where a wide range of acoustic conditions is required. The best systems provide good acoustics over a wider range of uses than would otherwise be possible, without the audience (or musicians) being aware that the sound that they hear is not due to the 'real' acoustic of the auditorium. These systems are seen as acoustically very advanced and are not commonly used in schools, but present a viable option where a large hall is to be used for both speech and music on a regular basis. These systems require loudspeakers in the auditorium side walls and ceilings, and should not be confused with the sound reinforcement system for speech (in this case a central cluster of loudspeakers over the forestage), although some electro-acoustic enhancement systems can also be used for speech reinforcement.

References

[1] Building Bulletin 86, Music Accommodation in Secondary Schools. DfEE, 1977. ISBN 0 11 271002 6.

[2] J Miller. Design standards for the sound insulation of music practice rooms. Acoustics Bulletin 18(6), Institute of Acoustics, 1993.

Acoustic design and equipment for pupils with special hearing requirements

SECTION 6

When considering classroom acoustics, children with a permanent hearing impairment have traditionally been treated as a special group, separate from the mainstream school population. However recent surveys of the school population show that about 75% of deaf children are educated in mainstream schools.

6.1 Children with listening difficulties

A recent survey by the British Association of Teachers of the Deaf (BATOD)[1] showed that about 75% of deaf children were being educated within mainstream schools. With the continuing trend towards inclusive education there is no reason to suppose that this proportion should do anything but increase.

In addition to the children with permanent hearing impairments there are large numbers of children within mainstream schools who have listening difficulties placing them in need of favourable acoustic conditions. These include children:

- with speech and language difficulties
- whose first language is not English
- with visual impairments
- with fluctuating conductive deafness
- with attention deficit hyperactivity disorders (ADHD)
- with central auditory processing difficulties.

Effort given to addressing the acoustic needs of the hearing impaired population also favours other groups whose needs for good acoustic conditions are not dealt with elsewhere in this document. Put together, the number of children falling into one or more of these categories could conceivably be a significant proportion within every mainstream classroom.

6.2 Children with hearing impairments and the acoustic environment

The majority of children with hearing impairments use speech and hearing as their main form of communication. The BATOD survey[1] indicated that 67% of children with hearing impairments were using an auditory-oral approach and a further 26% used an approach which combined sign with auditory-oral components. For these groups a poor acoustic environment can be a significant barrier to inclusion.

A hearing loss is typically described with reference to the audiogram. This is a graphical representation of an individual's threshold of hearing for a number of pure tones (typically measured at 250 Hz, 500 Hz, 1 kHz, 2 kHz, 4 kHz and 8 kHz) and presented to each ear using headphones. At face value, it suggests that the hearing impairment can be considered as a simple auditory filter and as such should predict a child's understanding of speech using traditional acoustic models. Although reliable, it says little about an individual's hearing for speech or the key skill of listening to speech with background noise. The audiogram is not a good predictor of educational outcome[2] and only a poor predictor of maximum speech recognition score[3]. Consequently, great care should be taken when considering the audiogram of a child as a predictor of the difficulties the child might have in a school environment.

At present there is little empirical data that specifically addresses the acoustic criteria required for the hearing impaired school population (see for example the review of the literature by Picard and Bradley[4]). What is currently available, however, suggests that the individual hearing needs of the hearing impaired child are likely to be more demanding than those of children with normal hearing. It would be helpful for the professional specifying classroom acoustics for a particular child to have available measures of the child's aided hearing and consequent acoustic requirements in terms of, for example, acceptable levels of

6 Acoustic design and equipment for pupils with special hearing requirements

Table 6.1: Recommendations of BATOD and ASHA for the acoustics of classrooms

Acoustic parameter	British Association of Teachers of the Deaf[5]	American Speech Language Hearing Association[6]
Unoccupied noise level	35 dB(A)	30 – 35 dB(A)
Reverberation time (unoccupied)	0.4 s across frequency range 125 Hz to 4 kHz	0.4 s
Signal to noise level	+20 dB across frequency range 125 Hz to 750 Hz +15 dB across frequency range 750 Hz to 4 kHz	≥ +15 dB

noise, desirable reverberation times and required signal to noise levels. However, such hearing measures are not routinely obtained.

Although it is not possible at present to provide definitive acoustic requirements for hearing impaired individuals, it is important for acousticians and architects to be aware of the recommendations published by specialist professional organisations. These include the British Association of Teachers of the Deaf[5] and the American Speech Language Hearing Association[6], see Table 6.1. Account has been taken of these recommendations in the setting of performance criteria in Section 1.

6.3 Hearing impairment and hearing aids

Modern hearing aids are designed to make speech audible to the listener without being uncomfortably loud[7]. They deal largely with the issue of audibility and are less able to address the issues of distortion that typically accompany a sensorineural hearing impairment.

One of the major challenges in the design of hearing aids is dealing with noise. Recent developments include the use of algorithms that attempt to enhance speech whilst reducing background noise, and better implementation of directional microphones. However, noise will continue to remain a significant obstacle to effective listening. Noise not only masks the amplified speech signal but also leaves a child tired from the effort required to listen. It is therefore essential that attention be given to creating a quiet classroom.

Sound insulation must be of a high standard, with the lowest background noise levels possible to ensure that a good signal to noise level is achieved. Typically a signal to noise level of +20 dB is considered desirable[5]. Short reverberation times are also critical in ensuring that sound does not build up when the class are working in groups. Care must also be taken to ensure that the level of low frequency noise is kept to a minimum. For many people with impaired hearing, low frequency noise can have a devastating impact on speech recognition, masking many important speech sounds in a manner that cannot be appreciated by those with normal hearing.

6.4 The speech signal and hearing aids

Speech, as a signal, is a critical factor in classroom listening and an important speech source is the teacher. Evidence has shown that teachers' voices are not always sufficiently powerful to deliver the necessary levels of speech required to ensure the best listening opportunities[8]. A growing body of evidence suggests that teachers are at above average risk from voice damage[9]. Few teachers have voice training and the vocal demands of teaching are probably underestimated.

Hearing aids are usually set up to amplify a 'typical' speech signal based on various measures of the long-term average speech spectrum recorded either at the ear of the speaker or at a distance of 1 m directly in front of the average speaker, as if in conversation. If the actual speech signal is weaker than average, perhaps because of distance, or is masked by babble or steady state background noise such as that from a classroom computer fan, then the hearing impaired listener will have increased difficulty. Listening to speech will become particularly effortful and challenging[10].

Children are not only required to listen to the teacher but also to other children. Children typically have less powerful speaking voices[8] and listening to their peers is frequently identified by children with hearing impairments as being difficult. One study suggests that 38% of a child's time in the classroom might be spent working in groups and 31% of the remaining time spent in mat work[11], both situations where listening to other children is important. There are no wholly satisfactory solutions to this. Technology and careful class management have a role to play but considerable attention needs to be paid to establishing low reverberation times and maintaining low ambient noise levels in order to reduce the auditory difficulties.

To minimise the challenges to hearing, use is often made of small acoustically treated rooms attached to mainstream classrooms in the primary school. These rooms are typically large enough for a group of four to eight children to work in. To allow supervision by the class teacher they will have a large window to allow a clear view into the classroom. The room will need to have a sufficient degree of sound insulation from the classroom to allow the children to talk to each other without being disturbed or disturbing the rest of the class. The favourable acoustic conditions and short distances between children and teacher, if present, ensure that communication is as easy as possible.

6.5 Listening demands within the classroom

Much of educational activity within classrooms revolves around speech. Some experts claim that 80% of all classroom activities require listening and speaking. It is important that within any room the acoustic characteristics allow for effective spoken language communication. The UK version of the Listening Inventories for Education[12] identifies the following listening demands within the classroom:

- listening to the teacher when s/he is facing away from the listener
- listening when the class is engaged in activities
- listening to the teacher while s/he is moving around the classroom
- listening when other children are answering questions
- listening when other adults are talking within the same room
- listening to peers when working in groups
- listening in situations with competing background noise from multimedia equipment.

A teacher should manage teaching in such a way as to ameliorate the challenges faced by a student with hearing difficulties. However, the better the acoustic conditions, the less challenging will be the situations described above.

6.6 Strategies developed to assist children with hearing and listening difficulties

Effective classroom management by the teacher is critical in ensuring that the children can have access to all that is spoken and there are many guidelines available for teachers (see for example publications by the Royal National Institute for the Deaf[13], the National Deaf Children's Society[14] and DfES[15]). Classroom management alone, however, cannot ensure that speech communication is sufficiently audible and intelligible if the classroom acoustics are not adequate, or if a child has a hearing or listening difficulty.

In order to ensure that children are able to hear the teacher and, to a lesser extent, their peers, a number of technological solutions have been developed, see Table 6.2. These solutions that work in tandem with the child's own hearing aids (if used) can be classified as either individual technology or whole class technology. In both these cases it is important to understand the underlying principles when specifying classroom acoustics.

Technology	Advantages	Disadvantages
Personal radio aids	Reduce the effect of the distance between speaker and listener Portable and convenient Particularly useful in situations where there is a poor signal to noise ratio at the position of the listener	Do not address the needs of group work directly Can require a high level of sophistication to gain maximum benefit Benefits can be lost if the child's personal hearing aid microphones are used in noisy environments
Classroom soundfield systems	Reduce the effect of the distance between the speaker and listener Inclusive technology Benefit to the teacher and the class Can ensure good signal to noise levels are maintained throughout the classroom	Do not address the needs of group work directly Poor classroom acoustics (eg high reverberation times or poor sound separation between neighbouring teaching areas) can limit the benefit of this technology
Personal soundfield amplification	Portable Addresses the issue of speaker to listener distance Can ensure favourable signal to noise levels for a particular listener or small group of listeners	Can be cumbersome to transport and manage Does not address the needs of group work directly
Auditory trainers and hard-wired systems	Provide excellent signal to noise levels Provide a high level of sound insulation Can be arranged to allow group work	Users are restricted in movement when using the device Can be heavy and uncomfortable to use Not an inclusive technology
Induction loop systems	Discreet and cheap Most hearing aids have a telecoil facility	Unpredictable acoustic response for the hearing aid user Spill over of signal into other rooms Do not deal with the needs of group work Susceptible to electromagnetic interference User normally isolated from environmental sounds

Table 6.2: Advantages and disadvantages of different technologies for aiding hearing and listening in the classroom

6.7 Individual technology

There are two main types of aid that can be used to assist children's hearing on an individual basis: radio aids that can be coupled to a child's hearing aids, and auditory trainers that are used with headphones.

6.7.1 Radio aids

Radio aids (also known as radio hearing aids or personal FM systems) are widely used by children with hearing impairments in schools. They help overcome causes of difficulty in a classroom situation by:

- providing a good signal to noise ratio
- reducing the impact of unhelpful reverberation
- effectively maintaining a constant distance between the speaker and the listener.

All radio aids have two main components: a transmitter and a receiver. The person who is speaking (usually the teacher) wears the transmitter. A microphone picks up their voice. Typically the microphone is omnidirectional and is attached to the lapel of the speaker, however there are head worn microphones available that help ensure a consistent transmitted signal to the child.

The sounds are transmitted by an FM radio signal to the receiver, which is worn by the child. The receiver converts the signal to a sound that the child can hear.

Radio aids are usually used in conjunction with the child's hearing aids. Most children use 'direct input' (also known as 'direct connection' or 'audio input') to the hearing aids using a lead. Direct input is a facility available on many behind-the-ear (post-aural) hearing aids and a smaller number of in-the-ear hearing aids.

Alternatively, the child can use an inductive neck loop - a small wire loop that can be worn over or under clothes. The loop is connected to a radio aid receiver usually worn around the waist or attached to a belt.

Direct input is generally recommended as preferable to the use of a neck loop for children in school. This is because the level of sound that a child hears using a neck loop can be variable and there is a risk of electromagnetic interference from nearby electrical equipment.

Radio aids are also beneficial for children who have cochlear implants. The radio aid receiver is connected to the child's implant processor using a dedicated lead.

Traditionally, radio aid receivers have been worn in a chest harness or on a belt. Recent developments include miniature radio aid receivers that connect directly to a hearing aid and are worn entirely behind-the-ear. Behind-the-ear hearing aids that include built-in radio aid receivers are also being manufactured.

Most radio aids can be set up so that the child will not only hear the voice of the speaker using the transmitter, but also environmental sounds such as their own voice and the voices of other children near to them. Radio aids can do this in a number of different ways and it is often necessary to strike a balance between allowing the child to hear the voices he or she needs to listen to and the impact of hearing unwanted background noise.

For the best listening condition the hearing aid user will normally be required to mute his or her microphone on the hearing aid and listen exclusively to the transmitted voice of the speaker. This is good for formal teaching situations but requires considerable skill on the part of the teacher to include the hearing impaired child in classroom discussion. This solution is less helpful for children engaged in group activity, where the child will need to work with a small group of peers.

Most radio aids are able to operate on a range of carrier frequencies. For example, each school class might have its own frequency so that there is no interference with a neighbouring class. In the UK, radio aid channels lie in the range 173.350 MHz to 177.150 MHz. Those channels in the range 173.350 MHz to 173.640 MHz are dedicated exclusively to use by radio aids. A licence is required to use radio aids operating on frequencies between 175.100 MHz and 177.150 MHz.

The sounds heard by a child using a radio aid will depend on the quality and correct use of their own hearing aids. The level of amplification is determined by the settings of the hearing aids, not the radio aid. Accepted procedures exist for setting up a radio aid to work with hearing aids (a process sometimes known as 'balancing').

A general principle is that if a child uses a hearing aid, then the child is also likely to find a radio aid helpful in many classroom situations.

Radio aids have often been seen as the solution to poor acoustics in the classroom. However, it must be noted that they only partially solve the problem; the solution must lie in addressing the issue from three directions:

- the class teacher and classroom management style
- technology that assists listening
- careful attention to classroom acoustics.

Current information about radio aids is available from a number of sources including the National Deaf Children's Society[16].

6.7.2 Auditory trainers and hard-wired systems

An auditory trainer is a powerful amplifier used with high-quality headphones. As a large, stand-alone piece of equipment, an auditory trainer can be designed without the restrictions of size that exist with typical behind-the-ear hearing aids, and a good quality high level sound output with extended low and high frequency range can be achieved.

Within the mainstream educational environment, auditory trainers are most likely to be used for short periods of individual work and speech therapy sessions. However, it is also possible to link several auditory trainers together for group work. In some schools for deaf children this equipment is permanently installed within a classroom. The teacher's voice is picked up by a microphone and the output is available at every desk. Each child wears headphones that are configured to meet their individual amplification requirements. The children may also wear microphones to enable everyone in the class to participate in discussions.

6.8 Whole class technology

The use of a personal system is sometimes essential for a hearing aid user to be able to succeed in a particular environment. There is, however, a trend to use the inclusive technology termed 'sound field amplification' to ensure that the signal level of the speech is delivered to all parts of the classroom at an appropriate level above the background noise. This technology is of benefit for all with listening difficulties in the classroom, not just the hearing aid user, and has particular benefits for classroom management and the voice of the class teacher.

It is important to note that whole class technology is not a substitute for remedying poor classroom acoustics. However, it can be particularly valuable in maintaining good signal to noise levels and improving classroom management. Soundfield amplification systems are also used in conjunction with personal radio aids. In situations where a deaf child is part of a mainstream class, advice should be sought from members of a relevant professional group (educational audiologist, clinical audiologist or teacher of the deaf) as to the most appropriate technology.

Figure 6.1: A simple schematic drawing of a soundfield system in a typical classroom

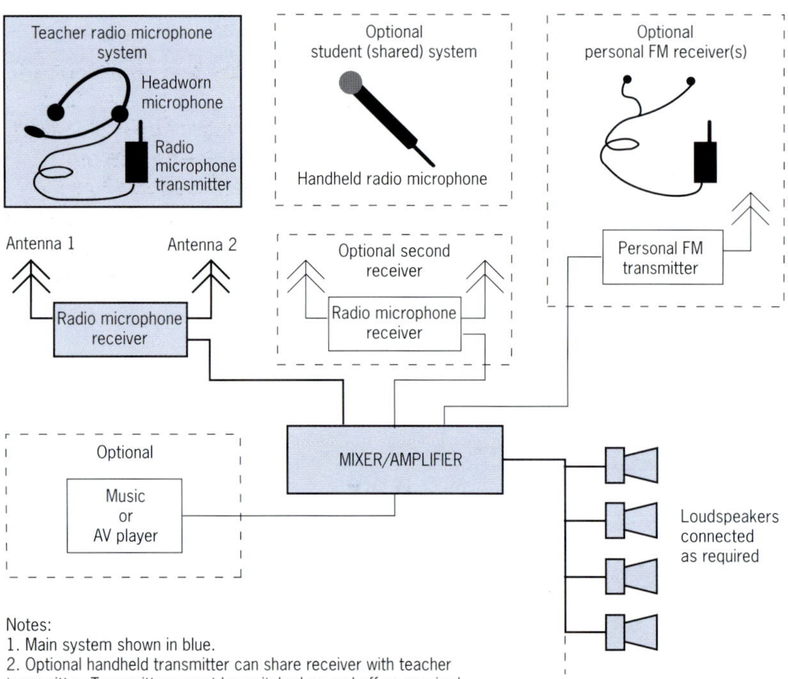

Notes:
1. Main system shown in blue.
2. Optional handheld transmitter can share receiver with teacher transmitter. Transmitters must be switched on and off as required.
3. Alternative second receiver allows simultaneous use of teacher and student transmitters.
4. Personal FM transmitter(s) for use by pupils with serious hearing impairment can be connected to output of system.
5. CD, cassette and/or video player can optionally play through the system.

6.8.1 Whole classroom soundfield systems

Soundfield systems provide distributed sound throughout a classroom. They use a wireless link between the microphone and amplifier which will operate on VHF, UHF radio or infra red frequencies. Soundfield systems have been shown to be beneficial for hearing children and children with a mild or temporary hearing loss. They will not by themselves usually provide sufficient improvement in signal-to-noise ratio for a child with a significant hearing loss, when a personal radio aid is also usually necessary.

A soundfield system is perhaps more widely known as a sound reinforcement system; the term 'soundfield' system originated from the field of Audiology and continues to be associated with classroom sound reinforcement systems. The technology has matured since it was first introduced into classrooms in the late 1970s in the USA, and has evolved to take into account new technologies and teaching management styles. Its benefits have been variously described as:

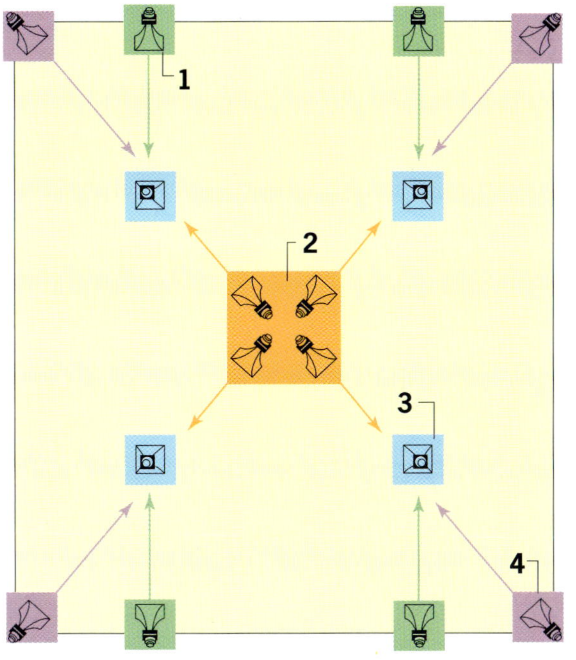

1 Speakers mounted one quarter of wall length from corners, mounted flush with wall, at 2 m height, directed to point on floor at angle of about 60°.
2 Speakers mounted at centre of room, 600 mm apart in square orientation, directed to the room corners.
3 Speakers mounted flush with ceiling, facing directly at floor, in centres of 4 quadrants of the ceiling.
4 Speakers mounted in room corners, directed to centre of room.

The four speaker layouts were investigated using a computer model to predict the interference effects due to phase changes of the speaker outputs.

Layout 1 causes some peaks in the room, but is likely to work reasonably well.

Layout 2 produces large interference effects and is therefore not recommended.

Layout 3 produces the most even coverage at all frequencies and the least loudspeaker interaction. It is therefore recommended for all rooms where a suspended ceiling exists. Loudspeakers should be selected to provide a wide and even coverage that is constant with frequency.

Layout 4 provides relatively even coverage of the room with interference effects which, although complex, appear relatively benign. Location of loudspeakers in room corners tends to produce a rise in the bass response of systems, which will usually require equalisation using system controls to produce optimum speech clarity and naturalness.

Layouts 1 and 4 using wall mounted loudspeakers are recommended where ceiling mounted units are not practical. For layout 1 speakers should be mounted at least 1 m from the side wall. Location of wall mounted speakers at least 2 m above the floor, and at least 600 mm below the ceiling is recommended. Brackets should keep the loudspeakers very close to the wall to minimise self-interference effects.

Figure 6.2: A plan of a classroom showing four alternative speaker layouts. *The speakers are drawn horn-shaped to show the directionality of the speaker output, although many modern speakers are flat*

- academic improvements for all class members
- more on task behaviour
- greater attentiveness
- improved understanding of instructions
- less repetition required from the teacher
- improved measures of speech recognition
- reduced voice strain and vocal fatigue for the teacher.

6.8.2 System overview

Figure 6.1 shows a simplified block diagram of a typical soundfield system. Each element shown can be a separate unit, or some of these can be combined into an integrated unit. The current trend is for manufacturers to create more integrated products, designed especially for classroom soundfield use. Typical arrangements of loudspeakers are shown in Figure 6.2.

Table 6.3 describes the various components of a soundfield system. A possible detailed specification is included in Appendix 8.

Where a soundfield system has not been designed specifically for the classroom it should be used for a trial period before being selected from the range available. The manufacturers and suppliers should all provide installation information including commissioning of installations and operating instructions, plus ongoing support. Large rooms or rooms that are unusually shaped will usually need specialist advice. Teachers must receive adequate training in using the systems.

6.8.3 Personal soundfield systems

A child who cannot physically wear a conventional hearing aid, who has a unilateral hearing loss, or has Central Auditory Processing Disorder or

Component	Requirements	Comments
Loudspeaker Wall mounted, ceiling mounted and flat panel speakers are used in schools.	The purpose should be to provide high quality distributed sound reinforcement throughout the whole classroom and over the whole speech frequency range. Selection of appropriate speakers should therefore address this requirement.	Often the location of loudspeakers is determined by the necessity to fit in with the current use of the classroom, when not installed as part of the original building work.
Microphone and transmitter Using infra red, UHF or VHF carrier frequencies and high quality headworn or lapel microphones. Radio system information is available at www.radio.gov.uk	This should be a high quality system which retains both the frequency and dynamic properties of speech. It is important that all teaching styles can be accommodated so a choice of microphones should be available. It is important that the transmitter can operate without interference from other systems or from public services.	In order to retain good dynamic range a compander system is typically required (see Figure 6.3). A head worn microphone can improve the consistency of the transmitted signal and help to prevent feedback that is present in systems that do not have feedback control technology. However teachers often like a choice of microphone and will use headworn, lapel or wrap around microphones depending on activity and personal preference. Battery life of at least one school day is essential for a transmitter if it is to be acceptable for school use.
Receiver Matched to the transmitter	Will provide a complementary system to the transmitter, avoiding interference or frequency dropout.	A compander technology and diversity system is particularly suitable for classroom use, ensuring good dynamic range and avoiding frequency dropout respectively. Some teaching situations require twin channel inputs, so that a pass around radio microphone can be used. Where infra red systems are being used separate additional receivers might be necessary to avoid 'blind spots'.
Amplifier	The amplifier should be correctly matched to the loudspeaker system. It should offer a wide flat frequency response which can be adjusted if necessary. It should allow for additional inputs from multimedia within the classroom, such as TV, computer and radio and outputs to radio systems.	Some schools might require an additional output facility for use by deaf children with personal FM systems. The amplifier is usually combined with the receiver unit.

Table 6.3: Components of a soundfield system

Attention Deficit Disorder, might use a portable soundfield system. Personal soundfield systems comprise a radio transmitter and microphone worn by the teacher and a small, portable unit for the child. The portable unit includes an FM receiver, amplifier and loudspeaker and is designed to be carried around school by the child and placed on the desk next to them. The sound of the teacher's voice is amplified and played through the loudspeaker.

6.8.4 Infra red technology

Infra red technology has been available for many years with little market presence. However, this technology has recently undergone considerable development and

Technology	Advantages	Disadvantages
Infra red Frequency range 2.3–2.5 MHz	Physically limited to enclosed room Allows equipment to be shared between rooms Wideband transmission Can be used with personal hearing aids using a neck loop (an induction loop worn round the neck)	Occasionally needs extra IR receivers in a room
Radio VHF narrowband 173.35–177.15 MHz	Reserved frequency bands for use in schools Many frequency bands available Equipment compatible across manufacturers	Poor signal quality when compared to wideband
Radio UHF wideband 790–865 MHz	Can allow a higher quality signal than narrow band equipment Many frequency bands available, although a site licence might be required	Not available for personal FM equipment

Table 6.4: Advantages and disadvantages of infra red and radio technologies

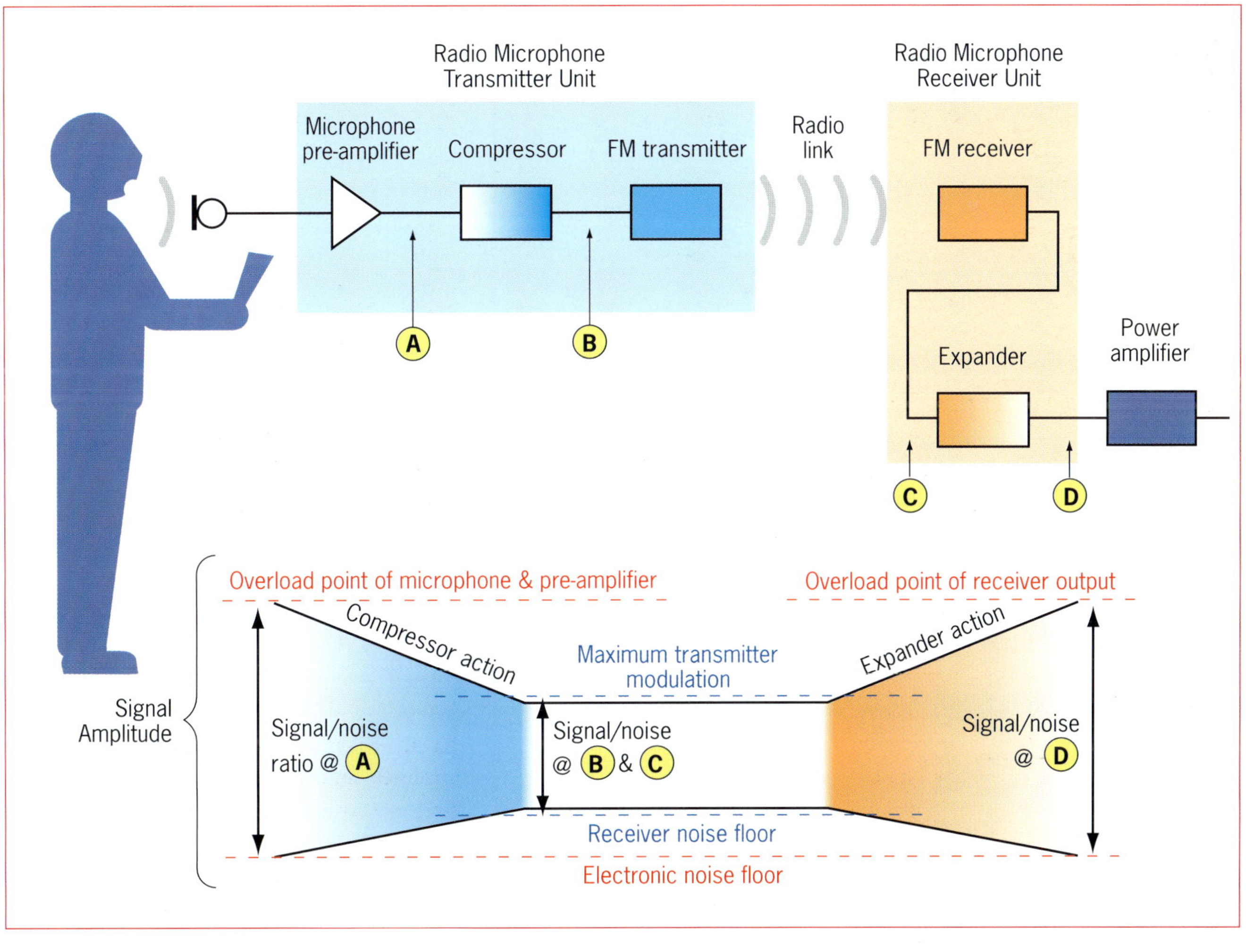

Figure 6.3: FM radio microphone system

EXPLANATION OF TECHNICAL TERMS

Matching loudspeakers and amplifier

Audio power amplifiers for sound reinforcement are made with two main types of outputs described as 'low impedance' and '100 V' or 'high impedance'. Similarly loudspeakers come in 4 or 8 ohms (low impedance) or 70 V or 100 V (high impedance).

Low impedance amplifiers and loudspeakers

If an amplifier is rated for 2, 4, 8 or 16 ohms, then it is a low impedance type. Care must be taken to ensure that the loudspeakers add up to a total load that is both within the amplifier's power rating (W or watts), and between its maximum and minimum load impedance range. Low impedance speakers, usually rated at 8 ohms for smaller types, have to be connected in a way that creates a total load within the range the amplifier is designed for. High impedance, 100 V or 70 V loudspeakers cannot be used satisfactorily. The advantage of low impedance systems is optimum audio performance, especially at low frequencies. Hi-fi loudspeakers are usually low impedance.

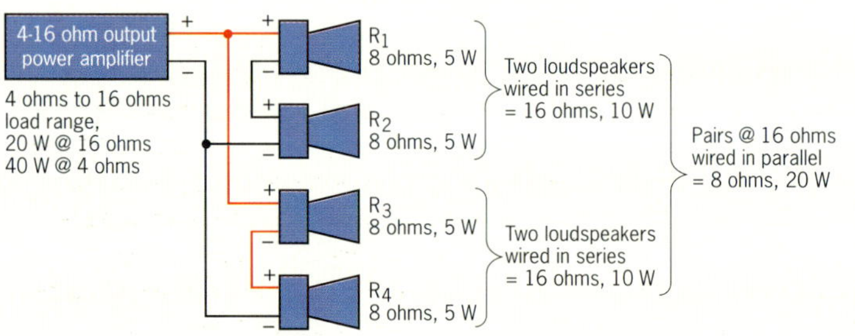

Calculating the load impedance

For loudspeakers wired in series – add up the individual impedances $R_{total} = R_1 + R_2 + \dots + R_N$

– add up the individual power $P_{total} = P_1 + P_2 + \dots + P_N$

For loudspeakers wired in parallel – add up reciprocals of the individual impedance

$$\frac{1}{R_{total}} = \frac{1}{R_1} + \frac{1}{R_2} + \dots + \frac{1}{R_N}$$

– add up the individual power $P_{total} = P_1 + P_2 + \dots + P_N$

In above example $R_1 + R_2 = 8 + 8 = 16$ for each series pair $= R_{1+2}, R_{3+4}$

Wiring the pairs in parallel gives $\frac{1}{R_1+R_2} + \frac{1}{R_3+R_4} = \frac{1}{16} + \frac{1}{16} = \frac{2}{16} = \frac{1}{8} = \frac{1}{R_{total}}$

Therefore $R_{total} = 8$

High impedance, 70 V or 100 V amplifiers and loudspeakers

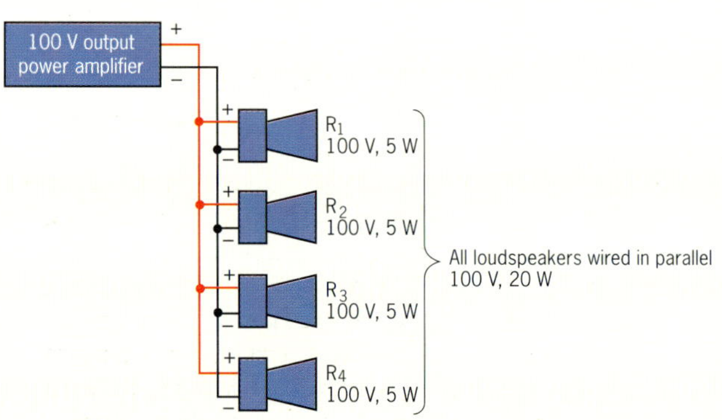

Calculating the load impedance

Impedance is taken care of automatically by the 100 V transformer in the system.

Total power is the sum of all devices connected.

is available to be used with many of the technologies identified within this section.

One of the major developments is the use of the 2.3 MHz and 2.5 MHz frequencies, allowing greater resistance to interference from fluorescent lighting and sunlight.

Table 6.4 compares the advantages and disadvantages of infra red and radio technologies.

6.8.5 Induction loop systems

Induction loop systems take advantage of the telecoil facility available with most hearing aids and cochlear implants. A telecoil is a small receiver capable of picking up audio frequency, electromagnetic signals. It is usually activated by setting a switch on the hearing aid to the "T" position. An induction loop system comprises a sound input (usually a microphone), an amplifier and a loop of cable which is run around the area in which the system is to be used. The loop generates an electromagnetic field which is picked up by the telecoil in the hearing aid. The hearing aid user will hear the sound while they are within the looped area.

Induction loop systems have many applications, from large-scale installations in theatres and cinemas to small, domestic products used to listen to the television. In the UK they are now rarely used in a classroom setting. Alternatives such as radio aids offer improved and more consistent sound quality and are less susceptible to interference. Induction loop systems can also be difficult to use in multiple applications, as the signal from one area can overspill into another.

In schools, induction loop or infra red hearing aid systems should be considered in large assembly rooms or halls. This is primarily for visitors to the school rather than for deaf pupils themselves, who would normally have their own assistive listening equipment. They should also be considered in performance spaces, meeting rooms and at reception area desks. In such situations the output from an existing PA system is often connected directly to the loop amplifier.

Pay phones in schools should have

inductive couplers (a form of induction loop).

Induction loop systems should be installed in accordance with British Standard BS7594. Their advantages and disadvantages are listed in Table 6.2.

6.8.6 Audio-visual equipment

Wherever possible, classroom equipment should be integrated with the assistive listening devices used by deaf children. For example, the audio output from audio-visual equipment, televisions and cassette recorders, can be connected to radio aid or soundfield transmitters. 'Direct input' leads are available to enable the audio output of computers or language laboratory equipment to be connected directly to a child's hearing aid.

6.8.7 Other assistive devices

There is a wide range of other devices that can be used by deaf children in school, besides those that primarily assist listening. These include subtitled and signed video, speech recognition software and text telecommunication devices, eg telephones.

For further details of these devices contact the professional or voluntary organisations listed at the end of this section. Furthermore, it is recommended to seek advice to ensure that all public spaces in a school meet the needs of deaf and hard of hearing people.

6.9 Special teaching accommodation

It is not the intention within this document to address the needs of special schools for deaf children. Specialist advice should always be sought from an educational audiologist or acoustician when designing or modifying accommodation for this particular purpose.

Many hearing impaired children attend mainstream schools with resource facilities, sometimes called 'units'. These contain specialised rooms that exceed the acoustic specifications for regular classrooms. Within these rooms, children are able to learn the language skills that might not be possible in a busy

EXPLANATION OF TECHNICAL TERMS CONTINUED

High impedance, 70 V or 100 V amplifiers and loudspeakers

If an amplifier is rated for 70 V or 100 V, then it is a high impedance amplifier. It will also have a power rating. High impedance loudspeakers, rated at 70 V or 100 V must be used. All loudspeakers should be either 70 V or 100 V. In this case the loudspeakers are simply wired in parallel and their individual power requirements are added up. Thus four 100 V loudspeakers rated at 5 W would be wired in parallel and will provide a 20 W load to the amplifier. External transformers can be added to low impedance loudspeakers to convert them for high impedance use. The advantage of this method is simple wiring. PA, paging and SFS loudspeakers are usually 100 V types in the UK.

Radio microphone system

Compander system (See Figure 6.3)

FM (frequency modulated) radio links provide a signal to noise ratio that is determined by the modulation bandwidth of the transmitter. Wider bandwidths allow fewer channels in a band of available frequencies, so regulations limit the bandwidth to two system types described as wideband FM and narrowband FM. Even wideband provides a limited signal to noise ratio of about 65 dB from real products. This is adequate if everything is perfectly adjusted so that a user's voice hits just below the maximum permitted signal level. However real users vary their voices, different users share systems AND they are often not correctly adjusted anyway. A compander system combines a compressor on the transmitter of the system, and an expander on the receiver. The two are matched in their action so that the result on the receiver output is very close to the original input signal. What happens is that a larger signal range of say 90 dB is compressed by 50% to fit into 45 dB. This allows for an improved safety margin in the transmitter so that it does not overload, and allows a wide working range that will tolerate user variations. At the receiver the 45 dB range is expanded back to 90 dB. This pushes the system noise down and the signal up. The result is a signal free from distortion due to overload and with a much reduced background noise when a soft talker is turned up at the receiver.

Diversity receiver

A FM radio microphone system emits a signal that has a fairly long wavelength. The waves can reflect from room surfaces and arrive at the receiver antenna in a way that causes the waves to cancel. The result is a 'dropout' which will be heard as a disappearance of the audio from the system. If the dropout is maintained, for example if the user is standing still in a location that produces a cancellation, the receiver can even hunt and locate an alternative signal to lock onto although this is uncommon. A diversity receiver provides two independent radio and audio paths, including two spaced antennae. The spacing minimises the risk that both antennae will receive a cancelled signal simultaneously. The unit will automatically and instantaneously select the stronger of the two signals to the audio output. While audio dropouts may be only slightly disturbing to a person with normal hearing, the hearing impaired child, especially one reliant upon a personal FM receiver, will get nothing and could therefore frequently lose the whole meaning or context of a piece of verbal information. Therefore, where possible, diversity receivers should be used.

mainstream classroom. They are also places where children can interact within a favourable acoustic environment.

It is not uncommon for these rooms to be used for 'reverse integration', where a small group of children from the mainstream work with the hearing impaired children. Occasionally this provision may be directly attached to a mainstream class in the form of a 'quiet room' leading from the classroom. In other situations the accommodation might be a separate room or even building. Teachers and support professionals might also use the areas for a range of activities involved in the audiological management of the hearing impaired child. Case Study 7.6 describes a junior school with a hearing impaired unit, now renamed as the RPD (Resource Provision for the Deaf). The characteristics of rooms in an RPD are:

- excellent sound insulation
- very short reverberation times
- very low ambient noise levels
- flexible space for individual and small group work
- good lighting
- storage facilities for audiological equipment.

Organisations

British Association of Audiological Scientists	http://www.baas.org.uk
British Association of Educational Audiologists	http://www.edaud.org.uk
British Association of Teachers of the Deaf	http://www.batod.org.uk
British Society of Audiology	http://www.thebsa.org.uk
National Deaf Children's Society	http://www.ndcs.org.uk
Royal National Institute of the Deaf	http://www.rnid.org.uk

Glossary

Term	Explanation
Natural-oral approach	An approach to the education of children with hearing impairments that seeks to promote the acquisition of spoken language using residual hearing.
Residual hearing	A term used to describe the hearing abilities that remain in the case of a hearing impairment.
Hearing aid	A battery powered device worn by an individual, either behind the ear or in the ear. A hearing aid will be selected and programmed to provide the maximum audibility of the speech signal consistent with an individual's residual hearing.
Cochlear implant	A special kind of hearing aid where the inner ear is directly stimulated electrically via an implanted electrode.
Central Auditory Processing Difficulty	A broad term used to describe listening difficulties, which are not due to the outer, middle or inner ear.
Radio aid	An assistive listening device, designed to provide an FM radio link between a transmitter (usually on the speaker) and the listener (coupled directly to the hearing aids).

6.10 Beyond the classroom

As far as possible children with hearing impairments should be included in all school activities. Improving listening conditions through better acoustics is a very important part of this, but not the only relevant factor. There are many others such as teaching style and context, staff training, deaf awareness issues, and a whole school approach to special educational needs.

Classrooms are not the only places where hearing impaired children interact. It is often overlooked in school design, but critical learning and interaction takes place outside the classroom, and if hearing impaired children are to be fully included, attention should be given to all areas of the school where the children might be expected to interact with others. These areas include rooms where aspects of the curriculum are delivered: libraries, assembly areas, sports halls, music rooms, ICT suites and gymnasia. In these areas the need for good speech communication is essential although constrained by the activities taking place.

Inclusion in most music activities requires good acoustic conditions, good planning and structuring of lessons, and the appropriate use of assistive listening devices.

Perhaps the most difficult areas for inclusion are large spaces such as assembly halls and sports halls. These areas require careful design and forethought.

In other areas, not used for delivering the curriculum, children still need to be able to interact verbally. These include the corridors, cloakrooms, medical rooms, school office, dining room, play areas and toilets. In these communal places important social interaction often takes place and if inclusion is to be effective, these areas need to be designed with the acoustic needs of the hearing impaired child and the child with listening difficulties in mind.

References

[1] M Eatough. Deaf Children and Teachers of the Deaf England, BATOD magazine, 2000.

[2] S Powers. S Gregory and E D Thoutenhoofd, The Educational Achievements of Deaf Children. DfES, 1998.

[3] J M Bamford, *et al.*. Pure tone audiograms from hearing-impaired children. II. Predicting speech-hearing from the audiogram. Br J Audiol, 15(1), 3-10, 1981.

[4] M Picard and J S Bradley. Revisiting speech interference in classrooms. Audiology, 40(5), 221-44, 2001.

[5] BATOD. Classroom Acoustics - Recommended standards. 2001.

[6] ASHA. Position Statement and guidelines for acoustics in educational settings. ASHA, 37(14), 15-19, 1995.

[7] S Gatehouse and K Robinson. Speech tests as a measure of auditory processing, in Speech Audiometry, Second Edition, M Martin, (Editor) Whurr: London, 1997.

[8] A Markides. Speech levels and speech-to-noise ratios. Br J Audiol, 20(2), 115-20, 1986.

[9] J A Mattiske. J M Oates and K M Greenwood. Vocal problems among teachers: a review of prevalence, causes, prevention, and treatment. J Voice, 12(4), 489-99, 1998.

[10] T Finitzo-Hieber and T W Tillman. Room acoustics effects on monosyllabic word discrimination ability for normal and hearing-impaired children. J Speech Hear Res, 21(3), 440-58, 1978.

[11] O Wilson *et al.*. Classroom Acoustics, Oticon Foundation in New Zealand: Wellington, 2002.

[12] D Canning. Listening Inventories For Education U.K., in LIFE UK, City University, London, 1999.

[13] RNID. Guidelines for mainstream teachers with deaf pupils in their class. Education guidelines project, RNID, 2001.

[14] National Deaf Children's Society. Deaf Friendly Schools – a Guide for Teachers and Governors, NDCS, 2001.

[15] DfES. Special Education Needs Code of Practice, DfES/581/2001.

[16] B Homer, R Vaughan and K Higgins. Radio Aids, NDCS, 2001.

[17] National Deaf Children's Society. Quality Standards in Education - England, NDCS, 1999.

Case studies

SECTION 7

This section contains ten case studies which illustrate some of the principles of the acoustic design of schools described in previous sections, and give examples of solutions to problems of poor acoustics in schools.

The school is situated at a considerable distance from the main road running through a large village in a quiet residential area. In the early 1990s, it was extended by adding seven new classbases and a new multi-purpose hall. Activities in the hall include assemblies, singing, concerts and physical education. The hall is of particular interest because it required remedial measures not long after completion to alleviate acoustic problems that were being experienced by teaching staff.

The new hall is adjacent to playing fields and background noise levels around the school are low. Therefore there is little disturbance to occupants of the hall from external noise.

The hall is built of conventional masonry cavity walls comprising 100 mm facing brick outerleaf, 50 mm cavity, and 140 mm blockwork inner leaf with a plaster finish. A plan and section of the hall are shown in Figure 7.1.1.

The roof has a hipped form and is constructed of steel trusses with 100 mm by 50 mm softwood rafters at 600 mm centres. It is covered with slates on battens and felt. The shallow pitched ceiling is formed from tongue and groove timber boards (119 mm by 19 mm), overlain with 150 mm thermally insulating mineral wool batts. The roof void increases from a height of 200 mm at the eaves to 2 m at the ridge.

Large external windows with opening lights are located in the north-east and south east walls with a row of smaller high level opening lights located in the external wall to the south-west, above the circulation corridor. The circulation corridor connects the hall to the main building at ground floor level via glazed doors in a glazed screen. The corridor also provides a useful acoustic buffer between the hall and the nearby classrooms and offices. External windows and doors are all thermally double-glazed. Internal doors and the glazed screen are of 6 mm glass.

Wall bars and similar apparatus are supported off the two long walls. The floor is of sprung timber strip to accommodate physical education, dancing, etc. The hall is naturally ventilated.

Figure 7.1.1: Plan and section of the new hall showing extent of remedial treatment

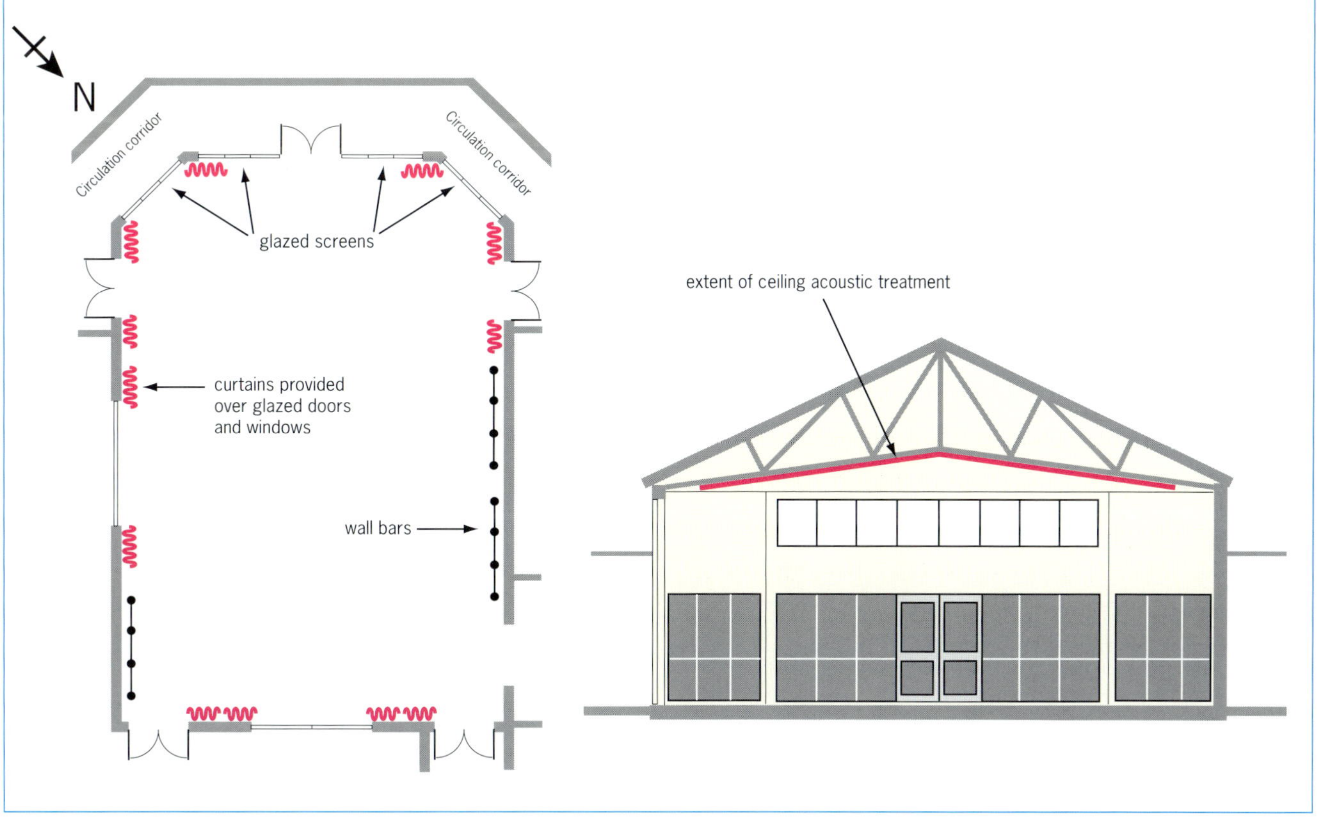

7.1 Case Study: Remedial work to a multi-purpose hall in a county primary school

The new hall suffered from:

- poor speech intelligibility, particularly with small groups of 30 or less
- distortion or colouration of speech
- unusually high background noise noise levels, eg from the shuffling of children's feet.

Teachers found that they could improve speech intelligibility slightly if they slowed down their normal rate of speech or addressed groups of pupils from a sidewall rather than near the centreline. In fact, speech from around the centreline of the hall appeared louder than normal and sounded coloured or distorted.

An acoustical assessment showed that speech was most distorted when both speaker and listener were near the centreline. Flutter echoes and enhanced reverberation were clearly evident and disturbing. When speaker and listener were both near a side wall the conditions were less severe although still poor.

The acoustical faults correlated well with the teachers' complaints. The majority of complaints stemmed from excessive reverberation, attributable to the predominantly hard surfaces in the hall. Both floor and ceiling were hard and acoustically reflective. Excessive reverberation caused consecutive syllables in speech to run into one another, reducing intelligibility.

This problem was compounded by the shape of the ceiling. It has a shallow pitch with hipped ends, similar to an inverted concave dish. Sound focused by the hard reflective ceiling onto the hard floor below and the resulting multiple reflections were detected as a longer reverberation time (RT) near the centreline. This effect caused sounds to appear louder than normal and coloured or distorted.

To rectify these faults, it was proposed that the ceiling should be made acoustically absorbent. This would reduce the RT to a level suitable for primary school uses and reduce the focusing effect.

Although it provided a solution in this case, it is not normally advisable for ceilings to be sound absorbing in rooms where good speech intelligibility is a requirement. If the size, shape and geometry of the space are right in the first place, then the ceiling should be reflective to sound. The reason for the success of the ceiling treatment in this case was the overriding need to make a substantial reduction in RT and the fact that the floor has a timber finish, which provides a useful reflection path in the absence of a comparable reflection from the ceiling.

The school wanted to retain the timber ceiling. Therefore the timber boards were taken down and a series of 20 mm by 200 mm slots were cut into them (see Figure 7.1.2) to give an open area of approximately 25%. A mineral fibre acoustic quilt, 25 mm thick, was laid directly over the slots in the ceiling void. The quilt was faced with an acoustically transparent black scrim on the hall side for aesthetic reasons. The existing layer of thermal insulation was replaced over the acoustic quilt. Figure 7.1.1 indicates the area of the ceiling that was treated. The acoustic treatment to the timber ceiling is considered to be in keeping with the appearance of the hall (see photograph, Figure 7.1.3).

In addition to the ceiling treatment, acoustically diffusing panels were recommended for the walls to distribute sound evenly around the hall and prevent flutter echoes. An example of a diffusing panel is shown in Figure 7.1.4. However, these panels were omitted due to lack of funds. As a result of this omission and the presence of an acoustically absorbent ceiling, there is a tendency for sound to

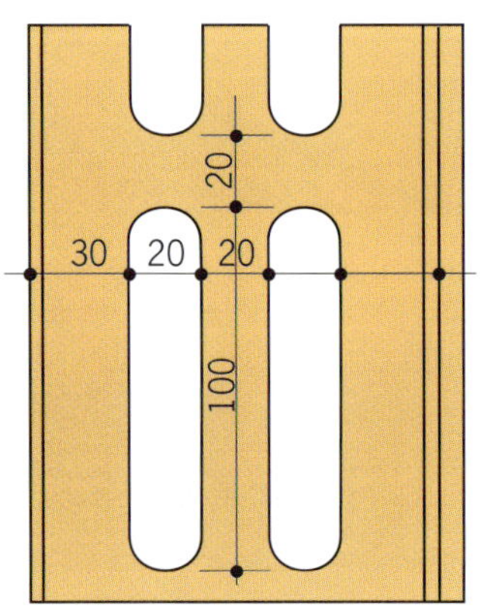

dimensions in millimetres

Figure 7.1.2: Detail of timber slats used to line the hall ceiling

Figure 7.1.3: The hall ceiling after acoustic treatment

reverberate around the hall in a horizontal plane, particularly when occupancy is high and the floor is obscured. Under certain conditions, this manifests itself as distracting flutter echoes between the hard parallel side walls. One teacher reported this effect as a disturbing 'ringing' noise whilst rehearsing music and dance with a small group of children at the south west side of the hall.

Following remedial acoustic treatment to the ceiling, the response from the teachers to the modified acoustics of the hall was very favourable and all reported a very noticeable improvement.

Speech intelligibility was found to be much improved when addressing both small and large groups of children, and noise from physical activities and children shuffling feet during assembly has been reduced to acceptable levels. Communication during physical education and similar noisy activities is easier, and accompanied by lower levels of background noise.

The reverberation time was measured in the same positions before and after the remedial work. Two sets of measurements were made; one with the source and receiver on the centreline of the hall and the other with the receiver positioned 2 m from a side wall. Measurements were made while the space was unoccupied. Curtains were pulled back to their normal bunched positions either side of internal and external doors and windows. This arrangement was considered to produce

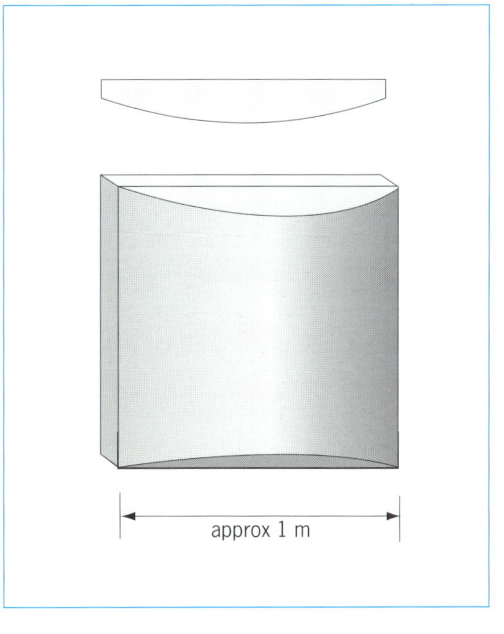

Figure 7.1.4: Example of acoustically diffusing panel

Case Study: Remedial work to a multi-purpose hall in a county primary school

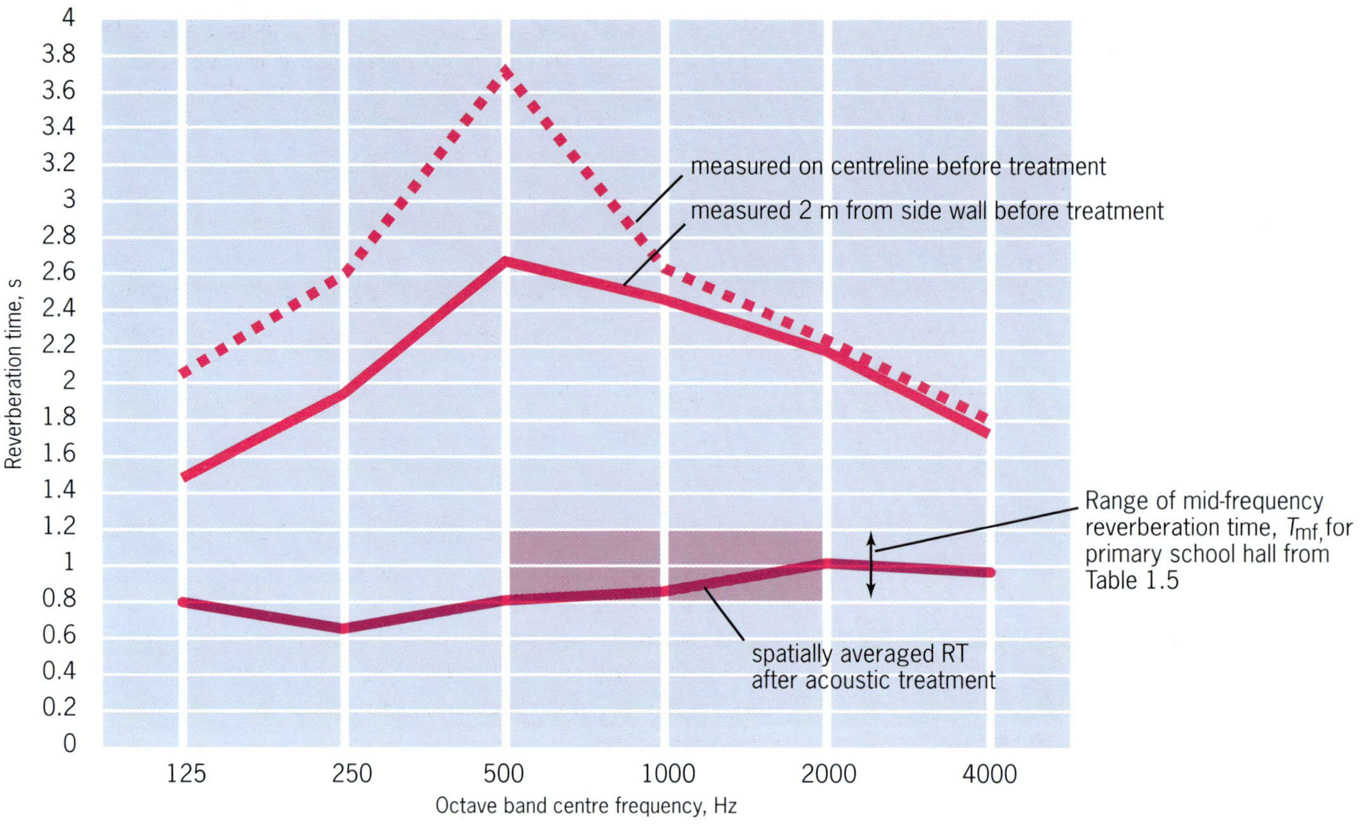

Figure 7.1.5: Measured reverberation time in the new hall before and after implementation of acoustic measures

the most reverberant condition likely to be encountered during every day small group activities.

Before remedial work, the measured T_{mf} was 2.8 seconds on the centreline but fell to 2.5 seconds along the side of the hall. Figure 7.1.5 shows the measured RT curves as a function of frequency. The T_{mf} after treatment is within the range specified in Table 1.5 for a primary school hall, which should be between 0.8 and 1.2 seconds.

Concerts and musical activities take place in less reverberant conditions than before, with substantial reductions in colourations and distortions. These conditions have been found to be satisfactory. The introduction of acoustic absorption into the ceiling of the new hall has been successful in providing acoustic conditions which are suited to primary school uses.

It is clear from this study that the acoustics of a hall are of fundamental importance in the effective functioning of this key space in a primary school. In many halls, hard wall and floor finishes will be necessary and the required acoustic absorption will need to be accommodated in the ceiling. Ideally, absorbent and reflective surfaces should be more or less evenly distributed on both walls and ceiling. This case study, where modification of the existing ceiling was complicated and costly, highlights the importance of considering the acoustic requirements at the design stage.

Case Study 7.2: An investigation into the acoustic conditions in three open-plan primary schools

7.2

An investigation of the acoustic conditions in three recently built open-plan primary schools was carried out. Sound insulation between classrooms and reverberation times and sound levels in unoccupied classrooms were measured.

The effect of noise from adjacent areas on speech intelligibility within the learning bases was assessed. The Speech Transmission Index (STI) was measured in the classrooms using Maximum Length Sequence (MLS) analysis equipment as described in BS EN 60268-16. In each case an artificial mouth, positioned where the teacher usually stood during lessons, was used to produce a reference signal which was received by a microphone at different positions within the room. Speech intelligibility was rated using the measured STI values.

7.2.1 School 1 (pupils aged 5 -11 years)

The layout of the school is shown in Figure 7.2.1(a). The walls are full height between the classrooms and the corridors, with teaching areas accessed via open arches from the corridors The two teaching areas on each side of the corridors are open plan, being separated only by a quiet/IT area. Measurements were conducted in the Yellow and Green team areas indicated. The layout of the Green team area with measurement positions is shown in Figure 7.2.1(b).

Measurement results

The measured classroom mid-frequency reverberation times (T_{mf}) are shown in Table 7.2.1.

The sound level ($L_{Aeq,10min}$) was measured in classroom Y1 when occupied during a typical interactive science lesson, and when unoccupied after the school day had finished. The sound level was also measured in the unoccupied Yellow team (Team 4) practical area indicated in Figure 7.2.1. The measured sound levels are shown in Table 7.2.2.

It should be noted that although the level in classroom Y1 was measured after the children had left the school, the corridor adjacent to the room was still occasionally used by those involved in after school activities. The sound level in the unoccupied practical area was measured with lessons being conducted in all the adjacent teaching rooms.

As the school was in use, 10 minutes was the longest practical time period for the measurements of indoor noise levels.

The speech transmission index (STI) was measured at 5 positions in the unoccupied room G3 with and without masking noise being generated in rooms G1 and G2. The position of the artificial mouth and the 5 microphone positions are shown in Figure 7.2.1. The masking noise had the same level as was measured during the science lesson in classroom Y1 and was shaped to give similar levels, in the third-octave frequency bands between 50 Hz and 5 kHz, as those measured.

Table 7.2.1: Classroom mid-frequency reverberation times

Room	Mid-frequency reverberation time (s)
Y1	0.4
G1	0.4
G2	0.4

Table 7.2.2: Sound levels in Yellow team area

	Classroom Y1 occupied	Classroom Y1 unoccupied	Practical area unoccupied
$L_{Aeq,10min}$ (dB)	62.7	42.4	54.2

Table 7.2.3: Average STI values in unoccupied room G3 with and without masking sound in rooms G1 and G2

Microphone position	No masking		Mask in room G1		Mask in room G2	
	STI	Rating	STI	Rating	STI	Rating
1	0.803	Excellent			0.639	Good
2	0.673	Good	0.654	Good	0.642	Good
3	0.761	Excellent	0.691	Good	0.426	Poor
4	0.739	Good			0.550	Fair
5	0.745	Good			0.555	Fair

Table 7.2.4: Measured sound insulation between classrooms

Rooms	$D_{nT(0.6s),w}$ (dB)
Y1 to Y4	16
G1 to G2	19
G1 to G3	24
G2 to G3	14

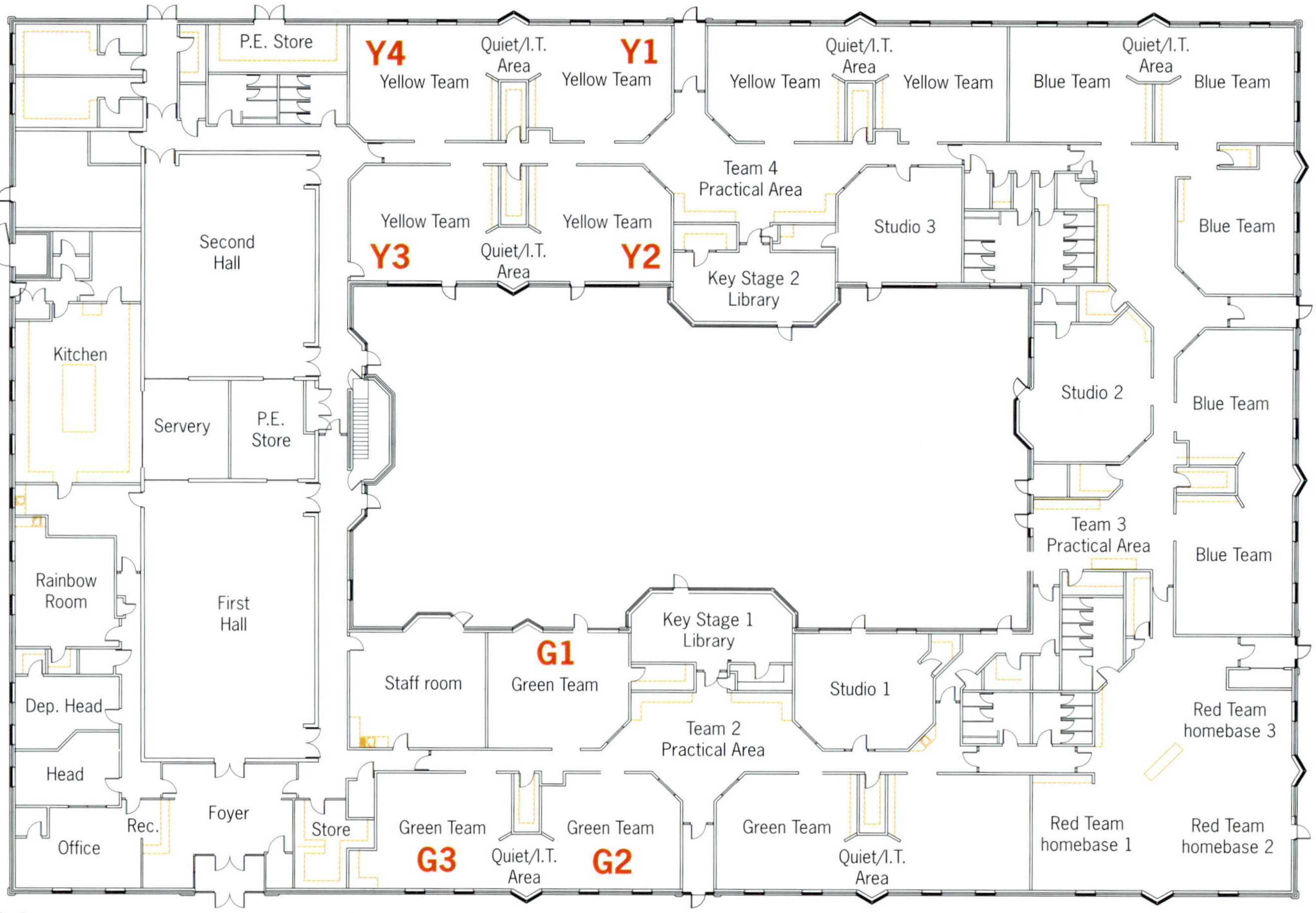

(a)

Figure 7.2.1:
School 1 layout
(a) Whole school
(b) Green team test area

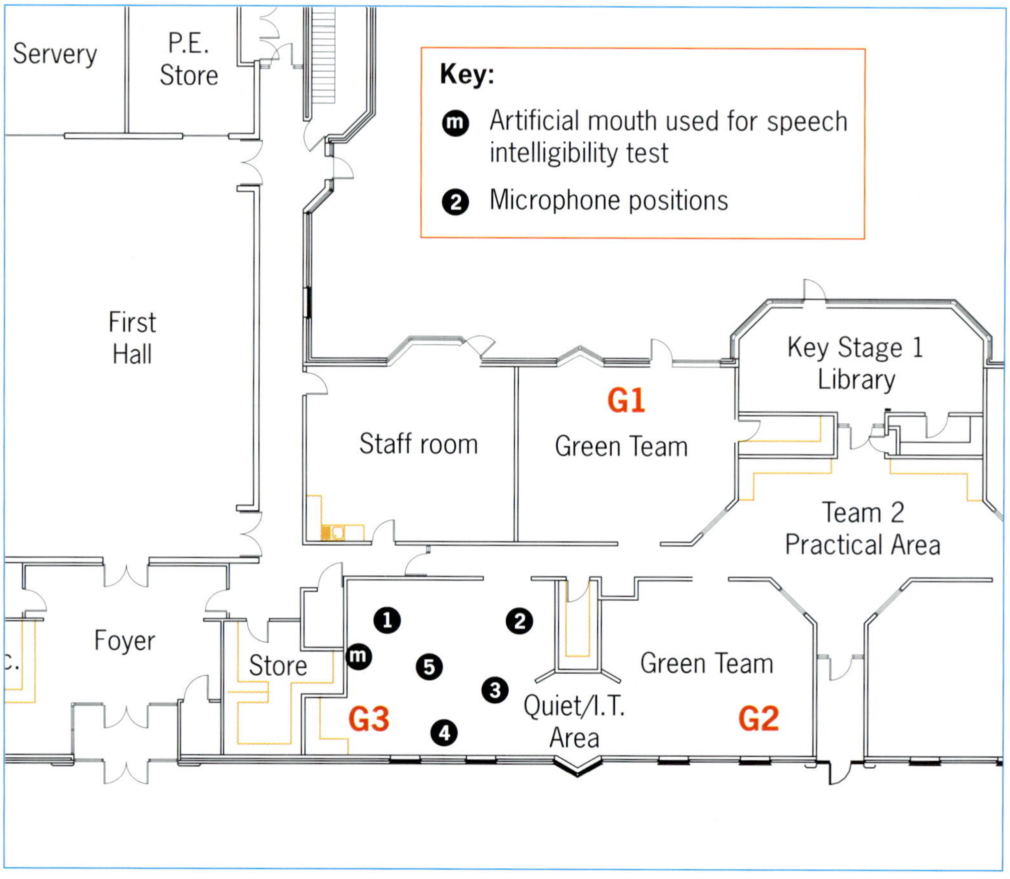

(b)

The STI measurement results are shown in Table 7.2.3.

The results of sound insulation measurements between classrooms are shown in Table 7.2.4.

Discussion

This school was selected for investigation primarily because it had been reported that the school's open-plan design worked well. The head gave the impression that he strongly favoured the open-plan layout and stated that he had been closely involved with the design process of the new school. However, other members of staff were less enthusiastic.

A team leader in the school stated that the open-plan design suited the teaching practices in the school although it had taken some time to get used to at first. Other teachers were forthright in their disapproval of the school's design and the restrictions that it imposed.

Of the teachers whose opinions were canvassed, the majority stated that they felt the open-plan design led to problems associated with disturbance. Timetabling was organised so that the activities in adjacent teaching areas produced similar levels of noise in order to avoid disturbance to pupils involved in quiet activities.

According to the teachers consulted, usually the arrangement was acceptable but problems could be caused if a teacher unfamiliar to the pupils was taking a class in an adjacent area. In such circumstances the usual strict enforcement of discipline on the children could be subverted leading to disturbance in adjacent areas.

The measured levels in the unoccupied Yellow team practical area and classroom Y1 were greater than those specified in Section 1. In the practical area, it can be assumed that the measured level was affected by sound from adjacent occupied classrooms. For example, it was noted that during measurements in the unoccupied practical area one of the teachers constantly reminded the children to work quietly by uttering the command "Shh" at regular intervals. At a different time, a teacher in classroom Y4 raised her voice sufficiently for her words to be heard clearly in classroom Y1. This was a result of her feeling the need to admonish a pupil for holding a conversation from the open corridor with one of her class members.

The behaviour of the teachers and pupils did not appear to be unusual and the strong impression was given that the day of the investigation was a typical school day.

The measurements of the Speech Transmission Index (STI) showed that speech intelligibility was reduced considerably during an interactive science lesson in classroom Y1. This was due to the increased sound level ($L_{\text{Aeq,10min}}$) during the lesson.

The mid-frequency reverberation time in each of the classrooms was 0.4 seconds, which is acceptable for classrooms for hearing impaired pupils. Because of this, in the absence of children and teachers, the measured STI rating varied between good and excellent in unoccupied classroom G3. However, when masking noise was generated in room G2, the STI rating was reduced to poor and fair in positions 3, 4 and 5. This suggests that, when the teacher is speaking to the class from the usual position, pupils sitting closest to room G2 are likely to experience more difficulty understanding the teacher's words than other pupils in the classroom due to noise emanating from room G2. The measurement of STI showed that noise generated in G1 had no significant measurable effect on speech intelligibility in room G3. This is likely to be due to the stagger between the entrances to rooms G1 and G2 on opposite sides of the corridor.

It should be noted that STI is an objective measurement of speech intelligibility, and cannot quantify disturbance to pupils. Disturbance may depend, for example, on whether pupils perceive sound generated in adjacent areas to be interesting or threatening.

7.2 Case Study: An investigation into the acoustic conditions in three open-plan primary schools

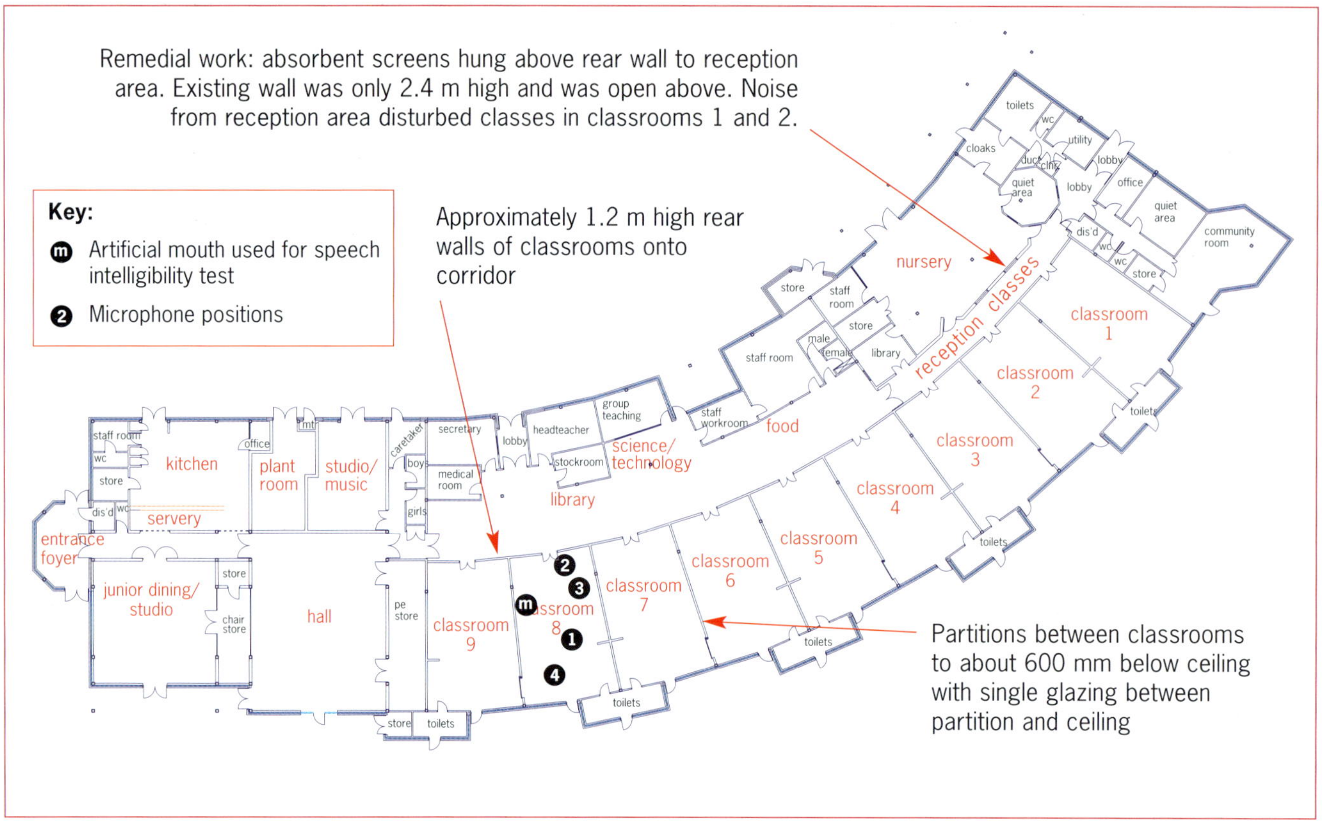

Figure 7.2.2: School 2 layout

7.2.2 School 2 (pupils aged 3 -7 years)

The school layout is shown in Figure 7.2.2. Classrooms 1 to 3 are for reception classes and no measurements were conducted in this area, where some remedial work had been carried out. Originally, the nursery wall onto the corridor was only 2.4 m high with a gap above. However, the disturbance to classrooms 1, 2 and 3 from the high noise levels resulting from normal nursery activities necessitated the closing off of the nursery area using full height acoustic panels above the existing nursery wall.

The walls separating the classrooms from the corridor area are approximately 1.2 m high. Some common resource areas, eg the library and the science/technology area, are located in the corridor space. It is possible to see over the low wall into the corridor and directly from one classroom to another in the vicinity of these walls. Although the height of the walls between classrooms increases towards the external wall of a classroom, at no point is there a continuous barrier from the floor to the ceiling between the classrooms. The gap above the partition wall provides a clear sound path, see Figure 7.2.3, and could with forethought have been easily closed off. However, it is doubtful if this would have made a big difference to the sound transmission given the low walls onto the corridor.

Figure 7.2.3: Partition wall between classrooms. Note the large gap above the glazed head of the partition.

Measurement results

Measurements were conducted in rooms 4 to 9.

Because the classrooms were identical in appearance, the mid-frequency reverberation time was measured only in unoccupied classroom 8, and was 0.5 seconds.

STI was measured in classroom 8.

First, measurements were conducted at the positions indicated in Figure 7.2.2, without any masking noise in adjacent areas. After this, white noise was generated as masking sound in room 9 to represent noise from an occupied classroom and STI was measured in positions 3 and 4 in room 8. Masking levels of 60 dB(A) and 70 dB(A) were used. White noise was used as masking sound because no pupils were in the school during the measurements. Therefore, a typical classroom sound spectrum could not be recorded and used as the masking signal. The artificial mouth was positioned 1m in front of the white board on the wall between rooms 8 and 9. as shown in Figure 7.2.2.

The average STI values measured are shown in Table 7.2.5.

Table 7.2.6 shows the measured airborne sound insulation between classrooms in terms of the weighted BB93 standardized level difference ($D_{nT(0.8s),w}$). Table 7.2.7 shows the measured sound levels in the classrooms with a sound source in classroom 9.

Microphone position	Masking level (dB(A))	STI	Rating
1		0.656	Good
2		0.616	Good
3		0.640	Good
4		0.588	Fair
3	60	0.459	Fair
3	70	0.263	Bad
4	60	0.541	Fair
4	70	0.430	Poor

Table 7.2.5: Average STI values in classroom 8 with and without different levels of masking noise in classroom 9

Discussion

Brief discussions were held with the head of the school and a few other teachers before and after measurements began. The head stated that she liked the open-plan design since it meant that pupils were accustomed to seeing her and she could enter classrooms without causing undue disturbance.

When the school was first used, problems with high noise levels had been experienced in the reception class area but these were alleviated by the addition of acoustically absorbent panels on the wall opposite the classrooms. No other adverse comments about the acoustics in the school were made by any of the teachers interviewed although one teacher did describe an unusual situation caused by the lack of acoustic isolation between classrooms.

The same story was being read to pupils in adjacent classrooms at the same time. The teacher said that she became aware that her colleague in the adjacent classroom was one or two words ahead of her in the story. She described the situation as being "like hearing an echo" and attempted to speed up her reading in order to synchronise the delivery to both classes.

Rooms	$D_{nT(0.8s),w}$ (dB)
7 and 8	13
9 and 4	28

Table 7.2.6: Measured sound insulation between classrooms

The design of the school means that acoustic isolation between classrooms and the area outside the classrooms would be expected to be low. The results of the measurements taken bear this out. 13 dB $D_{nT(0.6s),w}$ between classrooms 7 and 8 is a very low level of sound insulation. Indeed, 28 dB $D_{nT(0.6s),w}$ between classroom 9 and classroom 4 (which are

Table 7.2.7: Sound levels in classrooms 4, 5, 6, 7, 8 and 9 with sound source in classroom 9

Classroom	4	5	6	7	8	9
$L_{Aeq,2min}$ (dB)	68.2	69.5	72.8	75.8	81.4	96.1

not adjacent, see Figure 7.2.2) is significantly lower than the 45 dB between adjacent classrooms required in Table 1.2 of Section 1.

Comparison of STI values in a classroom with and without masking noise generated in an adjacent classroom demonstrates that there is a significant reduction in speech intelligibility due to the masking noise

The data in Table 7.2.5 show that the STI values and, consequently, speech intelligibility were reduced in the two positions used for the measurements when masking noise was generated in the adjacent classroom and when the level of the noise was increased. Position 4 was better screened from classroom 9 than position 3 where there was an almost uninterrupted path between the two rooms owing to the lower dividing partition at this point. The measurements show that speech intelligibility in position 3 is reduced by masking noise generated in room 9. The masking noise had less effect on STI in position 4 than in position 3. However, position 4 had the lowest STI value of the four measurement positions. This is largely due to the artificial mouth being directed into the classroom perpendicularly from the wall. Directing the artificial mouth towards position 4 would have increased the STI value at this position. Thus, unless the

Figure 7.2.4: School 3 layout showing recently added extensions

teacher is looking directly at a child at this position, the speed intelligibility will only be 'fair'.

Since the mid-frequency reverberation time measured in two of the classrooms was 0.5 seconds problems with speech intelligibility can be attributed to high ambient noise levels in the classrooms. Because the sound insulation between the rooms is so low, it is likely that noise generated in adjacent areas will contribute to the overall sound levels in the rooms.

7.2.3 School 3 (pupils aged 4 -8 years)

This is a recently built school which has been extended. The extensions accommodating rooms 1 to 8 are shown in Figure 7.2.4. Measurements were conducted in the original school building with children present and in the extensions both with and without the children present.

Measurement results

Table 7.2.8 shows mid-frequency reverberation times measured in the classrooms. Tables 7.2.9 and 7.2.10 show measured sound levels in occupied and unoccupied classrooms respectively.

The results from the measurements in Schools 1 and 2 demonstrate that STI values are reduced by noise from adjacent areas and that those positions closest to the noise are likely to be most affected. Therefore, STI was measured in only one position in two classrooms. In room 3, STI was measured with the curtains between rooms 3 and 4 open and closed. All measurements were conducted with the artificial mouth positioned where the teacher would usually stand, see Figure 7.2.4, and the receiving microphone was positioned 3 m in front of the artificial mouth. The results are shown in Table 7.2.11.

Discussion

The head in this school was strongly in favour of the open-plan design of the school for the following reasons:

- she felt that the staff worked better as a team
- she felt that the children worked better as a group
- she felt that open-plan design allowed more flexibility
- she felt that organising pupils in common teaching areas was "more natural", especially for those joining the reception class.

However, prior to this investigation, the head had contacted her local education authority due to problems encountered in the extensions to the school containing rooms 1 to 4 and 5 to 8. Here, difficulties had been encountered which resulted in 'acoustic curtains' being fitted to separate the classrooms from the communal areas 4 and 5. When the measurements were made, the curtains were temporarily removed from rooms 6 to 8.

The measurement results given in Table 7.2.8 show that the reverberation time in the original building is shorter than in the two extensions. They also show that the reverberation times in the rooms with curtains (rooms 3 and 4) are lower than

Room	Mid-frequency reverberation time (s)
9	0.4
6	0.7
5	0.9
3	0.6
4	0.6

Table 7.2.8: Classroom mid-frequency reverberation times

Room	Lesson type	$L_{Aeq,3min}$ (dB)
5	Project work	74.9
6	Literacy	69.7
4	Project work	69.3
3	Numeracy	69.8
10	Project work	66.2
9	Room empty	56.2

Table 7.2.9: Sound levels in occupied classrooms

Room	$L_{Aeq,3min}$ (dB)
5	35.4
4	32.8
3	31.8

Table 7.2.10: Sound levels in unoccupied classrooms

Table 7.2.11: Average STI values in unoccupied classrooms. Note: adjacent rooms were occupied during the measurements in Room 9.

Room	STI	Rating
9	0.570	Fair
3 (curtains open)	0.689	Good
3 (curtains closed)	0.693	Good

those in rooms without curtains (rooms 5 and 6), which exceed the values specified in Table1.5. These results suggest that the acoustic problems experienced by staff in the extensions to the original building can largely be attributed to the lack of sound absorption and the consequent relatively long reverberation times in these areas compared with the original school building. The original building had acoustically absorbent ceilings whereas the extension did not.

The results from the measurement of STI show that in unoccupied classroom 3 speech intelligibility was good with the curtains both closed and open. In room 9, the speech intelligibility rating was fair. Since the reverberation time in room 9 is 0.5 seconds, this lower rating can be attributed to noise generated by children in the adjacent areas at the time of the measurements.

The results shown in Table 7.2.12 show that the curtains reduce the sound transmission between classrooms, in addition to reducing reverberation times, although none of the sound insulation values measured meets the specification of 45 dB in Table 1.2 of Section 1.

Table 7.2.9 shows the sound levels recorded in different rooms whilst lessons were taking place. Pupils were engaged in project work in room 4 while in rooms 3 and 6, more formal literacy and numeracy lessons were being conducted.

Project work meant that the children were working in small groups and noise levels generated by their interaction during these activities were higher than would be expected in a formal lesson. For example, in the numeracy lesson in room 3, the teacher sat in one corner of the classroom with the children seated close to her. In this lesson the teacher spoke and the children responded when it was appropriate.

The level in room 5 was approximately 5 dB higher than the level measured during the literacy lesson in room 6. This part of the extension did not have curtains fitted between the rooms at the time of this investigation. In the other extension, where curtains were fitted, the levels measured in room 4 (project work) and room 3 (numeracy lesson) were virtually the same. The curtains between rooms 3 and 4 were drawn so that only a gap of around 300 mm remained between the curtain and the wall separating room 3 from 4.

A teacher in the school volunteered comments on teaching conditions in rooms 1 to 4. She said that she found it difficult to hear some softly spoken pupils due to the high noise levels in the classroom and that parents would inform her if and when their child had difficulty hearing what was being said in lessons.

The teacher was of the opinion that room 2 was worse to teach in than rooms 1 and 3 because of the low sound insulation of the walls separating the rooms and the consequent noise transmission from rooms 1 and 3. Unprompted she described the difficulties experienced when reading the same story to her class as the teacher in the adjacent room but being a sentence or two behind or in front of her colleague next door.

The teacher felt that the curtains had improved conditions in the classrooms. Cupboards had also been placed in the openings between rooms 1, 2, 3 and 4 in an attempt to improve sound insulation between the different teaching areas. In her opinion, the cupboards had been useful for this purpose.

Table 7.2.12: Measured sound insulation between classrooms

Rooms	Curtains	$D_{nT(0.6s),w}$ (dB)
3 to 4	open	10 dB
3 to 4	closed	15 dB
3 to 2	open	21 dB
3 to 2	closed	28 dB

7.2.4 Summary

In all the schools visited, the head teachers appeared to approve of the open-plan design in their school. Some teachers shared their head's enthusiasm for the design but others felt that problems caused by the transmission of sound between rooms were significant.

The measurement of STI in the schools demonstrated that speech intelligibility is reduced by noise generated in adjacent rooms. In all the open-plan schools, high ambient noise level was the most significant cause of low speech intelligibility.

From the few opinions canvassed in the schools it would appear that there are benefits to adopting an open-plan design. These appear to be that the design is favourable for team working, that it engenders a feeling of inclusion in the school and that it allows for a visually attractive environment. However, placing cupboards in spaces between rooms in order to increase isolation between them may detract from the original open-plan design.

From the results of this survey, it is difficult to justify the use of open-plan schools in terms of their acoustic environment. None of the schools met the requirements for sound insulation between classrooms contained in Building Bulletin 93. Although reverberation in classrooms was well controlled (apart from in the extension in School 3), ambient sound levels during teaching periods were too high for the measured STI values to indicate good speech intelligibility. As a consequence of the levels in the classrooms, both teachers and pupils would need to speak more loudly in order to be clearly understood.

In many open-plan teaching spaces it is difficult to achieve clear communication of speech between teacher and students.

For this reason, careful consideration should be given as to whether to include open-plan teaching spaces in a school. If open-plan areas are required then rigorous acoustic design is necessary to satisfy the performance standards in Section 1.

The primary school in Case Study 7.1 extended its facilities in the early 1990s by adding seven new classbases and a multi-purpose hall to the existing school. The new classbases were arranged in two open-plan areas, of three and four classbases respectively.

The teaching area with four classbases, numbered 4 to 7 in Figure 7.3.1, is an interesting example of the limitations of an open-plan environment. Acoustic problems were experienced by the teaching staff which subsequently led to the implementation of remedial measures. Visits were made to the school before and after the remedial work.

The new extension to the school is of conventional masonry cavity walls, comprising 100 mm facing brick outerleaf, 50 mm cavity and 140 mm blockwork inner leaf with a plaster finish.

The roof over the teaching area is made up of a combination of pitched sections with a tiled exterior and flat roof constructions with a felt finish. Each pair of adjacent classbases has a roof light located in a flat roof section. Windows are thermally double-glazed and openable.

Internal walls are generally constructed of either 100 mm or 140 mm lightweight blockwork. Surface finishes are generally hard and reflective except for the floor which is covered in a short pile carpet. The walls are plastered and have an emulsion paint finish. The ceiling in the open-plan teaching areas is constructed of 12.5 mm plasterboard with a painted skim finish.

The general features to note about the layout of the open-plan teaching areas are:

- All four teaching spaces are incorporated in an open-plan arrangement.
- The physical separation between classbases 4 and 5, and between classbases 6 and 7 is minimal which implies negligible acoustic separation. A small improvement in acoustic separation may be gained by strategic positioning of tall items of furniture, eg bookcases, between adjacent classbases.
- The physical separation between classbases 5 and 6 is only partial and is formed by the projection of the group room on one side and book shelving on the other.
- The teaching spaces are separated from potentially noise producing and noise sensitive spaces, eg other classrooms and the main hall, by a corridor. This arrangement is advantageous for reducing noise disturbance to or from other parts of the school.
- Toilets and services provided in the corner of each pair of bases are buffered from the teaching spaces by lobbied doors.

Teaching staff perceived the open-plan teaching area to be difficult to work in because of poor acoustics. They had two

Figure 7.3.1: Plan of open-plan teaching spaces before modifications

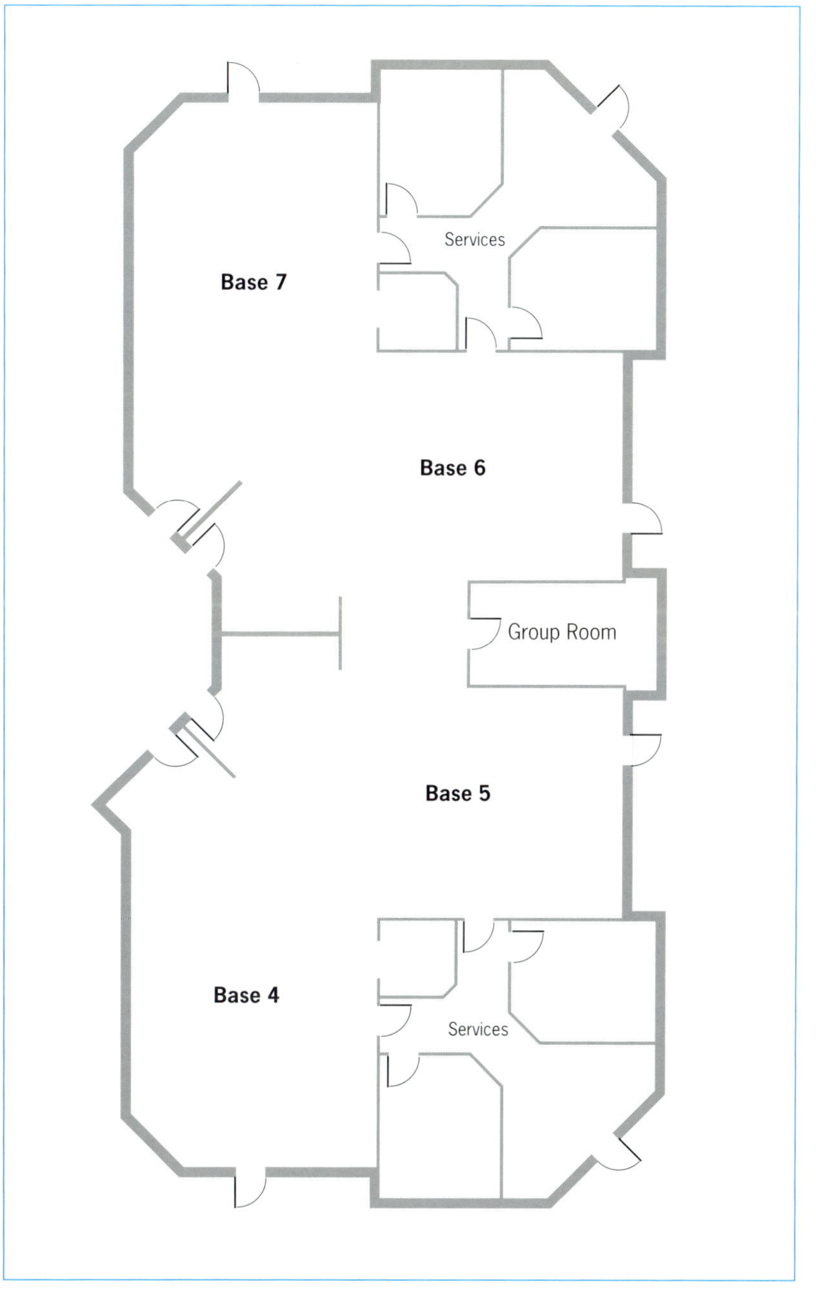

Case Study: Remedial work to an open-plan teaching area in a primary school

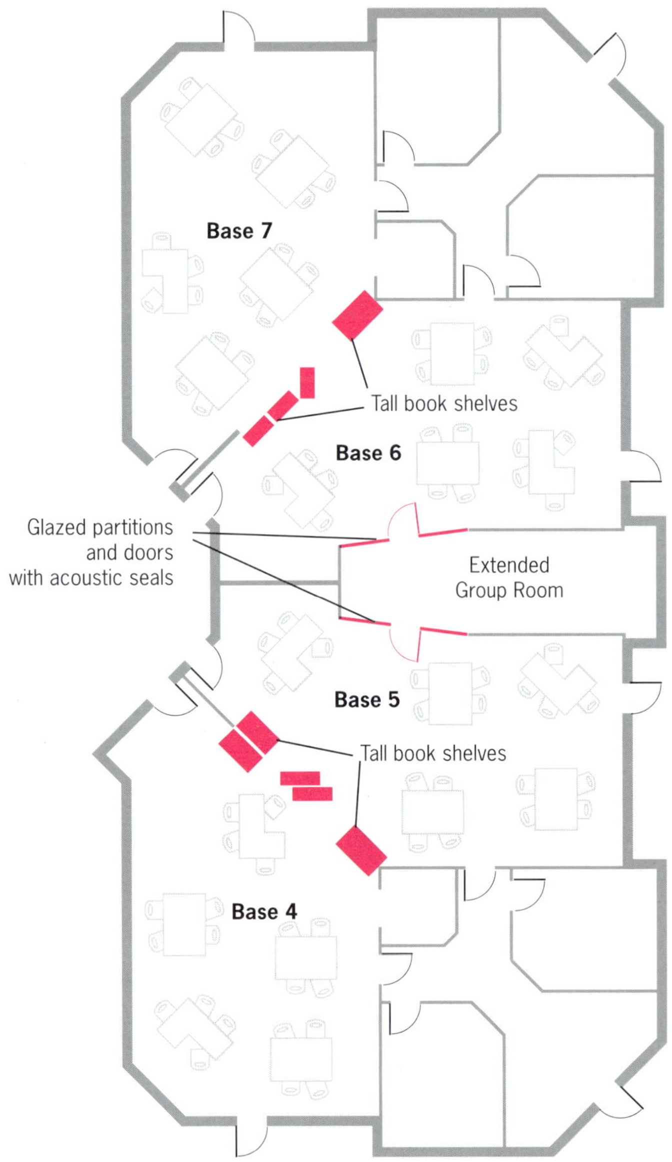

(a)

(b)

Figure 7.3.2: Arrangment of open-plan teaching spaces following modifications
(a) floor plan
(b) new glazed partition extending the group room

main complaints. Firstly, when teaching in a classbase, they could clearly hear teaching activities in other classbases, even the most distant ones, and they found this very disturbing and disruptive. Some teachers perceived this as a 'funnelling' of sound from one end of the open-plan teaching area to the other. Conditions were worst during normal table activities when a comparatively active and excited class returned to an adjacent classbase from a PE lesson.

Secondly, noise levels within a classbase during teaching activities were excessively high and adversely affected concentration and working ability.

Remedial measures were designed to improve the acoustic separation between classbases in order to reduce difficulties arising from mutual disturbance, and to reduce the build-up of noise levels during classroom activities to promote an improved teaching environment.

The acoustic separation between classbases was increased by installing a full height double-leaf glazed partition between the group room and bookshelves as shown in Figure 7.3.2. The glazing is 6 mm thick and doors have perimeter and threshold acoustic seals. This construction extends the size of the group room and forms an effective acoustic barrier between the two pairs of bases 4 and 5, and 6 and 7.

To provide further acoustic separation between the individual bases 4 and 5, and 6 and 7, several tall bookcases were positioned along the dividing line between these classbases. The acoustical separation provided by this type of partial barrier is, of course, considerably less effective than that provided by a full height partition.

Noise control

Noise levels during class were high because surfaces were hard and acoustically reflective with the exception of the carpeted floor. Acoustic absorption was added to reduce these noise levels. The ceiling was the most suitable area for treatment and acoustic tiles were applied over the whole of the ceiling in the open-plan teaching area. The precise absorption coefficient of the ceiling tiles is not known, but an absorption coefficient of 0.9 over the speech frequency range is normally needed to maximise the benefit of an acoustic ceiling. As well as controlling noise within the classbase, the ceiling treatment helps to reduce the propagation of sound from one classbase to another.

The teachers reported an immediate improvement in aural conditions with the installation of the partitions. They found that they were now only disturbed by the classbase immediately adjacent to them. By strategic location of items of tall furniture they were able to slightly reduce this remaining source of disturbance.

The acoustic ceiling, installed a few months later, was perceived by teachers to produce a small but significant reduction in the noise levels during class activities.

Acoustic measurements

The noise levels during class and the reverberation times of the spaces were measured. Measurements were also made to evaluate how well sound propagates from one classbase to another. The majority of measurements were made after the remedial treatment had been implemented although noise levels during class were also measured before treatment.

Activity noise

The noise levels were measured in the four classbases, before and after the remedial treatment, during typical table activities. Approximately 25 pupils were present with 1 or 2 teachers in each classbase. The octave band frequency spectrum for all measurements was consistent in shape and a typical sound level spectrum for classroom table activity before treatment is given in Table 7.3.1.

For typical table activities, the background noise levels prior to the acoustic modifications ranged from 67 dB(A) to 71 dB(A). Following the installation of the acoustic ceiling and partitions, noise levels ranged from 64 dB(A) to 69 dB(A), a reduction of 2 to 3 dB(A) which appears to be a small but significant subjective decrease.

Reverberation time

The reverberation time was measured in classbases 4 and 5. After remedial treatment, the unoccupied mid-frequency value was 0.4 seconds with a rise to 0.7 seconds at 125 Hz. The mid-frequency reverberation time, which will undoubtedly have dropped with the installation of the acoustically absorbent ceiling, is now generally below 0.6 seconds, as required for primary school classrooms in Table 1.5.

Sound insulation

The sound insulation was measured between classbases 5 and 6. A value for $D_{nT(0.6s),w}$ of 48 dB was obtained which meets the requirements between standard classrooms specified in Section 1.

Sound propagation

Before the partitioning of the room, simple tests showed that speech could easily be understood between extreme ends of the open-plan area even when there was no line of sight. Whilst the partitioning provided by extending the group room gives good sound separation between two pairs of teaching bases, the acoustic ceiling and physical obstructions, such as tall bookshelves, are the only means of achieving a degree of acoustic separation between the other adjacent bases.

To measure the sound propagation with distance across an adjacent pair of classbases, a broadband sound source was used to simulate the radiation of sound from a nominal teaching position and sound level measurements were made across the classbase and into the adjacent classbase. Figure 7.3.3 illustrates the three

	Octave band centre frequency (Hz)							
	63	125	250	500	1 k	2 k	4 k	8 k
Sound pressure level (dB)	56	60	65	69	68	62	57	54

Table 7.3.1: Typical measured activity noise levels in an open-plan classbase before remedial treatment

propagation paths that were investigated:

- from base 4 to base 5 with line of sight
- from base 4 to base 5 via an indirect path
- from base 5 to base 6 via the partitioning formed by extending the group room.

The results for the three paths are shown in Figures 7.3.4 (a) and (b).

Figure 7.3.3: Sound propagation paths investigated

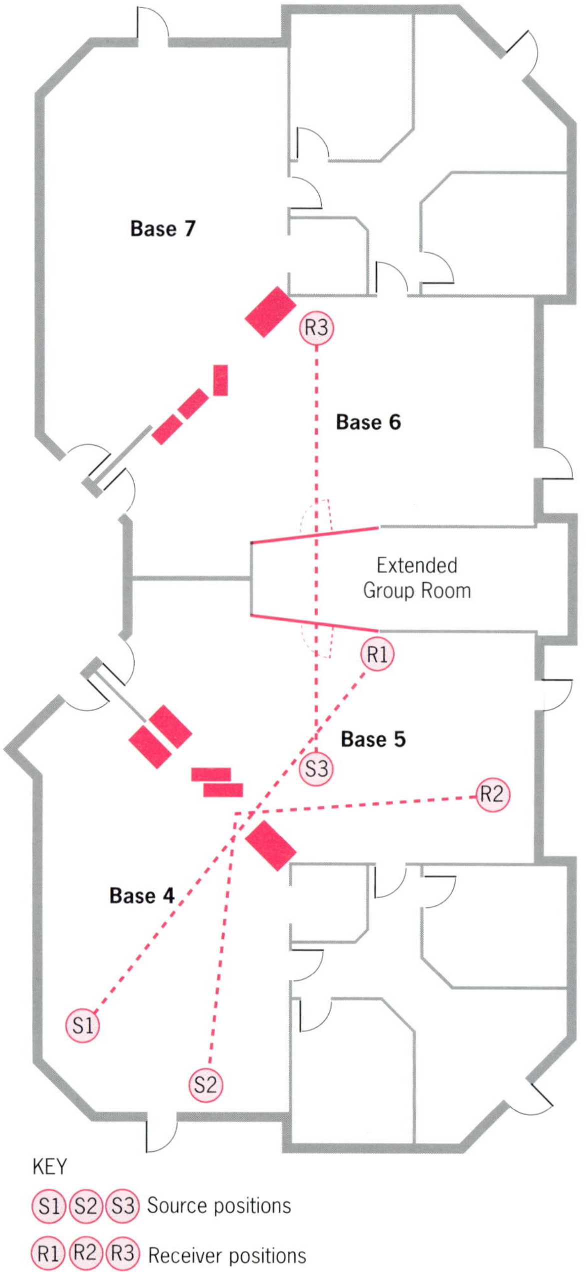

By comparing the two figures, it is evident that the reduction in sound level with distance between bases 4 and 5 is very modest compared with the large reduction between bases 5 and 6 (ie across the partition). This is clearly reflected by the subjective impressions of teachers who are disturbed by noise from an adjacent classbase on the same side of the partition but are not disturbed by classbases beyond the partition.

The erection of a physical barrier across the middle of the open-plan teaching area was clearly effective in improving conditions. It is important to note the constructional simplicity of this barrier and its acoustical effectiveness in reducing sound transmission. This was achieved by using two partitions with a large air cavity in between (the extended group room). A single partition would have needed to be substantially heavier with more elaborate acoustical detailing.

The new partition did not solve all the problems of sound transmission since classbases 4 and 5, and classbases 6 and 7 are still open to each other and some mutual disturbance is still occurring. This has been reduced by partial barriers but can not be effectively eliminated without a complete barrier.

The acoustic ceiling treatment is beneficial in reducing noise levels but did not result in a dramatic effect since the classbases were already carpeted and furnished.

Conclusions

The effect of mutual disturbance in open-plan teaching areas is clearly illustrated in this case study and relatively simple remedial measures have been shown to work moderately well.

Before embarking on the design of an open-plan teaching area, serious consideration should be given as to whether the advantages of the open-plan arrangement outweigh the serious inherent acoustic disadvantages.

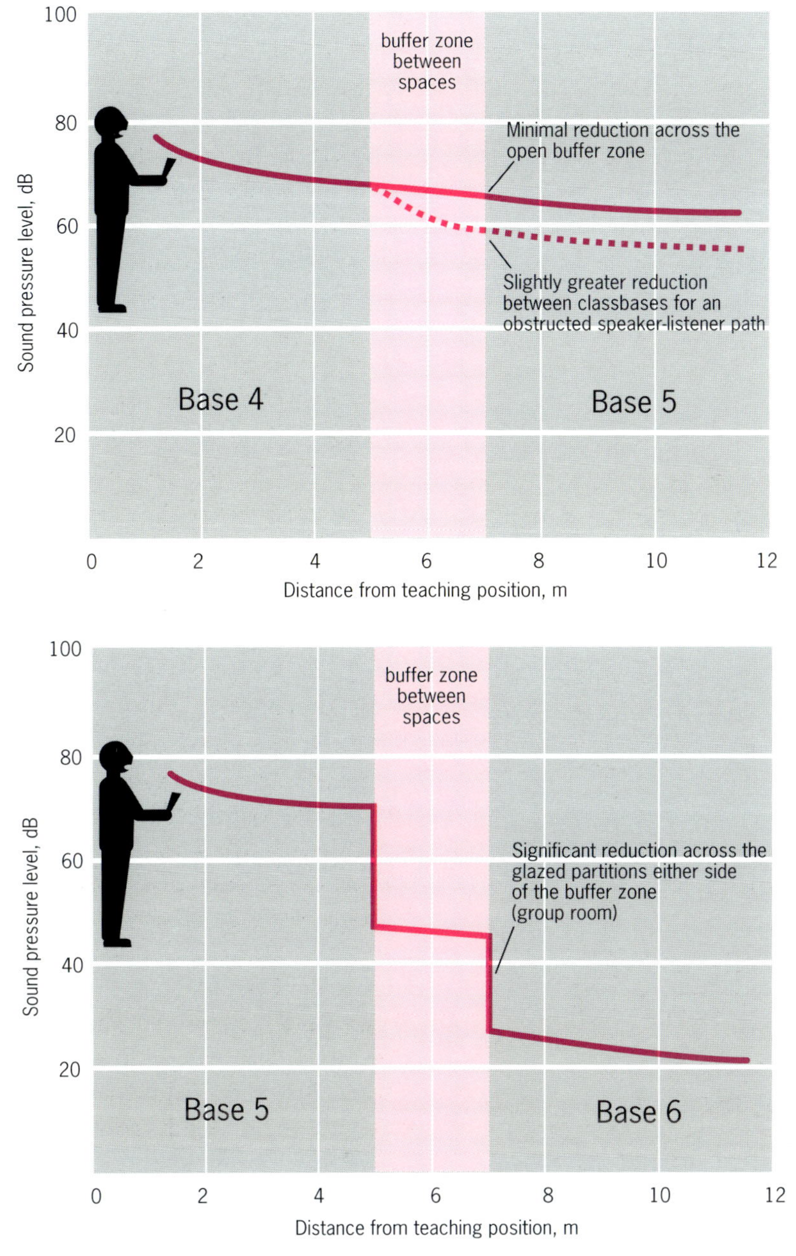

Figure 7.3.4: Sound propagation from one classbase to another

(a) without full height partition

(b) with full height double partition

Existing school buildings may have spaces that are less than ideal and compromises have to be made during remodelling. A design and technology (D&T) workshop was converted into music accommodation for an 11–16 comprehensive school with 600 pupils on its roll. Figure 7.4.1 shows plans of the original workshop and the conversion.

The floor area of the conversion is 96 m^2, including an adjacent 13 m^2 space with independent access.

The original workshop was built in 1954 using a prefabricated, reinforced concrete system of modular design having concrete roof panels and double-skin walls; there is a wood block floor. The south-east and north-west facades of the building were fully glazed from a sill height of about 1.0 m. The ceiling height in the main space was 3.3 m. The size of the main space was suitable for a music room but there were some disadvantages with the accommodation:

- Existing floor and ceiling surfaces were hard, resulting in an unacceptably long reverberation time of 2 seconds. Standing waves and flutter echoes were likely due to parallel walls and hard surfaces.
- The north-west wall abutted the

Figure 7.4.1:
(a) Plan of the original workshop
(b) Plan showing conversion to music accommodation

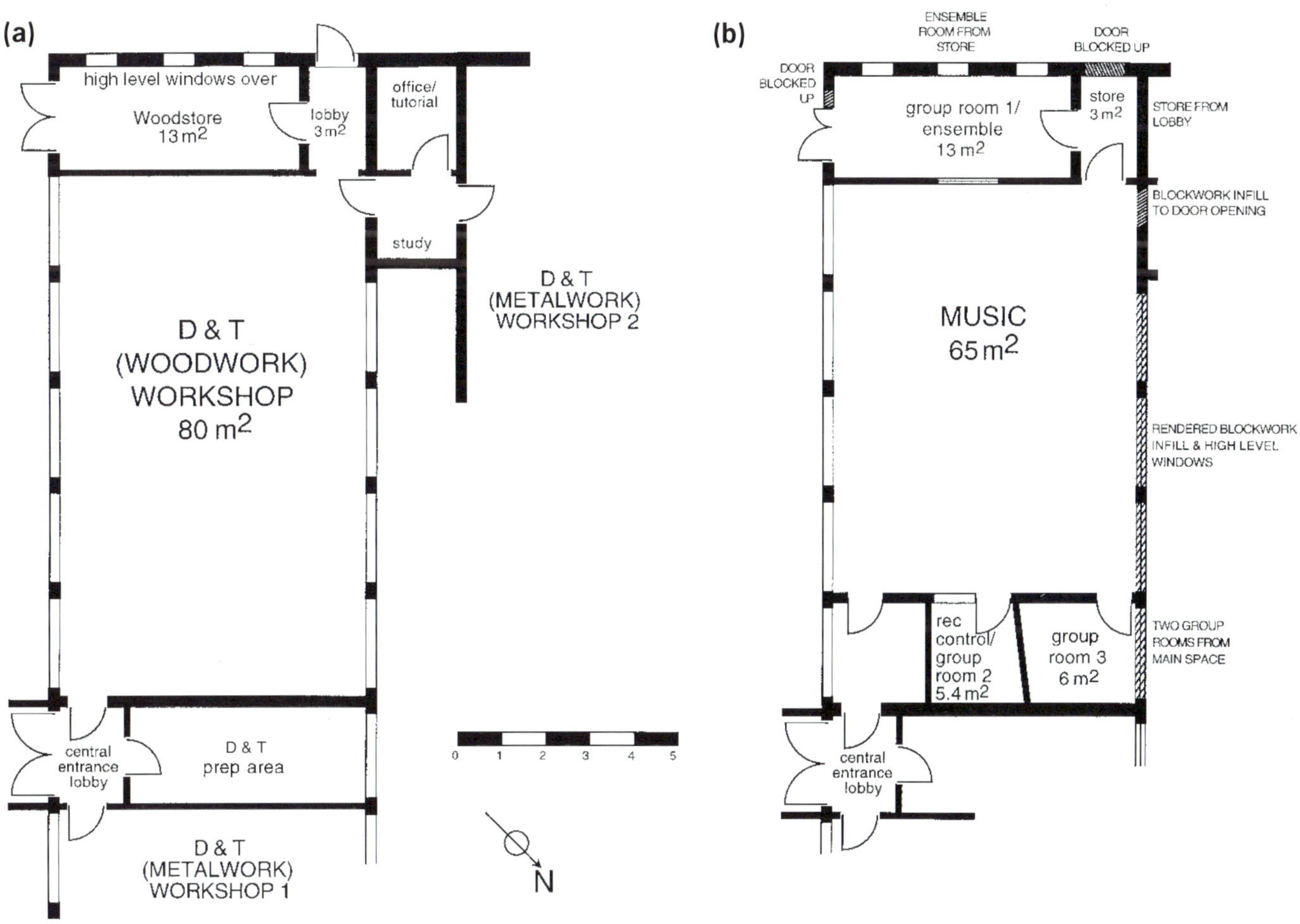

school playing field. The extent of glazing was excessive and considered undesirable from a security point of view on a side with potential for intrusion.

- The school playground, a potential source of noise, is adjacent to the south-east wall.
- A second design and technology workshop is adjacent to the space (although an entrance lobby and store provide a buffer between the teaching spaces).
- The building is free-standing and circulation is external which results in an excessive number of entrances.
- The reverberation time of the space was too long for a music room.

The adaptation

Structural alterations were kept to a minimum in order to constrain costs and maximise available funds for acoustic treatments and finishes. Within the existing area, it was possible to provide a music room of 65 m^2, three group rooms and a store, see Figure 7.4.1(b). Performances to an audience or large-scale rehearsals take place in the school hall. The largest group room (or ensemble room) is converted from the existing store and can be accessed separately, if necessary, to avoid disturbing classes. The dimensions of this space are not ideal as proportions are long and narrow but compromise has been accepted and the wall treatment is designed to optimise room responses. An entrance lobby houses coats and bags and provides additional sound insulation between the main space and the adjacent D&T room.

The sound insulation of the music room was a priority. The key aspects of the acoustic treatment are shown in Figure 7.4.2, and described below.

Figure 7.4.2: Plan showing acoustic treatments

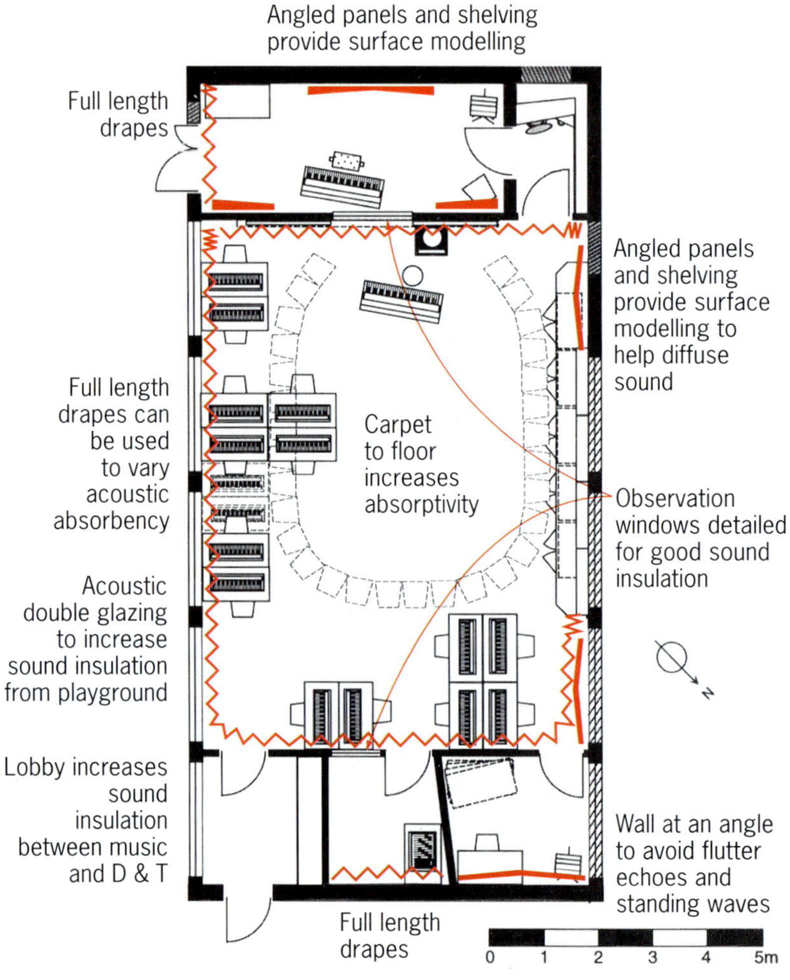

Construction

In order to improve security, glazing to the north-west wall was removed and the opening was infilled up to two-thirds of its height with rendered blockwork. Medium density block (1500 kg/m^3) was used to give appropriate sound insulation. The top third of each panel was thermally and acoustically double-glazed with bottom-hung openable fanlights.

Angled panels of medium density particle board were fixed to studding on the inside face of the north-west wall of the main space. These help to prevent standing waves between parallel side walls and can provide much needed display space. The panels are without fabric covering since this would compromise the high frequency response. Panels are omitted where there are shelves as these have an equivalent acoustic effect. Angled panels are also used in the group rooms.

Secondary acoustic glazing was added to the windows to the south-east (playground) side, as two sliding panels. This allows access for maintenance and to open casements or fanlights. Solar reflective film was added to the outside of the existing fenestration to reduce solar

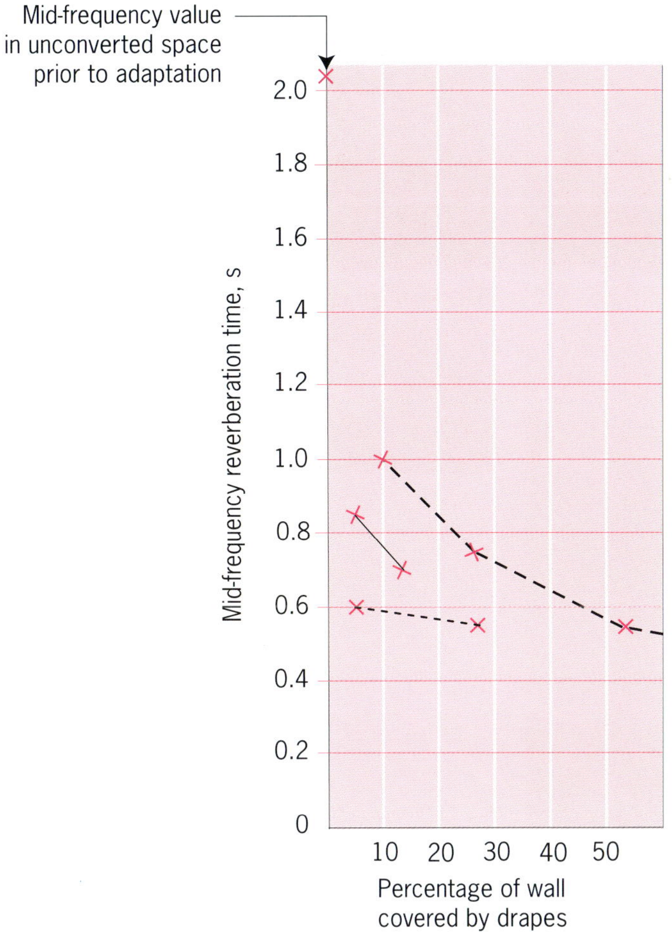

Position of drapes	Percentage of wall area covered
CLASSROOM	
bunched	10%
rear wall	27%
rear and side	53%
ENSEMBLE ROOM	
bunched	5%
one wall	14%
GROUP ROOM	
bunched	5%
one wall	28%

Key
Classroom (dashed line)
Ensemble room (solid line)
Group room (dotted line)

Figure 7.4.3: Graph showing effects of drapes on reverberation times in classroom, ensemble room and group room

gain. In a new building, an alternative solution may be to incorporate a fixed sunshade or 'brise soleil' at the eaves soffit.

Internal doors into the music classroom, the adjacent D&T space and the ensemble room were upgraded to heavy solid core doors with double seals all round including threshold seals. Doors to the two group rooms have vision panels for supervision with 10 mm glass. Acoustically double-glazed observation windows are formed in the partition walls to the ensemble room and one of the group rooms.

Fixtures and finishes

The existing plasterboard ceiling finish was retained; the existing wood block floor to the main space was also retained and carpeted for acoustic reasons. The ensemble room floor has a basic underlay and corded carpet, so that the finish is not too acoustically absorptive.

Wall finishes in the main space are supplemented by heavy drapes of at least 0.5 kg/m^2 at 200% gather, providing acoustic variability and control. The curtain track is ceiling mounted along three sides of the room providing maximum flexibility allowing curtains to be positioned to suit the configuration of the musical activity. This is useful in a school where one classroom serves a number of functions.

Curtains are also provided in group rooms. In the ensemble room, they are positioned at the south-east end of the space, screening the doorway or bunched in the corner as required.

On completion, acoustic measurements were taken in the music classroom, ensemble room and a group room (all when unoccupied). Resulting mid-frequency reverberation times are depicted in Figure 7.4.3. This graph shows that measured values are in accordance with Table 1.5 of Section 1, and demonstrates the potential of providing acoustic variability using drapes.

In the 65 m^2 classroom it can be seen

that curtains can be very effective in reducing mid-frequency reverberation time. Because of the number of variables combining to affect the reverberation time in a room including volume, the weight and location of curtains, surface finishes and furniture, the results shown here are indicative only. The graph shows that the ensemble room at 13 m^2 has a measured RT of 0.7 to 0.8 seconds, within the range given in Table 1.5 of 0.6 to 1.2 seconds for an ensemble room.

The background noise level in the unoccupied music classroom measured whilst adjacent classes were in session was 29 dB $L_{Aeq,\ 1hr}$. This suggests that the indoor ambient noise level is less than the required level of 35 dB $L_{Aeq,\ 30min}$ given in Table 1.1 of Section 1.

The music department at a school with 650 pupils between the ages of 11 and 18 was replaced. The new self-contained suite comprises a large music room, three music practice rooms, an ensemble room and ancillary accommodation.

The school is located in a quiet rural district with low ambient outdoor noise levels. The music block is several metres away from other buildings, which ensures that noise egress to other parts of the school is minimised.

The building is constructed of masonry with an external leaf of brickwork, an insulated cavity and internal leaf and walls of blockwork, some of which are plastered. The density of the blockwork is not known but ideally it should be the highest available, ie 2000 kg/m^3. The tiled roof has an internal sheathing of plywood which benefits sound insulation.

A full height blockwork crosswall, up to the roof soffit, separates the large music room from the rest of the building. The music practice rooms also have full height walls.

Windows are double-glazed and can be opened. Doors are generally hollow core with basic seals giving around 20 dB R_w for the doorsets.

The music suite is a good example of how to control noise transmission between rooms, and thus reduce disturbance, by careful planning of the room layout, see Figure 7.5.1. The key features are:

- The large music room is separated from other music rooms by a corridor and storage areas.
- The ensemble room with its associated recording/control room is also separated from other rooms by a corridor.

Music classroom

The geometry of the large music room is good, with a rectangular plan shape and a fairly steeply pitched ceiling, see Figure 7.5.2. The light fittings and recessed roof lights provide some useful modelling to break up and diffuse the sound.

Two large encased purlins, projecting down from the plane of the ceiling cause a minor localised problem. Small sound colourations occur when musicians play in the area underneath the main roof beams. It is possible that these are caused by strong reflections from the junction between the roof beam and ceiling as shown in Figure 7.5.2. Additional localised measurements would be necessary to investigate this effect. A solution in this particular case would be to treat one side of the beam with absorbent material, as indicated.

As a general principle, it is useful to incorporate elements into a ceiling to provide diffusion and hence uniformity in

Figure 7.5.1: Plan of music department (furnished)

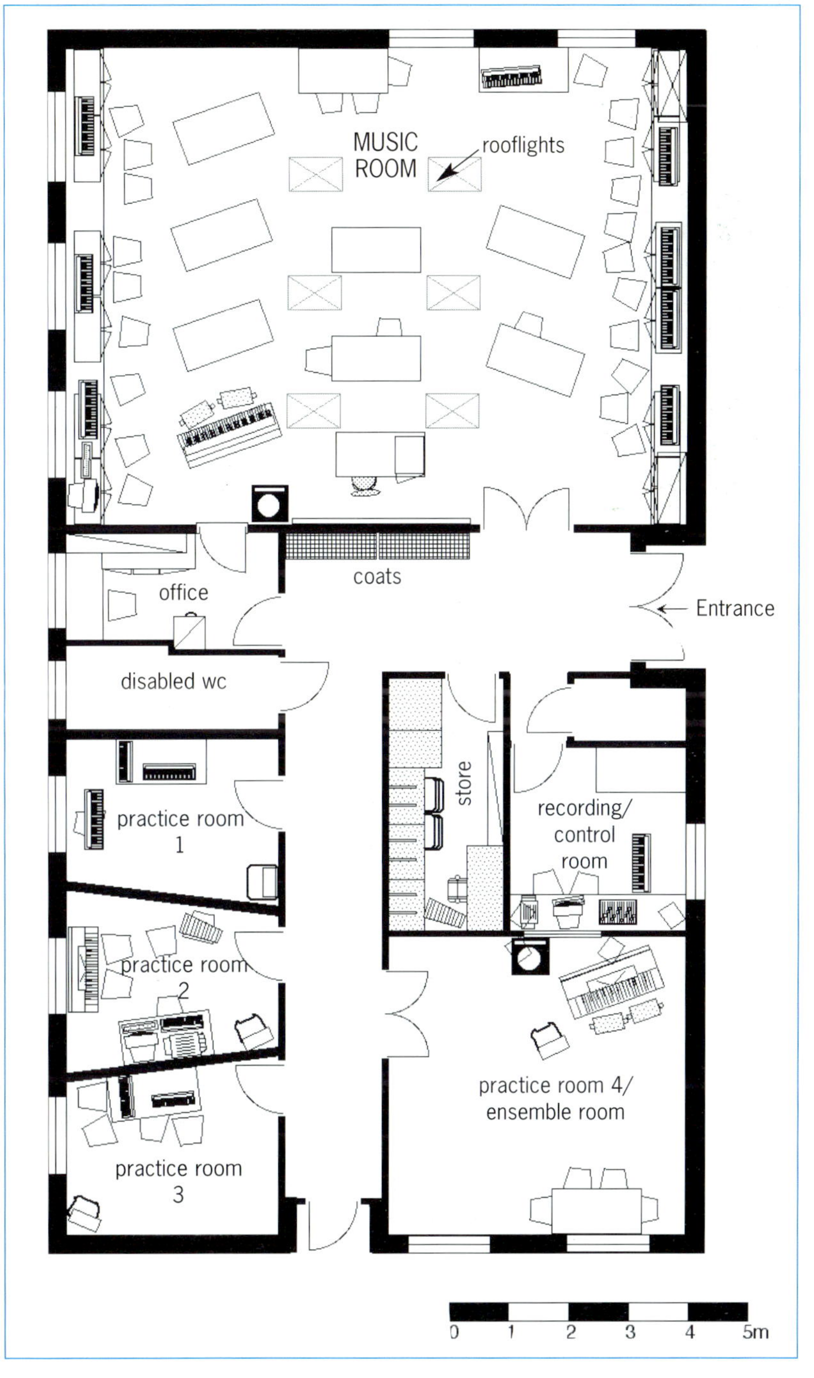

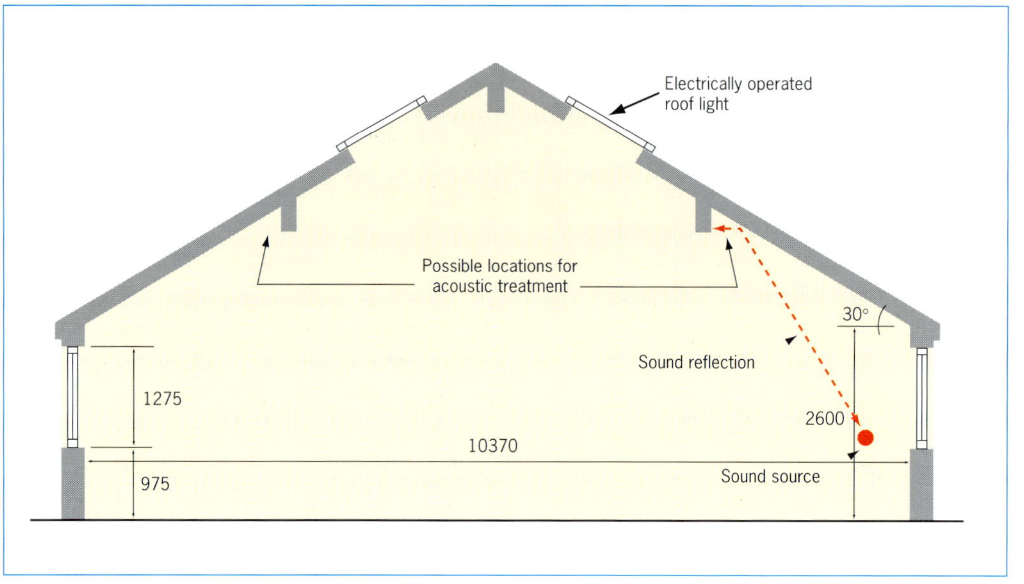

Figure 7.5.2: Section through music room

the sound field. For effective diffusion, projections of 0.3 m to 0.5 m are necessary. However, such projections should be distributed over the whole ceiling area; a single large projection can lead to a prominent and potentially disturbing reflection, as in this case.

Surface finishes are generally hard and reflective except for the floor which is covered with a short pile carpet. In detail, the walls are of plastered blockwork with an emulsion paint finish and the ceiling is of plasterboard with a plaster skim finish. This combination of hard and soft finishes ensures that the reverberation is sufficiently long for music performance and adequate for teaching.

No provision has been made for varying the acoustics, eg by use of heavy curtains. This would be desirable but not essential.

The measured mid-frequency reverberation time (RT), with 25 children and 4 adults present, was 1.0 seconds with a rise to 1.5 seconds at 125 Hz. The full RT curve as a function of frequency is shown in Figure 7.5.3.

This RT is within the range for ensemble rooms specified in Table 1.5 of Section 1.

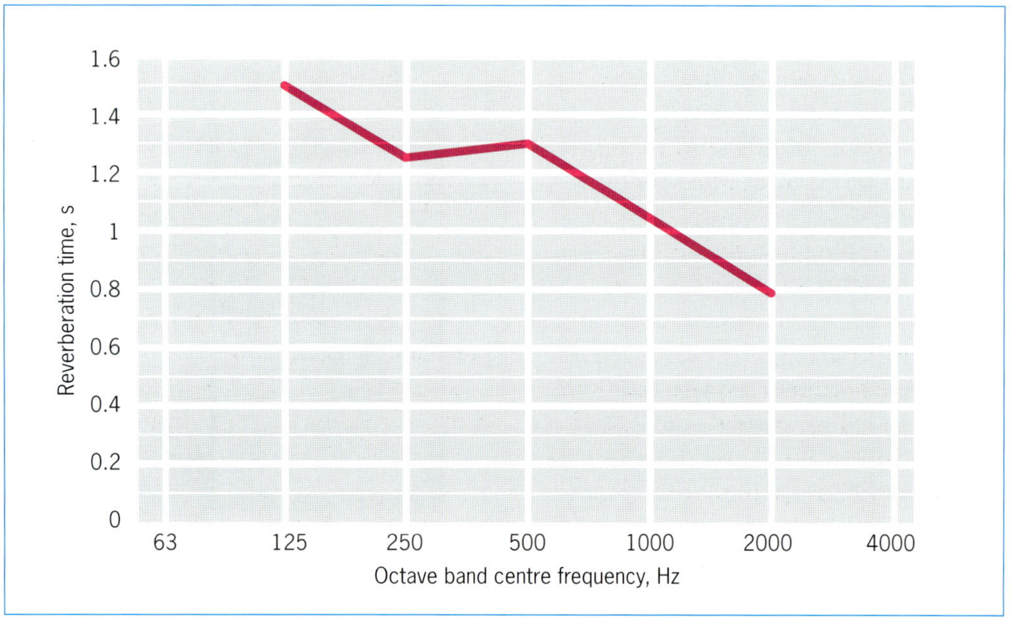

Figure 7.5.3: Measured reverberation time in the music room

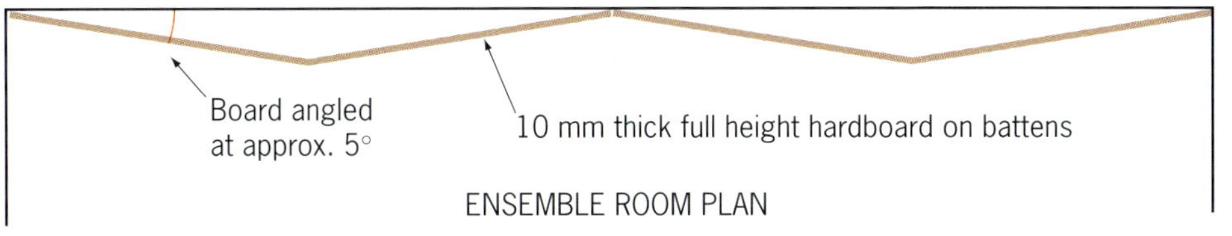

Figure 7.5.4: Simple acoustically diffusing elements

Practice rooms

All practice rooms include one pair of non-parallel walls which reduces the possibility of sound colouration from standing waves.

The practice room volumes are of the order of 20 m^3 and can accommodate up to 5 pupils working on composition. Ceiling heights are 2.6 m which is lower than desirable but acceptable.

The measured mid-frequency RT in one of the practice rooms was 0.4 seconds with a rise to 0.9 seconds at 125 Hz. This is at the lower limit recommended for practice rooms which results in a 'dry' sound but is nevertheless satisfactory. The moderate rise in bass frequencies is generally a welcome feature giving fullness of tone to certain instruments.

A combination of acoustically reflective and absorbent finishes has been used. The walls are of blockwork with a paint finish, the floor is carpeted with short pile carpet and the ceiling is treated with acoustic tiles. Although the selection of materials has resulted in acceptable reverberation times, the distribution of absorption is concentrated on the ceiling and floor which tends to emphasize sound reflections in the horizontal plane; an undesirable effect that has been noted by users. This problem could be overcome by redistributing a proportion of the absorption onto the walls, eg by installing absorbent wall panels and replacing some absorbent ceiling tiles with reflective ones.

Ensemble room

The ensemble room is square which could give rise to strong standing waves and hence possible colouration. However, one wall has simple angled reflective panels acting as diffusing elements which work effectively to counteract this. Surface finishes in this room are the same as in the practice rooms. Again the distribution of absorption is uneven although the sound field is rendered more diffuse by the installation of the angled panels on one wall, see Figure 7.5.4. Treatment of these panels with hessian is not desirable since it could reduce 'brilliance' of the sound as there is already sufficient high frequency absorption in the carpet and ceiling.

The measured mid-frequency RT was 0.4 seconds with a rise to 0.6 seconds at 125 Hz. This is lower than the range given in Table 1.5. A longer mid-frequency RT could have been achieved by reducing the ceiling absorption. There would then have been scope for having acoustic variability using curtains.

Recording/control room

The control room is also square but without the benefit of diffusing elements. This could give rise to standing waves although these were not evident. The shelving and equipment probably provide sufficient diffusion to avoid sound colouration. A facing wall could be treated with absorbing material to assist in preventing this.

This room has the same surface finishes as the practice rooms and is generally suitable for music practice and composing as well as monitoring and recording sound from the ensemble room. (Monitoring is normally done in a very dead acoustic although a suitable compromise has been struck here between practicing and monitoring).

There is good visual communication with the ensemble room through an acoustic double-glazed window.

Sound insulation

The measured sound level difference between two practice rooms in octave

Table 7.5.1: Measured level difference between two practice rooms

	Octave band centre frequency (Hz)						
	63	125	250	500	1 k	2 k	4 k
Level difference *D* (dB)	22	27	34	46	50	52	58

Table 7.5.2: Measured indoor ambient noise levels in practice room

	Octave band centre frequency (Hz)				
	125	250	500	1 k	2 k
Measured sound pressure level (dB)	38	33	26	21	25

bands is shown in Table 7.5.1. This equates to a weighted BB93 standardized level difference of 44 dB $D_{nT(0.8s),w}$.

The sound insulation between practice room 2 and the adjacent corridor was limited by the poor sound insulation of the doorset between them. The weighted BB93 standardized level difference between the ensemble room (practice room 4) and practice room 2 was 47 dB $D_{nT(0.8s),w}$.

Indoor ambient noise

The indoor ambient noise level was measured in practice room 3 during a period when people were moving around the building but no significant musical activity was taking place. The results are shown in Table 7.5.2. This equates to a single figure value of approximately 30 dB(A). It means that there is little masking of intrusive noise from adjoining spaces. Therefore, separation of sensitive spaces by storerooms and corridors is particularly important.

Table 7.5.3 compares the subjective assessments of the acoustic quality of the spaces with the acoustic measurements.

Discussion

One of the key issues relating to acoustics in music accommodation is sound transmission between different rooms which may cause disturbance to music practice and teaching.

Clearly, the layout of the building provides good separation between main rooms or groups of rooms. The measured $D_{nT(0.8s),w}$ from the ensemble room to practice room 2 was 47 dB.

However, the situation is more complicated because disturbance to a musician or teacher is also a function of the indoor ambient noise in the room they are playing/teaching in: the higher the indoor ambient noise (if relatively steady) the more masking of external sounds occurs and hence the lower the disturbance from external noise.

The sound insulation was significantly reduced by transmission through doorsets. The installed doors are hollow core with poor seals around the perimeter and threshold. The two sets of double doors in the music suite do not have effective seals at the meeting stiles. This is a common problem with double doors but can be overcome by careful detailing with rigid fixings at the meeting stiles. It might have been better to use unequal paired doors instead of double doors. The small leaf can then usually be bolted shut making the seal much more effective than on a normal double door. A wide single door is also a possibility.

Upgrading seals to proper acoustic seals and replacing doors by the solid core type would improve performance close to that required. (Preventing doors from squeaking would also be beneficial in reducing disturbance.)

A second key issue is the acoustic characteristics of each of the different types of space.

The large music room has a very good geometry for providing a diffuse sound field. The 1.0 seconds T_{mf} is sufficiently long to provide fullness of tone but short enough to maintain clarity which is an important quality in music teaching. The rise in RT at bass frequencies is beneficial in terms of adding ‘warmth’ to the acoustic characteristic.

The music practice rooms also have an

Table 7.5.3: Comparison of subjective and objective assessments

	Subjective impressions of the acoustic character of the space	Acoustic measurements		
		Mid-frequency reverberation time, T_{mf} (s)	**Airborne sound insulation $D_{nT(T_{mf,max}),w}$ (dB)**	**Indoor ambient noise level L_{Aeq} (dB)**
Music room	Moderately reverberant. Diffuse sound field. Adequate loudness	1.0		
Practice room (Tutorial 2)	Modest reverberation. Non-diffuse sound field. Reverberation concentrated in horizontal plane. Can hear instruments playing in adjacent rooms but just tolerable.	0.4	44 between practice rooms 2 and 3	29
Ensemble room	Modest reverberation. Adequate loudness. Disturbance from practice rooms only during quiet moments.		47 between the ensemble room and practice rooms	
Corridor	Can clearly hear piano from practice room and clarinet from ensemble room.			

appropriate geometry in plan, namely two non-parallel walls. However, absorption is not evenly distributed on the room surfaces which prevents a sufficiently diffuse sound field.

The ensemble room appears to be much favoured by teachers and musicians alike who feel comfortable with its size. A square plan shape is always problematic in terms of standing waves although this has been mitigated by using simple and effective diffusing panels.

The associated control room includes a well designed acoustic double-glazed viewing window and the two spaces together form a good quality recording suite. The window consists of 2 x 6 mm glass panes with a 100 mm air space.

Conclusions

The music block is exemplary in most respects in terms of its fitness for teaching and practicing music at secondary school level. The acoustic design is generally good but there are some minor shortcomings such as inappropriate selection of doorsets and door seals, and poor distribution of absorption in practice rooms.

In summary, the main points to note about the acoustic design of the music block are:

- location of the building at a distance from other buildings
- separation of large music room and groups of rooms by full height walls, corridors and buffer zones
- selection of a simple rectangular plan shape for the large music room
- selection of non-parallel walls for practice rooms
- use of simple angled wall panels to provide sound diffusion in the ensemble room
- use of solid, acoustically reflective materials for walls and ceilings in the large music room to ensure sufficiently long reverberation.

This case study describes a junior school and hearing impaired unit which provide an inclusive environment for hearing impaired children who are educated through a natural aural approach. The children attached to the unit all have a 'significant' hearing loss and abilities that fall within the 'average' range. The guiding principle that underlies their placement within the school is that they should be allowed to make best use of their residual hearing. The children have full access to the national curriculum and are members of a mainstream class. Children also have the use of a specialist teaching resource facility as described below.

Characteristics of the school

The junior school is of average size with about 230 children aged between 7 and 11. Sixteen children are included specifically within the resource provision for deaf pupils, although this number includes children currently attending the infant school and is liable to fluctuation depending on the unpredictable changes in the size of the hearing impaired population.

Accommodation

The school was built in the late 1950s and is set away from the road in a quiet location. The school has been pleasantly decorated throughout. Some attention has been given to reducing internal noise by carpeting classrooms and some corridors. Most of the ceilings have some degree of acoustic treatment. There are no open-plan classrooms within the junior school. It is the intention of the school to further improve the acoustics of the classrooms and a report on sound treatment has been commissioned.

There are 8 classrooms of similar size. In addition there is a dedicated ICT space, a drama room, music room and a large hall. A library has been established in one of the larger corridors.

Attached to the main building by a covered walkway is a building formerly called the hearing impaired unit, but now renamed as the RPD (resource provision for the deaf). This has extensive sound treatment and the main teaching room is situated so that it does not face the playground.

This case study focuses on two rooms: the main teaching space in the RPD (marked as RPD on the plan) and a

Figure 7.6.1: School room layout

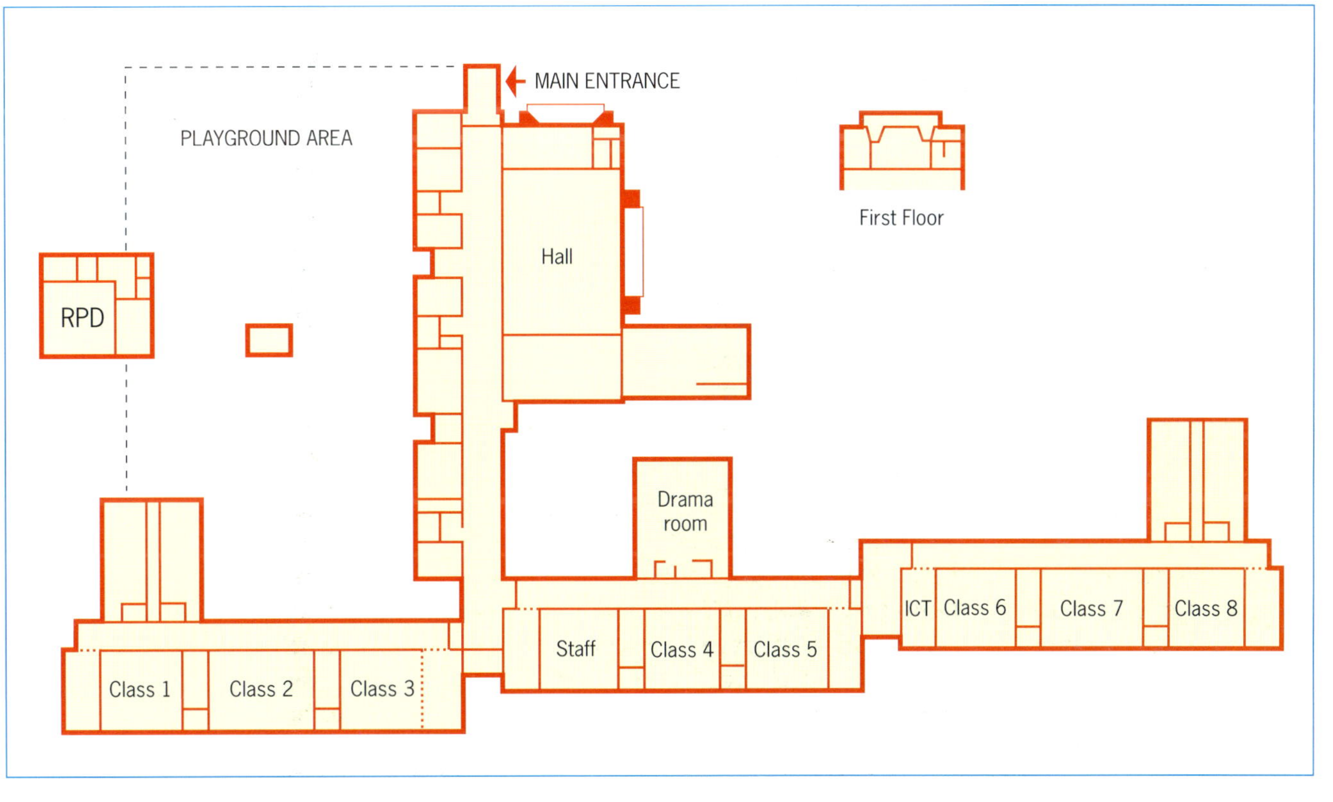

Figure 7.6.2: 'Circle time' in Class 4

typical classroom within the school (Class 4 on the plan).

Figure 7.6.2 shows the children all facing each other during circle time. The hearing impaired child has been placed next to the teacher to ensure that she can hear the teacher well and see all contributions. Figure 7.6.3 shows the layout of the room and the positions of the children during circle time. The teacher is wearing a radio transmitter that transmits her voice directly to the child's hearing aids and to a classroom soundfield amplification system. This will ensure that the teacher does not have to raise her voice and distort her speech unhelpfully. All children benefit and as a consequence are better able to participate.

Figure 7.6.3: Class 4 layout

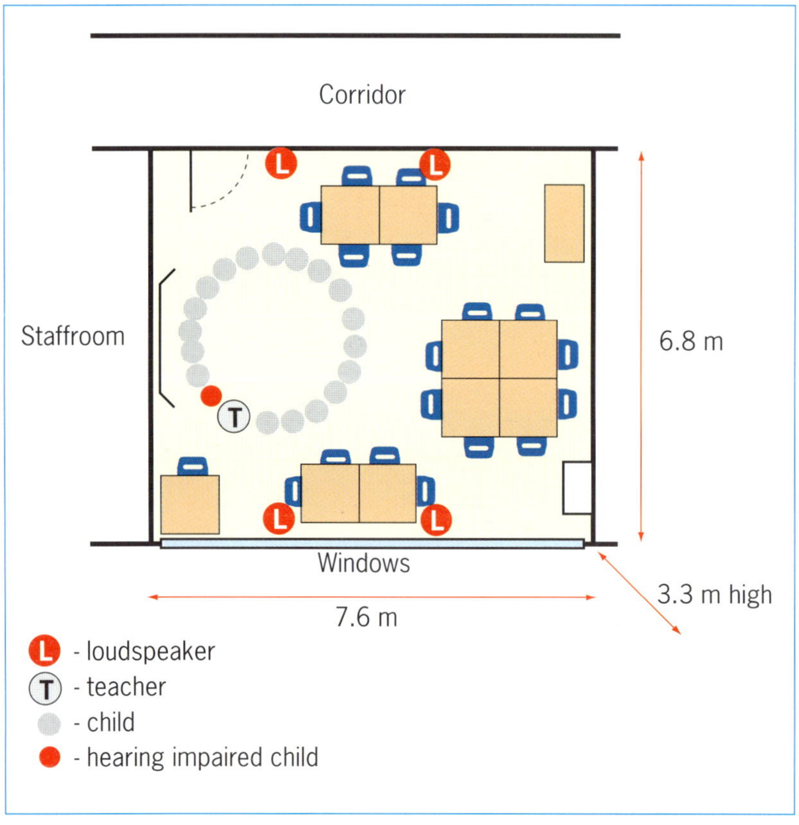

Acoustic and behavioural measures

A number of acoustic and behavioural measures have been obtained in order to present an account of the acoustic environment of the classroom. These measures include:

- listening inventories for education (LIFE UK, see Section 6.5)
- sound level during school day (1 minute average dB(A))
- short term sound level measurements (2 minute runs at 6 time intervals)
- room acoustic measures.

LIFE UK is a protocol for evaluating listening abilities of children. Application of the protocol indicates that the class are able to hear the teacher and each other well most of the time, see Figure 7.6.4. The hearing impaired child has a similar profile with the exception of several

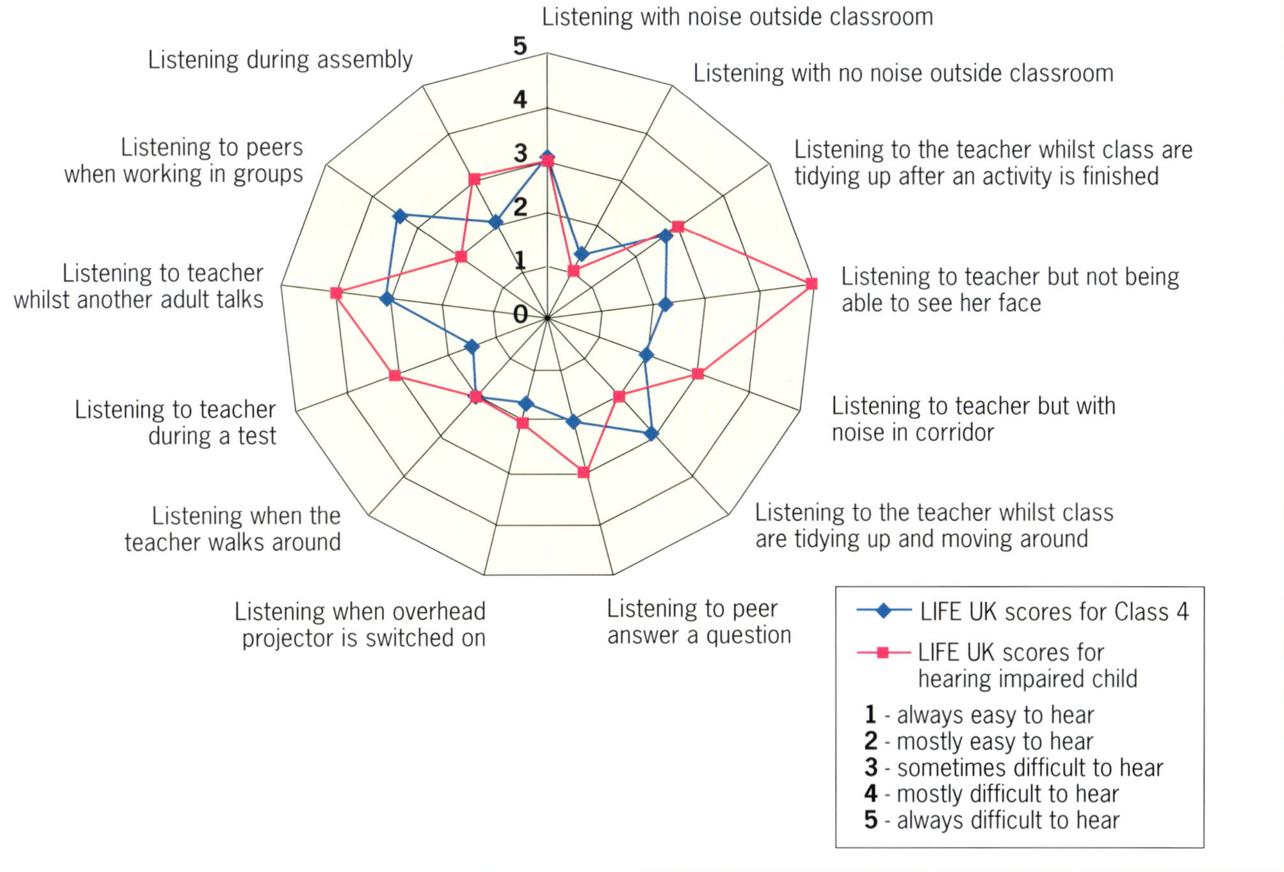

Figure 7.6.4: LIFE UK scores for Class 4

critical areas, primarily the child indicates that she needs to be able to see the teacher's face in order to understand what is being said. This is consistent with the benefits offered by lip-reading in less than ideal listening conditions. This can be addressed through the teacher modifying her teaching style. Other areas where the hearing impaired child finds greater difficulty include listening to her peers answer questions; listening when there is another adult talking; and listening when there is intrusive noise, for example from the corridor. The child indicates that she is making satisfactory use of the personal radio system and classroom amplification system to overcome many of the potential obstacles to hearing effectively.

Figure 7.6.5 shows a chart obtained using a noise logging dosimeter placed at the front of the classroom and out of the reach of the children. The chart presents the one minute history of the sound level obtained between 11.30 am and 15.28 pm during a typical school day. 'A' represents the class quietly engaged in group work.

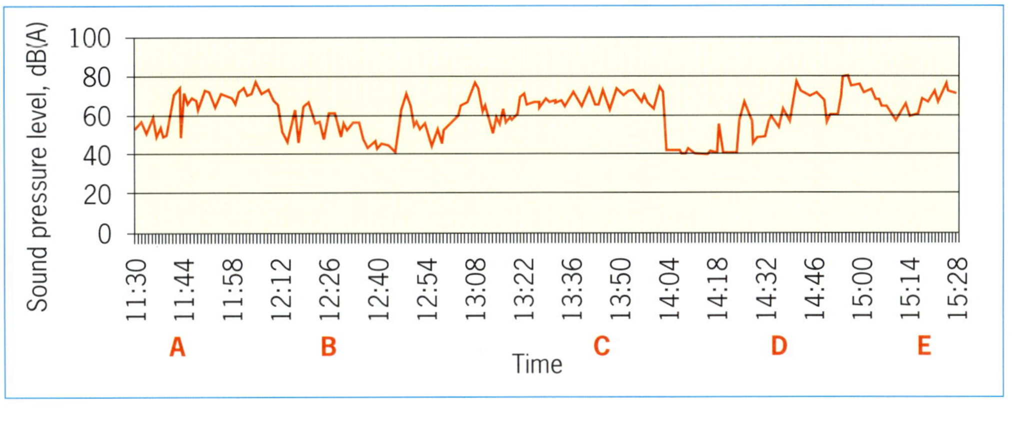

Figure 7.6.5: Sound levels in Class 4 during the day

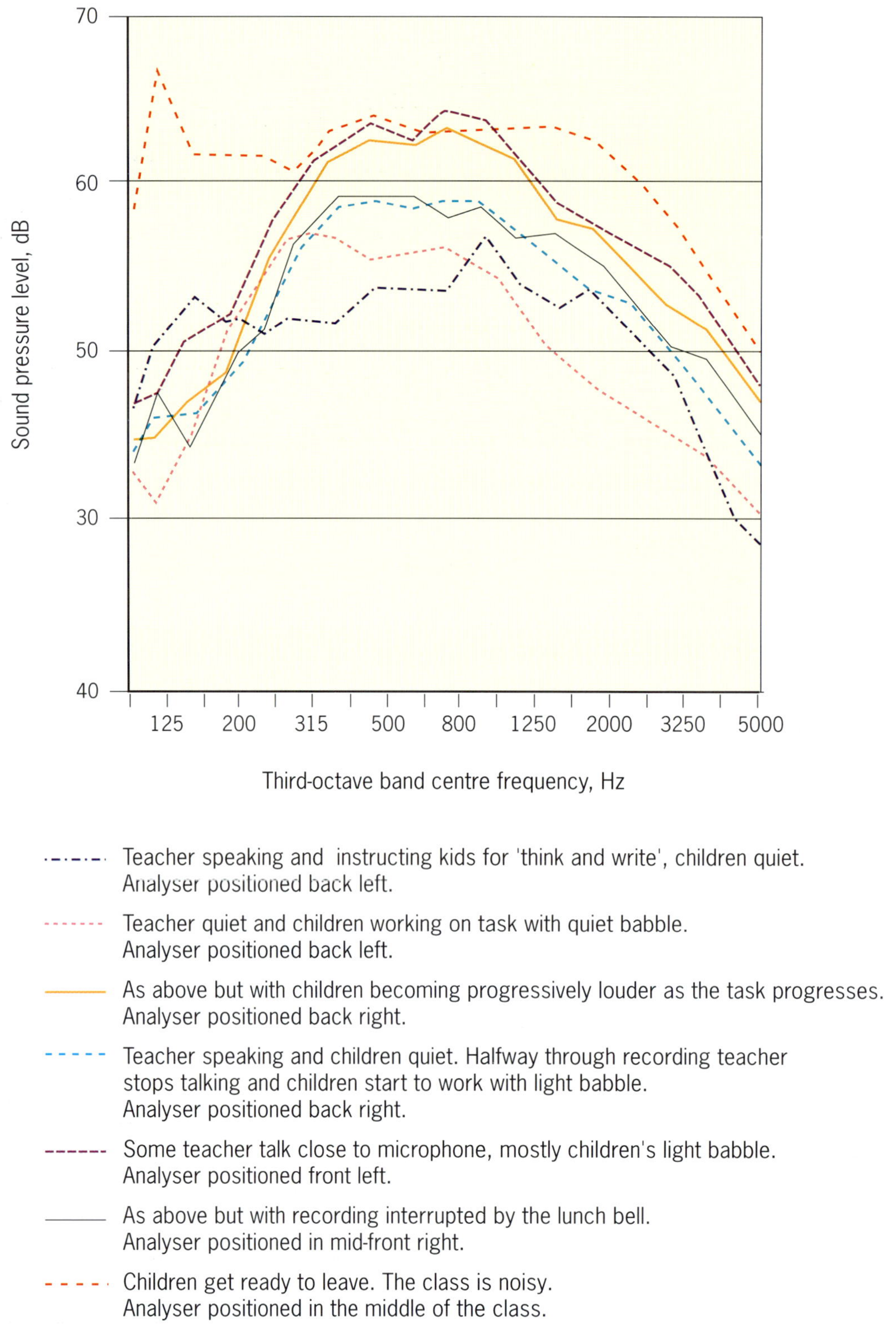

Figure 7.6.6: Frequency spectra for various classroom activities

'B' is the lunch break. During the period marked 'C' the class are again engaged in quiet group work; the end of period 'C' coincides with a break. During the period marked 'D' the sound level gradually rises while the children take part in a carefully controlled circle time discussion. A classroom soundfield system is used by the class teacher and a personal radio FM system is used by the one hearing impaired child who uses a hearing aid. Period 'E' represents the end of the school day and the sound level rises as children and adults use the room informally.

Figure 7.6.6 shows the third-octave band frequency analysis for some of the classroom activities.

Room acoustic measures

Ambient noise level

Measured levels with the classroom empty were in the range 32-36 dB L_{Aeq}, but this was after the end of school so did not include noise from other areas of the school. Noise from other areas was not perceived by the teachers as a problem.

Sound level changes due to use of soundfield system

Measurements of the sound pressure level and L_{Aeq} did not show changes that could be definitely attributed to the use of the system. The teacher being measured had, as judged subjectively, an exceptionally powerful voice, and it is quite possible that she was able to monitor the acoustic impact on the class and adjust her speaking level accordingly.

It is worth noting that the system is not purely an amplification system, it exists to distribute the sound from the teacher's voice evenly about the classroom. Simultaneous acoustic measures would have been useful to indicate the extent to which this was achieved.

Subjectively, there was an increase in clarity at mid and high frequencies. The increase in clarity does not imply a pleasant quality of sound and it was felt that the sound from the speakers was rather harsh. This could be a function of the frequency response of the speakers or the adjustment of the system.

Room acoustics assessment, Class 4

The measured reverberation times (RTs) are shown in Figure 7.6.7. The T_{mf} is above the value specified in Table 1.5. In addition, detailed analysis of the measured impulse responses showed flutter echoes between the parallel, reflective walls. These were not at such a level as to be annoying but they probably reduce speech clarity in the room. The room has predominantly reflective wall surfaces and although the ceiling and the carpets provide some absorption, more absorption on the walls would reduce or eliminate the flutter echoes as well as reducing the RTs to acceptable levels.

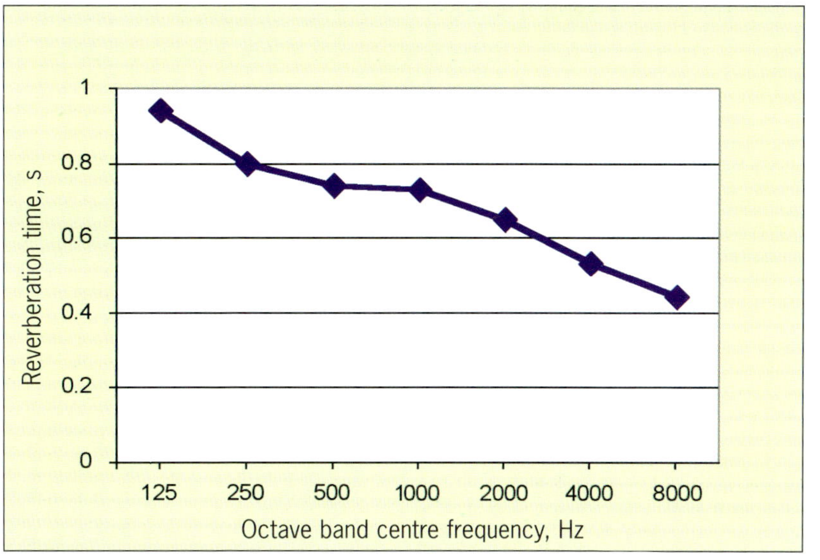

Figure 7.6.7: Reverberation times in unoccupied Class 4

Room acoustics assessment, RPD room

The measured RTs are shown in Figure 7.6.8. As expected for an RPD room, the T_{mf} is lower than the value of 0.4 seconds given in Table 1.5 for classrooms designed specifically for use by hearing impaired pupils. Furthermore, the RT across the frequency range is lower than 0.4 seconds as recommended in Table 6.1. There are no apparent flutter echoes or other

Figure 7.6.8: Reverberation times in unoccupied RPD room

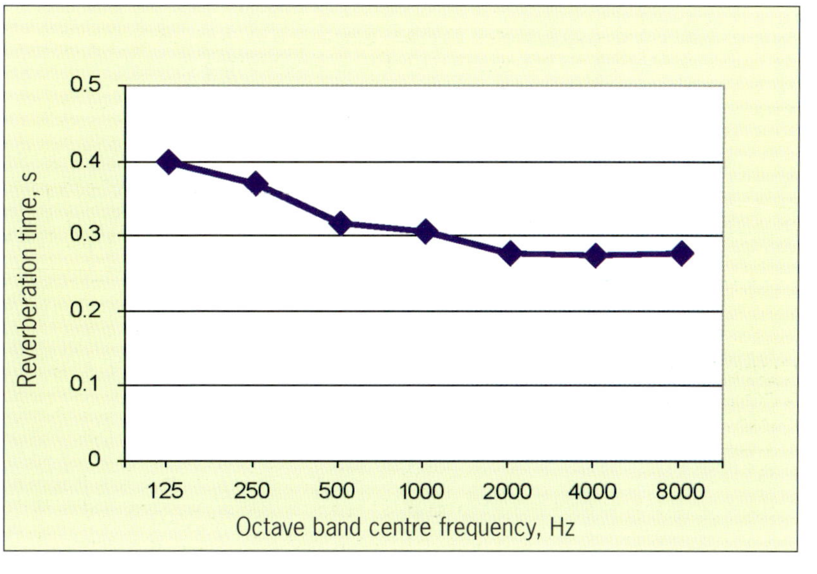

problems and no complaints of acoustic problems in this room, which would be considered to be very well designed acoustically.

Teaching resource base

The RPD is separated from the main school by a short covered walkway. There are two rooms and the entrance lobby outside the rooms is large enough to provide a space for small group work. The larger room shown in Figures 7.6.9 and 7.6.10 is used for teaching larger groups. The whole building has extensive sound treatment, ensuring that the environment has little reverberation. The building is set in the playground, but is only used for teaching purposes outside playtime. The largest space faces away from the playground. The windows are not double-glazed and there is no air conditioning, however the setting is very quiet and the rooms are large.

Figure 7.6.9: Teaching resource base

Figure 7.6.10: Teaching resource base room layout

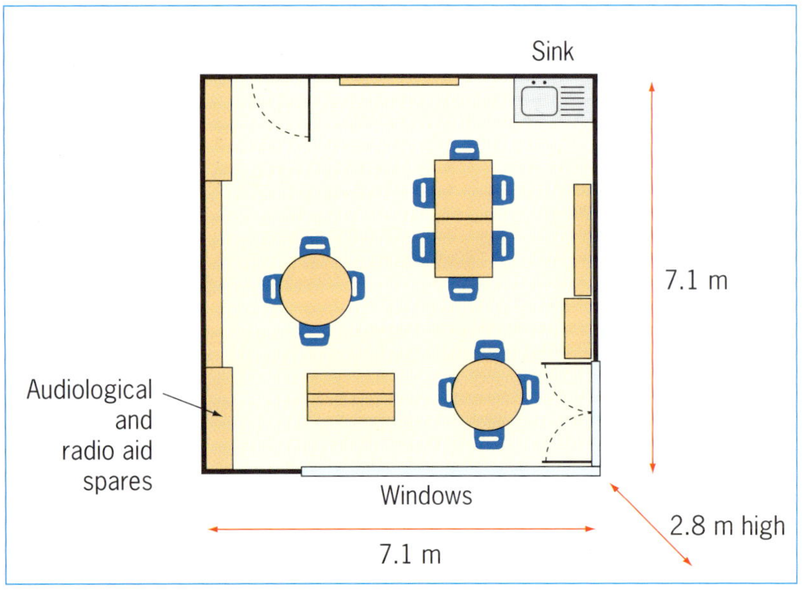

Strengths of the school

A review of the school shows that there has been considerable investment in ensuring that the school is one that reduces acoustical barriers to learning for hearing and hearing impaired children alike. The key features are:

- carpeting to reduce noise in corridors and classroom noise caused by movement
- attaching rubber ends to chairs and tables to reduce movement noise
- maximising lighting, and where appropriate using blinds, so that children and teachers are visible but not silhouetted against the light, thereby ensuring that lip-reading is effective
- using personal radio systems for the hearing impaired children to limit the effects of distance from the teacher
- using a soundfield system, which provides benefit to the hearing impaired child directly by increasing the strength and naturalness of the speech signal, and indirectly by modifying classroom behaviour in a positive manner
- making use of expertise in the in-service training of staff throughout the school
- providing an acoustically well-specified area for supporting those special needs of hearing impaired children that cannot be met within the mainstream classroom.

Future developments at the school

The school is about to undergo major roof repairs. As part of the process the school will take the opportunity to upgrade the acoustic treatment within the classrooms, seeking to lower the reverberation times. This will assist in reducing noise build up during critical learning times of group work and class discussion. The lower reverberation times will enable the soundfield system to work more effectively, and possibly enable the school to use ceiling mounted speaker systems for future installations.

This case study describes the acoustics of an all-age special school for hearing impaired pupils. The school is located on two sites. The primary aged pupils attend a primary special school for hearing impaired children and the secondary age pupils attend a special unit within a mainstream secondary school about one mile away from the primary school.

The primary school, the secondary special unit and the audiology room in the primary school are described separately.

The primary special school

The primary school is a school in which only severely hearing impaired pupils are taught. It was founded in 1975 and caters for up to 115 children between the ages of 3 and 11. The school consists of nine teaching classrooms and a nursery as well as a hall, a dining room and more informal open areas which are used for activities such as art and cookery. There is also an audiology room which is discussed in more detail later.

Pupils are taught in small groups by a teacher aided by a classroom assistant. The teachers wear radio transmitters and all pupils wear radio hearing aids so that they can make use of their residual hearing. Sign language (accompanied by speech) is used for teaching.

Classes were observed to gain an insight into the use of the school and measurements of background noise, reverberation time and sound insulation were carried out.

Location

The school is located on the outskirts of a city, a considerable distance away from the main road, in a residential area.

Layout and construction

A part plan of the school showing the classrooms for Year 2 and Year 3W is shown in Figure 7.7.1. Good internal space planning has generally ensured that noise sensitive areas have not been placed immediately adjacent to noise producing areas, thus avoiding the need for high performance sound insulating constructions. The dining room and main hall are separated from the nearest teaching rooms by an area which is used for activities such as art. Where there are two adjacent classrooms, storage areas have been created between them to act as a buffer zone. The school is single storey, so impact noise from footfalls and chairs being moved above is not an issue. The partitions are blockwork; the doors are timber hollowcore doors with no seals. The roof construction is not known, but the quiet location means that ingress of external noise is not problematic. All classrooms are naturally ventilated.

The classrooms for Year 2 and Year 3W are adjacent but are separated by store rooms. Both rooms are entered through a common area which is used for art and craft work, for storing teaching aids and as an area in which the classroom assistants can prepare teaching material. This common area is a useful acoustic buffer

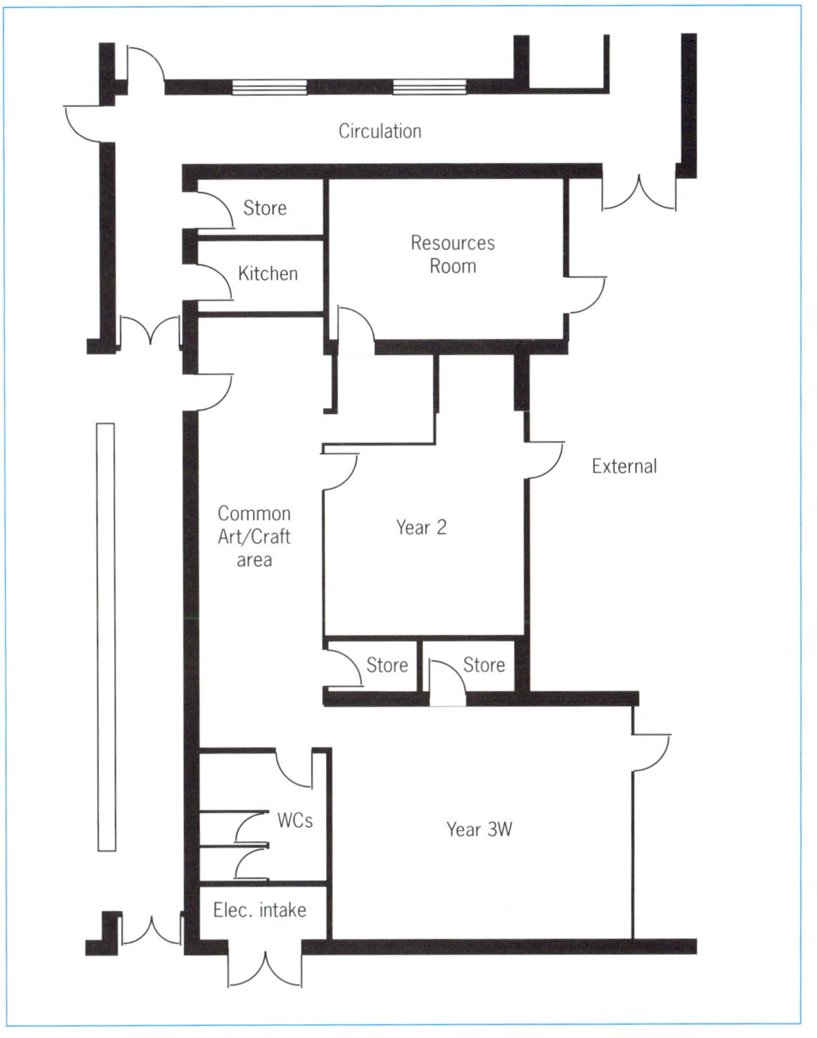

Figure 7.7.1: Part plan of primary school showing Year 2 and Year 3W classrooms

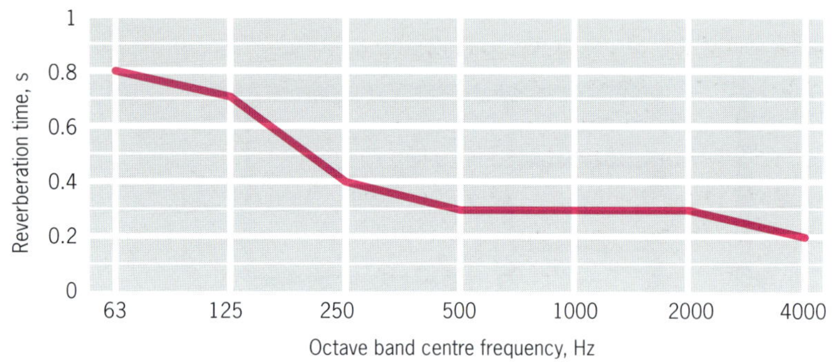

Figure 7.7.2: Measured reverberation time in Year 3W classroom

zone separating the classrooms from the corridor. Despite the fact that the two classrooms have been designed so that noise from one does not disturb teaching in the other, classes were being taught with the doors open between each classroom and the common area. Noise from one classroom was thus clearly audible in the classroom next door.

Surface finishes

In classrooms, surface finishes have been used to control reverberation times. All classrooms have thin carpet on the floors. The pitched classroom ceilings are covered in mineral fibre tiles; these extend down to cover the walls at high level (from the ceiling down to the height of the tops of the doors). The walls have a painted plaster finish with hardboard pinboards dispersed around them.

The amount of absorption provided ensures that the reverberation time is sufficiently short to provide good conditions for speech.

The measured unoccupied mid-frequency RT in the classroom for Year 3W was 0.3 seconds with a rise to 0.7 seconds at 125 Hz. The full spectrum is shown in Figure 7.7.2.

Sound insulation

Sound insulation measurements were carried out between the Year 2 and Year 3W classrooms because these represented a 'worst case' sound transmission configuration, no other two classrooms being located so close together.

The sound insulation between the Year 2 classroom and the common art/craft area via the Year 2 classroom door, which was closed, was also measured.

The BB93 standardized weighted sound level differences, $D_{nT(0.4s),w}$ between Year 2 and Year 3W classrooms and between Year 2 classroom and the common art/craft area were as follows:

Year 2 classroom to Year 3W classroom:

$$D_{nT(0.4s),w} = 53 \text{ dB}$$

Year 2 classroom to common area:

$$D_{nT(0.4s),w} = 18 \text{ dB}$$

Ambient noise levels during lessons

Measurements of ambient noise were made during a desk-based lesson in the Year 2 classroom. A teacher, two classroom assistants and five children were present. Maximum sound levels of 85 dB L_{Amax} were measured. The equivalent continuous sound level was 65 dB L_{Aeq}. Although there was some noise made by the pupils trying to talk, the dominant noise source was due to the teacher talking to the classroom assistants. Noise from the Year 3W classroom was also audible.

Measurements of ambient noise were also made during a physical education lesson in the main hall. The class consisted of a teacher, two assistants and approximately 10 children. Noise levels were very similar to those measured in the Year 2 class; namely a maximum sound level of 84 dB L_{Amax} and an equivalent continuous sound level of 65 dB L_{Aeq}.

Unocupied noise levels

Noise levels were measured in the Year 2 and Year 3W classrooms during a time when the rooms were unoccupied. These results are shown in Table 7.7.1.

The measured values were 40 dB L_{Aeq} in the Year 2 classroom and 29 dB L_{Aeq} in the Year 3W classroom. The dominant noise source in the Year 2 classroom was from the fan on a computer. In Year 3W there was no computer on, but at high frequencies noise from a fluorescent light fitting was dominant.

Discussion

A fundamental issue in the design of rooms for teaching hearing impaired children is the level of background noise which should be allowed. Background noise is amplified by hearing aids and reduces the signal to noise ratio of speech, reducing the effectiveness of the pupils' residual hearing.

The quiet location of the primary school and the absence of mechanical ventilation in the building ensures that indoor ambient noise levels in classrooms (29 dB L_{Aeq} in classrooms without computers) are low. This is lower than the recommended maximum indoor ambient noise level specified in Table 1.1 for classrooms for teaching severely hearing impaired pupils (see also Table 6.1 of Section 6).

A potential disadvantage of low background noise levels is that there is little masking of intrusive noise, so good sound insulation is essential. The layout of the school has been designed to tackle this by not locating noise-sensitive rooms adjacent to noise-producing rooms.

The sound insulation between the Year 2 and Year 3W classrooms of 53 dB $D_{nT(0.4s),w}$ meets the performance standard in Table 1.2. This would be exceeded between other classrooms in the school which are further apart than the Year 2 and Year 3W classrooms. If teaching were to be carried out with the doors between rooms shut, there would be little risk of noise from one classroom disturbing the class in the adjacent room.

The measured sound insulation of 18 dB $D_{nT(0.4s),w}$ between the Year 2 classroom and the common area is poor and implies that the door is a weak sound insulating element. If a class was being taught in the Year 2 or Year 3W classroom while a separate teaching activity was going on in the common area, then it is highly likely that noise from one would disturb the other. This was not perceived to be a problem by the teaching staff, as the classes were being taught with the doors open. Effective frame and perimeter seals would improve the performance of the doors slightly, but the door constructions would need to be changed to incorporate solid cores in order to significantly improve the sound insulation.

The control of reverberation time is vital, firstly to ensure that speech is intelligible and secondly to prevent an excessive build up of reverberant noise which can impair speech discrimination. The measured classroom mid-frequency reverberation time of 0.3 seconds meets the performance standards in Table 1.5.

It is often recommended that classroom ceilings are sound absorptive around the perimeter but reflective in the centre to aid propagation of the teacher's speech to the rear of the classroom. In this school, the classrooms were very small and pupils sit near the teacher because of the small numbers in each class, so sound propagation to the back of a large classroom is not an issue. In larger classrooms for teaching hearing impaired children, however, a central sound reflective ceiling zone may be advantageous.

Key design points to note are:

- quiet site location, away from any major noise sources such as roads, railways and industrial premises
- separation of classrooms by buffer zones such as store rooms, corridors and lobbies
- use of carpet and sound absorptive ceiling tiles in all classrooms to control reverberation times.

	Octave band centre frequency (Hz)							
	63	125	250	500	1 k	2 k	4 k	8 k
Year 2 L_{Aeq} (dB)	38	33	44	37	33	29	23	20
Year 3W L_{Aeq} (dB)	39	28	22	17	17	17	25	25

Table 7.7.1: Measured unoccupied noise levels in classrooms

7.7 Case Study: An all-age special school for hearing impaired children

A special unit in a mainstream secondary school

The special unit for hearing impaired pupils is located in a refurbished block of the mainstream secondary school and consists of six teaching classrooms, the resources area, a speech therapy room and an office for the headteacher of the unit.

When in the special school unit, pupils are taught in small class groups. Pupils are encouraged to communicate using sign language, lip-reading, speech and residual hearing with the help of hearing aids. For 30% of the time, hearing impaired pupils are taught in integrated classes in the mainstream school.

The special school unit forms an interesting comparison with the primary age part of the same school which is located in a nearby primary school, and is some 20 years older.

The unit was visited only four months after it was opened. During the visit, discussions were held with the headmaster of the hearing impaired unit to obtain his opinions on its acoustics. Measurements of background noise, reverberation time and sound insulation were carried out.

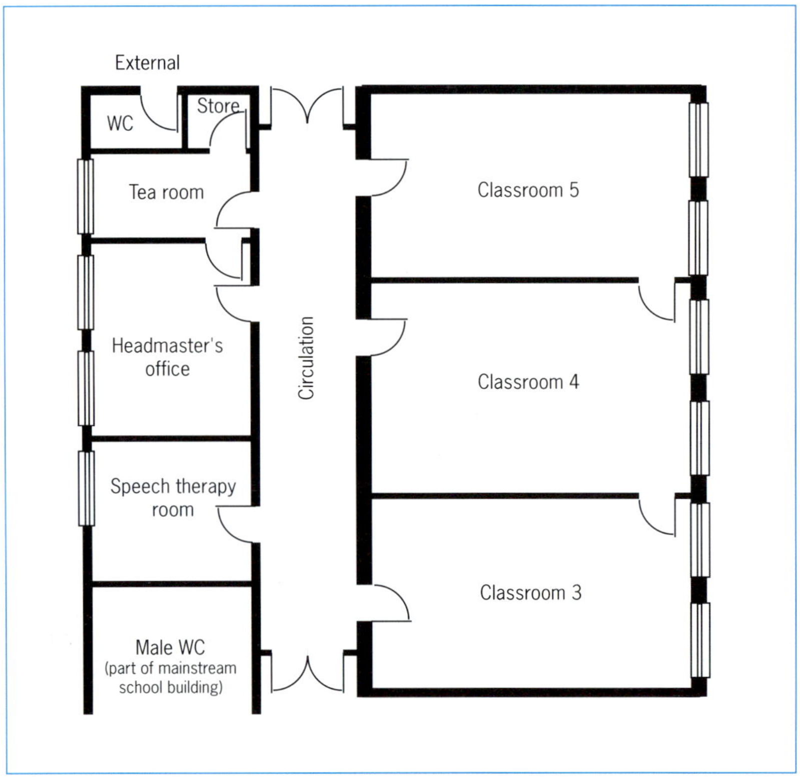

Figure 7.7.3: Ground floor plan of secondary school unit for hearing impaired pupils

Location

The secondary unit is surrounded on three sides by open countryside, and is far from any main roads. The unit is at one end of the school, so that the potential for noise break-out from other classrooms is minimised. Background noise levels on the site are low.

Layout and construction

The special school accommodation is divided between the ground and second floors, with mainstream accommodation in-between on the first floor. Figure 7.7.3 shows the plan of the ground floor.

Unlike the primary school, classrooms are located immediately adjacent to each other with no non-sensitive buffer zones in-between. Partitions between adjacent classrooms are studwork and consist of one layer of 12.5 mm plasterboard and one layer of 19 mm plasterboard on each side of a 48 mm stud. The partition between the headmaster's office and the speech therapy room is built of staggered 70 mm studs with two layers of 15 mm plasterboard on each side and mineral wool in the cavity. The partitions have been built to full height up to structural slab level. The headmaster complained, however, that there were gaps at the partition heads which reduced the sound insulation of the partitions, meaning that sound from one classroom could often be heard in the adjacent room. Examination of the partition heads revealed that pipe penetrations of the partitions had not always been properly sealed.

In many cases adjacent classrooms have connecting doors. All the doors in the special school are single solid core timber doors of 40 - 50 mm thickness with wiper seals acting on a raised timber threshold and compression frame seals. External windows are double-glazed with a deep cavity (approximately 200 mm) and are openable via a sliding casement mechanism.

The party walls between the specialist school and the adjoining mainstream school accommodation are masonry.

The first floor mainstream classrooms have been carpeted to reduce impact noise transmission to the ground floor classrooms, although the carpet has only a thin pile and does not appear to have underlay beneath it. Floor slabs are of concrete. No mechanical ventilation is provided.

Surface finishes

All the classrooms have thin pile carpets and mineral fibre suspended ceilings. The plasterboard walls have a sound reflective finish. Pinboards on the walls are timber, backed by an airspace, and provide some control of low frequency reverberation times. The speech therapy room also has a thin carpet and a suspended mineral fibre tile ceiling. The amount of absorption provided ensures that the reverberation time is sufficiently short to provide good conditions for speech.

Reverberation time

The measured unoccupied mid-frequency RT of classroom 4 was 0.4 seconds with a small rise to 0.5 seconds at 125 Hz.

The measured unoccupied mid-frequency RT of the speech therapy room was 0.3 seconds with a flat spectrum down to 125 Hz.

Sound insulation

Sound insulation measurements were carried out between classrooms 4 and 5 which are horizontally adjacent.

The weighted BB93 standardized level difference between classrooms 4 and 5 was 34 dB $D_{nT(0.4s),w}$.

The sound insulation between several other areas was also measured and the following weighted BB93 standardized level differences obtained:

- classroom 5 to mainstream classroom directly above: $D_{nT(0.8s),w} = 48$ dB
- headmaster's office to speech therapy room: $D_{nT(0.4s),w} = 47$ dB
- male toilets to speech therapy room: $D_{nT(0.4s),w} = 52$ dB

Unoccupied noise levels

Noise levels were measured in classroom 4 and the speech therapy room during a time when the rooms were unoccupied, but when there were staff elsewhere in the building. The noise spectra are shown in Table 7.7.2.

The corresponding indoor ambient noise levels are 26 dB L_{Aeq} in classroom 4 and 19 dB L_{Aeq} in the speech therapy room. The dominant noise sources in classroom 4 were a faint buzzing noise from the radiator and from fluorescent light fittings. Talking in other classrooms in the unit was just audible. The main noise sources in the speech therapy room were a clock ticking and a fluorescent light fitting buzzing. The headmaster's voice as he talked on the telephone in his office next door was clearly audible although the words were not intelligible. It should be noted that at high frequencies the reported octave band noise levels in the speech therapy room were due to electrical noise in the sound level meter; actual noise levels were probably lower.

Discussion

Measured noise levels in a typical classroom and the speech therapy room were very low (26 dB L_{Aeq} and 19 dB L_{Aeq} respectively) and meet the performance standards specified in Table 1.1. This is appropriate in rooms in which hearing impaired pupils are taught, to ensure good speech signal to noise levels. There was no unpleasant tonal content in the frequency spectra.

Very low unoccupied ambient noise levels mean that any extraneous noise intrusion will be especially audible. The site location and high performance external windows ensure that noise ingress from outside does not cause problems. The teaching staff have, however,

	Octave band centre frequency (Hz)							
L_{eq} (dB)	63	125	250	500	1 k	2 k	4 k	8 k
Classroom 4	34	30	29	18	22	13	12	13
Speech therapy room	28	23	15	12	12	11	11	12

Table 7.7.2: Measured unoccupied noise levels in Classroom 4 and the speech therapy room

complained about the sound transmission between horizontally adjacent rooms. The sound insulation appears to be of a lower standard than they had expected in a new purpose-built unit. These subjective comments are borne out by the results of the objective sound insulation measurements. The $D_{nT(0.4s),w}$ of 34 dB measured between classrooms 4 and 5 is lower than that required for classrooms in mainstream schools. Where background noise levels are low, hearing impaired pupils cannot discriminate between intrusive noise and speech as easily as pupils with full hearing, and a higher standard of sound insulation is needed. A minimum $D_{nT(0.4s),w}$ value of 50 dB is required, see Table 1.2.

Measurements showed that the sound insulation performance of the partition did not rise at high frequencies as would normally be expected. This confirms the existence of small gaps which were found at the partition heads. Notwithstanding this, the mid-frequency level difference across the partition is poor (between 30 dB and 35 dB). This indicates that the studwork partition selected was not of a sufficiently high performance. A partition with staggered studs, increased plasterboard thicknesses and mineral wool in the cavity would provide a higher standard of sound insulation. The overall sound insulation performance between adjacent classrooms is, however, ultimately limited by the communicating door. Although the doors are of a very high standard (this is discussed further below) they are still a weak sound insulation element. Whilst this may not be a serious problem between classrooms and the corridor, the presence of doors between classrooms is inconsistent with the requirement for a high standard of sound insulation. Connecting doors are not recommended.

The sound insulation measured between the headmaster's office and the speech therapy room was 47 dB $D_{nT(0.4s),w}$. This is below the performance standard in Table 1.2 for sound insulation between an office and a speech therapy room. The headmaster had complained that he was sometimes disturbed by noise from the speech therapy room and whilst in the unoccupied room the headmaster's voice was audible but not intelligible. This level of privacy means that although the headmaster's conversations would remain confidential, intrusive noise may disturb the concentration of both the headmaster and of users of the speech therapy room. A higher standard of studwork wall construction between rooms may have been considered to be impracticable in the special school design. An alternative solution would have been to locate non-sensitive acoustic buffer zones, such as storage areas, between the headmaster's office and other rooms.

A value of 48 dB $D_{nT(0.4s),w}$ was measured from one of the ground floor classrooms for hearing impaired pupils to the mainstream classroom directly above it on the first floor. This is an appropriate standard of sound insulation for the mainstream classroom and no complaints have been made by the teaching staff.

Visual inspection of the doorsets confirmed that they were of suitable quality and likely to meet the 30 dB R_w sound insulation specification for doorsets in Table 1.3.

Reverberation times in the classrooms are well controlled due to the provision of acoustic absorption on the floors and ceilings. The mid-frequency RT of 0.4 seconds meets the performance standards in Table 1.5. The wooden wall panels help to control the RT at low frequencies, on which hearing impaired people often rely for information. The teaching staff judged the classroom acoustics to be satisfactory.

The RT in the speech therapy room is also well controlled due to the carpet and mineral fibre suspended ceiling. The mid-frequency value of 0.3 seconds meets the performance standards in Table 1.5.

Conclusions

The acoustic design of the special school unit is good, in terms of room acoustics and unoccupied noise levels, although there are some deficiencies in the sound insulation provided by the internal wall constructions.

Key points to note are:

- The site is in a quiet location, away from any major noise sources such as roads, railways and industrial premises.
- Communicating doors between adjacent classrooms limit the sound insulation that can be achieved and are inconsistent with the need for low levels of intrusive noise.
- Partitions are full height, but poor workmanship has resulted in small gaps at partition heads.
- Sound transmission problems between the headmaster's office and the speech therapy room could have been avoided by better space planning.
- Use of carpet and sound absorptive ceiling tiles in all classrooms and the speech therapy room helps to control mid-frequency reverberation times.
- Wooden pinboards backed by an airspace help to control low frequency reverberation times.
- First floor classrooms are carpeted which reduces impact noise transmission. However, there was no underlay which would have reduced the impact transmission further.

Audiology room

In the primary school for hearing impaired children there is an audiology facility, which consists of a technician's room and an audiometric test room.

The tests carried out in the audiometric test room are generally carried out in the ambient acoustic field rather than using headphones. Activities range from testing hearing saturation levels and hearing aid discomfort (during which high noise levels of up to 90 dB(A) are generated in the room) to testing for speech discrimination against background noise, which requires low ambient noise levels.

Measurements were carried out of indoor ambient noise, sound insulation and reverberation time in the audiology suite. In addition, a discussion was held with the audiologist who uses the suite to obtain his opinion of the suitability of the acoustics.

Layout and construction

The location of the audiology suite within the primary school is shown in Figure 7.7.4. The audiometric test room is entered directly from the corridor. The test room also has an external wall and a window onto an enclosed courtyard.

The walls of the audiometric test room are a single skin of 100 mm thick blockwork of an unknown density. The single leaf doors into the technician's room and the corridor are a hollowcore timber construction with no frame or threshold seals. Noise from the corridor was clearly audible in the test room. There is a fixed double-glazed window between the test room and the technician's room which incorporates a deep acoustic cavity between the panes of glass. The external window which looks onto the courtyard is single-glazed and is openable.

The roof construction is not known, but the quiet site location means that ingress of external noise is not problematic. There is no mechanical ventilation system.

Figure 7.7.4: Plan of audiology facility

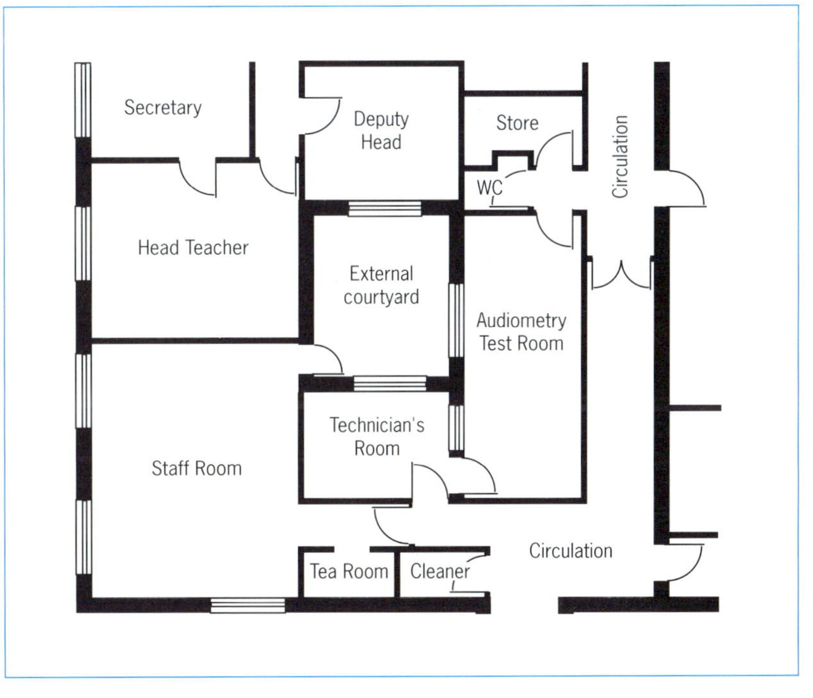

7.7 Case Study: An all-age special school for hearing impaired children

Surface finishes

The audiometric test room is carpeted and has a mineral fibre suspended ceiling. Apart from where there are windows or doors, the entire wall area is also lined with mineral fibre tiles. Subjectively, the room is very dead acoustically.

Audiologist's opinion

The audiologist finds intrusive noise very disturbing to his work, particularly when he is carrying out tests of speech discrimination against background noise. This means that the times when he can carry out certain measurements are determined by possible activity in the corridor. When there is no activity in the corridor, the background noise levels in the room are sufficiently low for his tests.

Conversely, when loud noises are generated in the audiometric test room (for example when hearing aid discomfort is being tested), these can clearly be heard in the corridor, although this does not cause disturbance to teaching.

The audiologist did not express any dissatisfaction with the internal room acoustics, but noted that achieving low ambient noise levels should be a first priority when designing audiology facilities and that good room acoustics were worthless without sufficiently low background noise levels or good enough sound insulation.

[1] 'Acoustics: Audiology', Health Technical Memorandum 2045, HMSO, 1996.

Acoustic measurements

Reverberation time

The unoccupied mid-frequency reverberation time, T_{mf}, in the audiometric test room was 0.2 seconds with a small rise to 0.3 seconds at 125 Hz. The RT across the frequency spectrum is shown in Figure 7.7.5.

Sound insulation

The sound insulation measured between the technician's room and the audiometry test room was 36 dB $D_{nT(0.4s),w}$.

Unoccupied noise levels

Noise levels were measured in the audiometry test room when the room was unoccupied. The results are shown in Table 7.7.3.

The noise level corresponds to an A-weighted sound pressure level of 21 dB L_{Aeq}. The dominant noise source was water running through the radiator. Voices in the corridor outside were audible. It should be noted that at high frequencies the reported octave band noise levels in the audiometry test room were due to electrical noise in the measurement system; actual noise levels would have been lower.

Discussion

Guidance for the acoustic design of audiology facilities in hospital audiology departments is given in Health Technical Memorandum 2045 "Acoustics: Audiology"[1], but the suite in the school is used for educational audiology and as such is provided by the county in which the school is situated rather than by the Health Service.The guidance includes maximum permissible ambient sound pressure levels in third-octave bands and reverberation times for audiometric test rooms, depending on the audiometric tests which will be carried out in the rooms. Because the use of each audiometric facility is specialised, the end users of any facility should be consulted and reference made to HTM 2045[1] before the acoustic design of a facility is undertaken.

In this test room the background noise

Figure 7.7.5: Measured reverberation time in audiometry test room

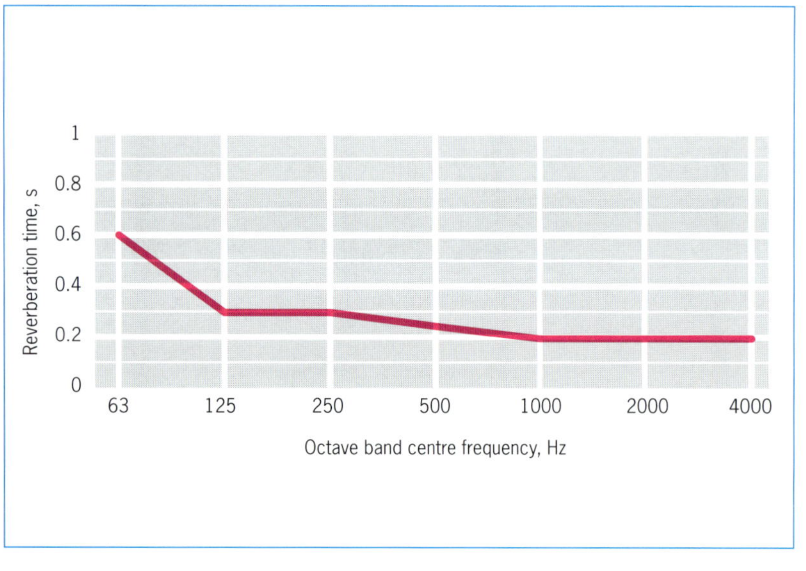

L_{eq} (dB)	Octave band centre frequency (Hz) 63	125	250	500	1 k	2 k	4 k	8 k
Audiometry test room	35	24	21	17	12	10	11	13

Table 7.7.3: Measured noise levels in audiometry test room, unoccupied

levels are sufficiently low for the audiologist to carry out his tests. The background noise spectrum does not contain any unpleasant tones, due to the quiet nature of the school site.

The limited sound insulation afforded by the single-leaf masonry wall and the poor quality single door mean that intrusive noise levels in the test room are high when there is activity in the corridor. The high intrusive noise levels disrupt the audiologist's work. An appropriate sound insulation performance for the wall between the test room and the corridor would be very dependent on the specific requirements of the audiologist and the school, but it is likely that a double-leaf masonry wall construction plastered on both sides (each leaf at least 415 kg/m^2 including plaster) would be the minimum required. The door from the corridor into the test room is a weak sound insulation element and would limit the performance of any upgraded wall construction. The best solution would be to allow entry to the test room only via the staff room and technician's room. Failing this, a lobbied door arrangement would be required.

HTM 2045 recommends that reverberation times at all frequencies between 125 Hz and 4 kHz are between 0.2 seconds and 0.25 seconds in audiology test rooms. The measured reverberation times are generally within this range, given the accuracy of the on site measurements. Thus the reverberation time in the test room has been well controlled by the selection of surface finishes. Recommendations are also made for reverberation times in third-octave bands from 31.5 Hz to 100 Hz. Due to the small size of the test room, reverberation times could not be measured accurately at these low frequencies.

Conclusions

Although background noise levels are low and the reverberation time is well controlled, the poor sound insulation means that the test room is unsatisfactory for its purpose.

Key points to note are:

- The site is in a quiet location, away from any major noise sources such as roads, railways and industrial premises, so background noise levels are low.
- The audiometric test room is poorly located adjacent to a noisy corridor.
- The 100 mm blockwork wall between the test room and the corridor is inadequate for controlling noise intrusion.
- The single door between the test room and the corridor is a weak sound insulation element.
- The reverberation time is well controlled by the use of carpet, a mineral fibre tile suspended ceiling system and mineral fibre tiles on all the walls.

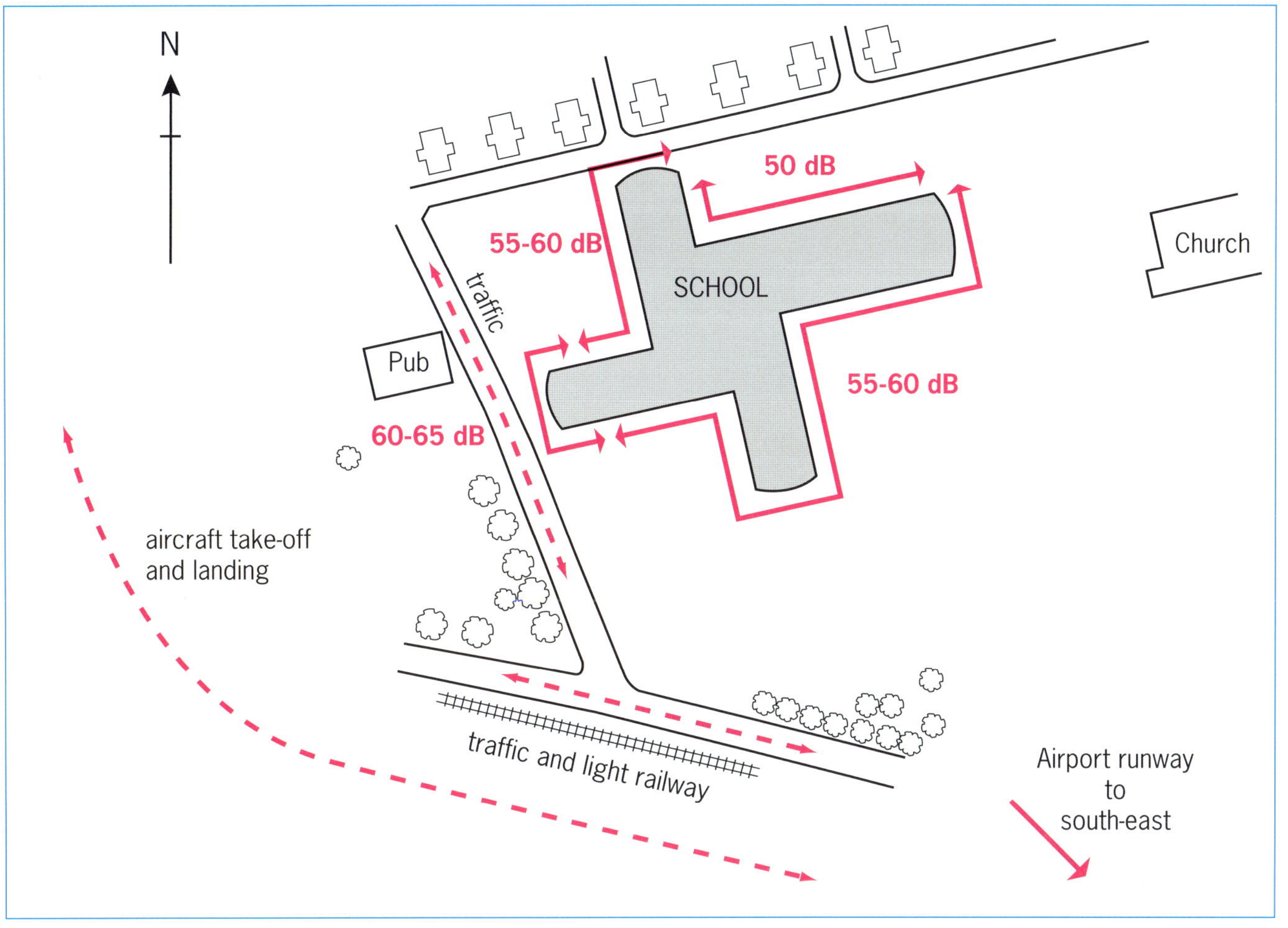

Figure 7.8.1: Site plan showing external noise levels, L_{Aeq}

Figure 7.8.1 shows a site plan, based on the noise survey carried out at the start of the project. The high external noise levels are generated by low-flying aircraft and traffic on nearby busy roads. One option would have been to acoustically seal the building envelope and mechanically ventilate the building. However, this was too expensive for the available budget. The design team also wished to reduce lifetime costs and opted for a naturally ventilated building which would maintain the same internal noise levels.

Being an inclusive school, the design had to accommodate pupils and other members of the community with hearing problems. The target for background noise was set at 35 dB(A). At the same time the design had to provide fresh air at a rate of up to 8 litres per second for each of the usual number of occupants. This equates to approximately 4.5 air changes per hour in both ground and first floor classrooms.

Standard products such as attenuated trickle ventilators inserted into window openings, as are often used in housing, would not have achieved the required air flow rates. Alternative purpose designed systems were therefore required.

Classrooms are naturally ventilated by means of inlet vents under the external windows and passive stacks located at high level at the rear of the rooms, adjacent to the central corridors. The inlet louvres duct air into the classrooms via grilles just inside the perimeter convector grilles. These inlet grilles are controlled by classroom users by easy to operate openable flaps covering the grilles.

Both inlets and outlets are acoustically insulated to prevent the entry of external noise.

Depending on the prevailing weather, wind driven or temperature driven ventilation provides sufficient fresh air.

- The more windy the weather, the greater the pressure difference across the building envelope and the greater the air

7.8 Case Study: Acoustic design of building envelope and classrooms at a new secondary school

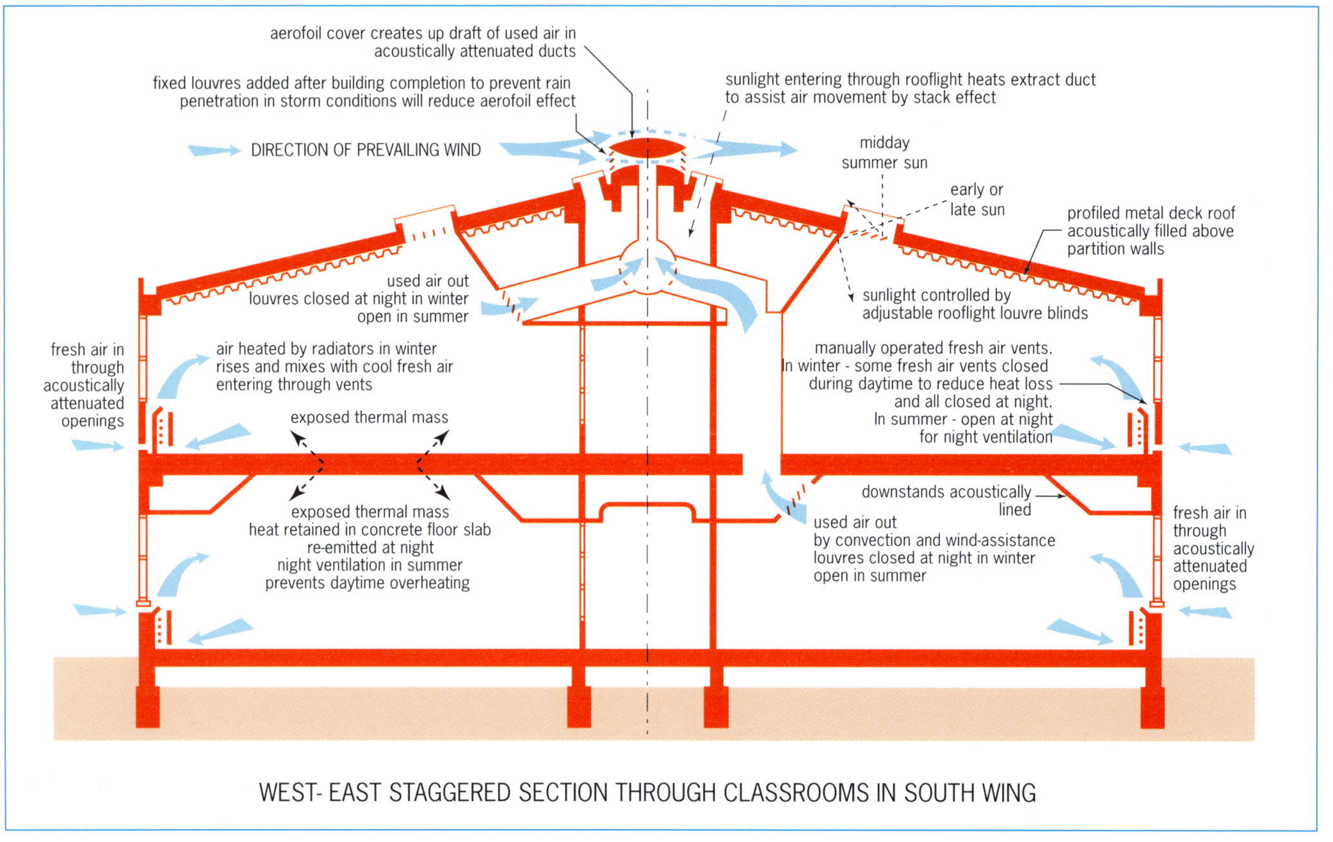

Figure 7.8.2: Schematic diagram of ventilation paths through two storey section of building

Figure 7.8.3: Ground floor air vents

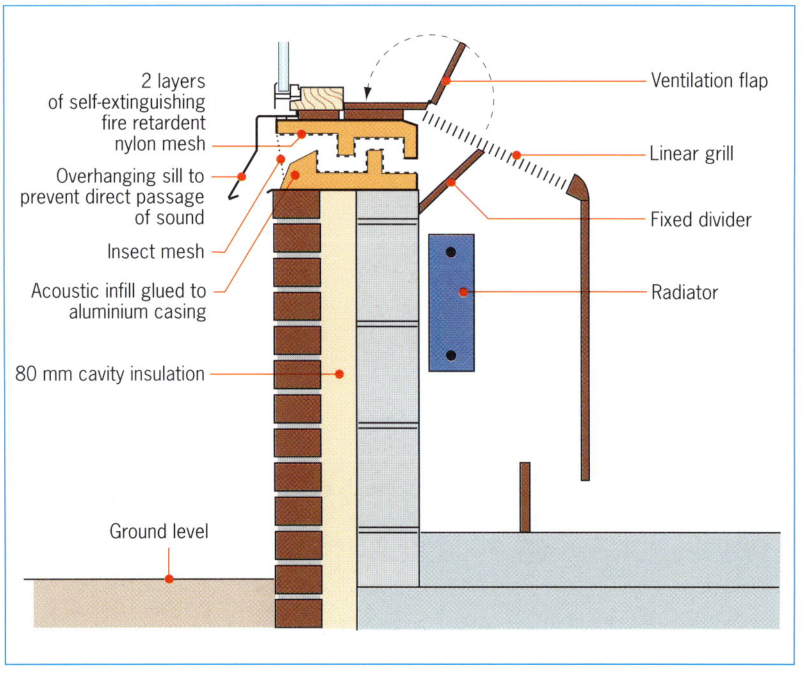

movement in the ducts.

- The temperature difference when the internal spaces are warmer than outside, as in winter, drives the stack effect ventilation causing air to rise up the central ducts.
- The central ducts which leave the back of the classrooms join into a combined duct over the first floor corridor which then rises to the outlet at roof level. The passive stack effect is enhanced by providing roof glazing over the combined section of duct which is painted black and encased over a drop ceiling area in the corridor. Solar gain raises the air temperatures in the top section of ductwork causing the air to rise. This is particularly effective in hot weather.
- An aerofoil is positioned at the duct outlet to enhance the wind-driven stack effect. The problem of wind-blown rain in storm conditions led to modification of the aerofoils to incorporate louvres beneath the aerofoil sections. This will probably have made the aerofoils on their own considerably less effective.
- The windows are openable and are designed to increase the maximum possible ventilation rate so that when the wind and stack driving forces are small there will still be adequate ventilation, although this will obviously let in some ambient noise.

The ventilation system is completely under the control of the occupants in individual spaces, who can open and close flaps over the inlets below the windows and high level adjustable louvres at the back of the classrooms, controlled by a short pole.

Ground floor vents (Figure 7.8.3) and first floor vents (Figure 7.8.4) are of different design. These proprietary/purpose designed external vents on the window walls are acoustically insulated.

The airborne sound insulation of prototype ground floor and first floor ventilators was tested in the laboratory. The resulting element-normalized level differences in octave bands and the resulting $D_{n,e,w}$ values are shown in Table 7.8.1. As a result of these tests, the ground floor vent design was modified to improve its performance. This included the addition of an overhanging sill and extended internal nibs of sound absorbing material. The final design, shown in Figure 7.8.3, appears as effective acoustically after installation as the first floor vents.

The passive stacks are acoustically lined which prevents cross-talk between classrooms which share the same discharge ductwork. Four classrooms are ventilated via a common extract duct.

Air flow tests were carried out in typical classrooms. These showed that on a typical spring day, with a moderate wind (10-15 kph), with all the vent flaps and louvres open, air entered at between 0.8 and 1.6 m/s depending on location within the building, and left through the high level louvres at between 0.3 and 0.7 m/s, again depending on location. This corresponds to a fresh air rate of 5.3 air changes per hour.

Metal deck roof

The roof structure, from outside in, is as follows:

- 0.9 mm gauge stucco embossed aluminium external covering
- 120 mm (compressed to 110 mm) thermal insulation
- 30 mm acoustic insulation
- vapour control layer
- 0.9 mm gauge polyester powder coated steel (internal support decking).

There is no void within the roof except between the profiles of the support decking. The profile voids are filled at partition lines with inserts of acoustically absorbent material.

There is some flanking transmission through the continuous profiled steel roof construction, which reduces the sound insulation between rooms.

Concrete floor

At first floor level, the floor finish on the precast concrete floor is a steel mesh reinforced sand/cement screed on 50 mm thick acoustic mineral wool board, which prevents the transmission of impact sound to the ground floor rooms below.

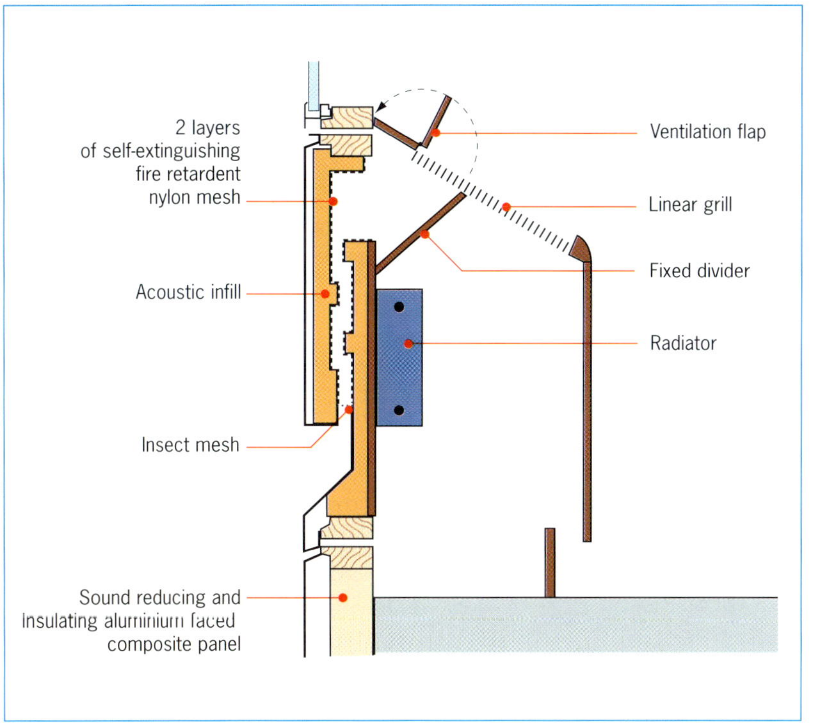

Figure 7.8.4: First floor air vents

Table 7.8.1: Element-normalized level differences for prototype ground floor and first floor vents

	Octave band centre frequency (Hz)								
$D_{n,e}$ (dB)	63	125	250	500	1 k	2 k	4 k	8 k	$D_{n,e,w}$ (dB)
Ground floor vent	28.0	21.6	22.5	25.8	40.8	57.9	54.0	53.9	33
First floor vent	24.5	19.6	22.2	28.4	42.2	50.8	53.4	53.0	34

Internal partitions

Internal partitioning generally uses two layers of 12.5 mm plasterboard each side of metal studs, with quilt in the cavity, giving a construction width of 200 mm. Laboratory test results for this form of construction indicate a sound reduction index of 52 dB R_W. Performance on site is usually at least 5 dB less than this. Tests carried out on site show a considerable variation in performance, from 44 dB to 38 dB D_W. The walls therefore meet the standard of 38 dB for classrooms given in Building Bulletin 87, that was required at the time of the design. However some fail to meet the 43 dB D_W design target for the project, which is lower than the present standard given in Table 1.2. There are a number of causes of the performance loss. On the first floor, flanking sound carried through the lightweight roof was seen to be a contributory factor, despite the use of fillers in the profile voids. However, the best performance was also recorded in one of the first floor rooms. A reduction in insulation was found in some ground floor rooms where partitions directly abut the precast concrete first floor. The conclusion was that variability in construction standards, rather than detailing, was the key factor.

Generally, staff and pupils at the school do not consider noise between spaces to be a problem, and are happy with the acoustics.

Classrooms are provided with acoustic treatment in specific ceiling areas to bring the mid-frequency reverberation times below 0.8 seconds. At first floor level, this is provided at the rear of the rooms, over the sloping soffit which encloses services and also helps to reflect daylight from the roof lights into the back of the room. On the ground floor, suspended ceilings run along both sides of classrooms to enclose services and cut off indirect sound paths. In the centre, the first floor precast concrete floor slab is exposed to absorb re-radiated solar radiation and to reinforce direct speech sound paths.

The site plan shows that the new extension is adjacent to a heavily polluted inner city road. The road is a very busy two lane highway and is the main source of noise in the area.

Acoustic design

The acoustic design was based on the noise limits recommended in Building Bulletin 87(BB87) of 40 dB $L_{Aeq,1h}$ in classrooms and not more than 50 dB $L_{Aeq,1h}$ in a gymnasium. These values are now superseded by the performance standards in Section 1 of Building Bulletin 93.

Noise survey

Noise surveys were carried out on site before and after the completion of the new extension. The aim was to establish the external noise levels and use these data to calculate the required sound insulation for the building envelope.

The measured free-field external noise level was 70 dB L_{Aeq}. The major source of noise was road traffic on the very busy main road.

The rooms with most exposure to road traffic noise are:

Ground Floor: Gymnasium

First Floor: Language classrooms 1, 2 and 3

Second Floor: Mathematics classrooms 2, 3, 4 and 5 and ICT rooms 1 and 2.

Sound insulation of the building envelope

The noise levels to which various parts of the building envelope would be exposed were calculated by extrapolation from the baseline noise measurements according to the *Calculation of Road Traffic Noise*. Design calculations of internal noise levels were made on an iterative basis to determine required acoustic specification of the windows, the roof and the wind scoop system so that background noise levels given as guidance in BB87 would not be exceeded.

The building envelope comprised:

- walls: part brick/block cavity, part blockwork with a terracotta tile rain screen and mineral fibre in the cavity
- windows: double glazing incorporating 10 mm and 6.4 mm laminated glass
- main roof: proprietary double skin steel roofing system (38 dB R_w)
- mansard roof: proprietary roofing system, supplemented by an internal plasterboard lining with mineral fibre infill
- roof lights: double glazing incorporating 4 mm glass.

Recommendations were given for the attenuation of external noise through the wind scoop system. It was recognised that new measures to attenuate external noise might affect the airflow characteristics and therefore any suggestions would need to be confirmed by the manufacturer.

The manufacturer of the wind scoop system arranged for acoustic tests to be undertaken in a UKAS test laboratory. A number of different internal lining treatments were tested. The results are summarised in Table 7.9.1.

Initial calculations for the classrooms indicated that a 5 m length of lined duct would provide sufficient attenuation to reduce the internal noise to approximately 40 dB L_{Aeq}. For classrooms on the second floor, the wind scoop ducts were not long enough and it was necessary to increase the attenuation by fitting an additional attenuator. For classrooms on the lower floors the length of the wind scoop duct was sufficient and no additional attenuator was required. The proposed duct details are summarised in Table 7.9.2.

7.9 Case Study: Acoustically attenuated passive stack ventilation of an extension to an inner city secondary school

Table 7.9.1: Laboratory measurement data for the airborne sound insulation of the wind scoop system

Test element	$D_{n,e,w}$ (C;C_{tr}) (dB)
620 mm x 620 mm square hole	13 (0;0)
Vent, unlined with dampers open	16 (0;0)
Vent, unlined with dampers closed	26 (0;-1)
Vent lined with acoustic tile, dampers open	26 (0;-3)
Vent lined with acoustic tile, dampers closed	35 (-1;-5)
Vent lined with open cell foam, dampers open	26 (0;-3)
Vent lined with open cell foam, dampers closed	35 (-1;-4)
Vent lined with open cell foam, linear ceiling grille fitted	27 (-1;-4)

Table 7.9.2: BB87 background noise levels and proposed duct details

Classroom	Floor	BB87 background noise levels $L_{Aeq,1hr}$ (dB)	Treatment
Mathematics 2, 3, 4 and 5	2	40	Internal acoustic lining plus 500 mm attenuator
ICT 1 and 2	2	40	Internal acoustic lining plus 1800 mm attenuator
Languages 1, 2 and 3	1	40	Internal acoustic lining
Gymnasium	Ground	50	Thermal insulation only to blockwork ducts

Post-completion measurements of indoor ambient noise levels

Following completion of the building, measurements of the indoor ambient noise levels were carried out at a number of different locations in each room and averaged. Simultaneous measurements were taken of the free-field external noise level which was 70.6 dB $L_{Aeq,5h}$ and within 1 dB of the level measured prior to development. The measurement results are summarised in Table 7.9.3 where they are compared with the design targets and predicted values.

The measured results in the Gymnasium, Languages 1, Mathematics 2 and Mathematics 3 were found to meet the design limits. The failure to meet the internal noise limit in Languages 3, ICT 1 and ICT 2 can be explained by the factors noted in the comments column, that is, excessive noise transmission via unsealed window frames and the noise from computer fans in the operational ICT rooms.

Internal noise measurements were also

Table 7.9.3: Comparison between the BB87 background noise levels, calculated and measured indoor ambient noise levels

Room	BB87 background noise levels $L_{Aeq,1h}$ (dB)	Calculated $L_{Aeq,1h}$ (dB)	Measured $L_{Aeq,1h}$ (dB)	Comments
Gym	50	45	42	Significant flanking transmission around escape door
Languages 1	40	37	38	
Languages 3	40	40	43	Some window frames not yet sealed
Mathematics 2	40	39	38	
Mathematics 3	40	40	38	
Mathematics 4	40	41	41	
ICT 1	40	42	44	Computer noise present
ICT 2	40	33	42	Computer noise present

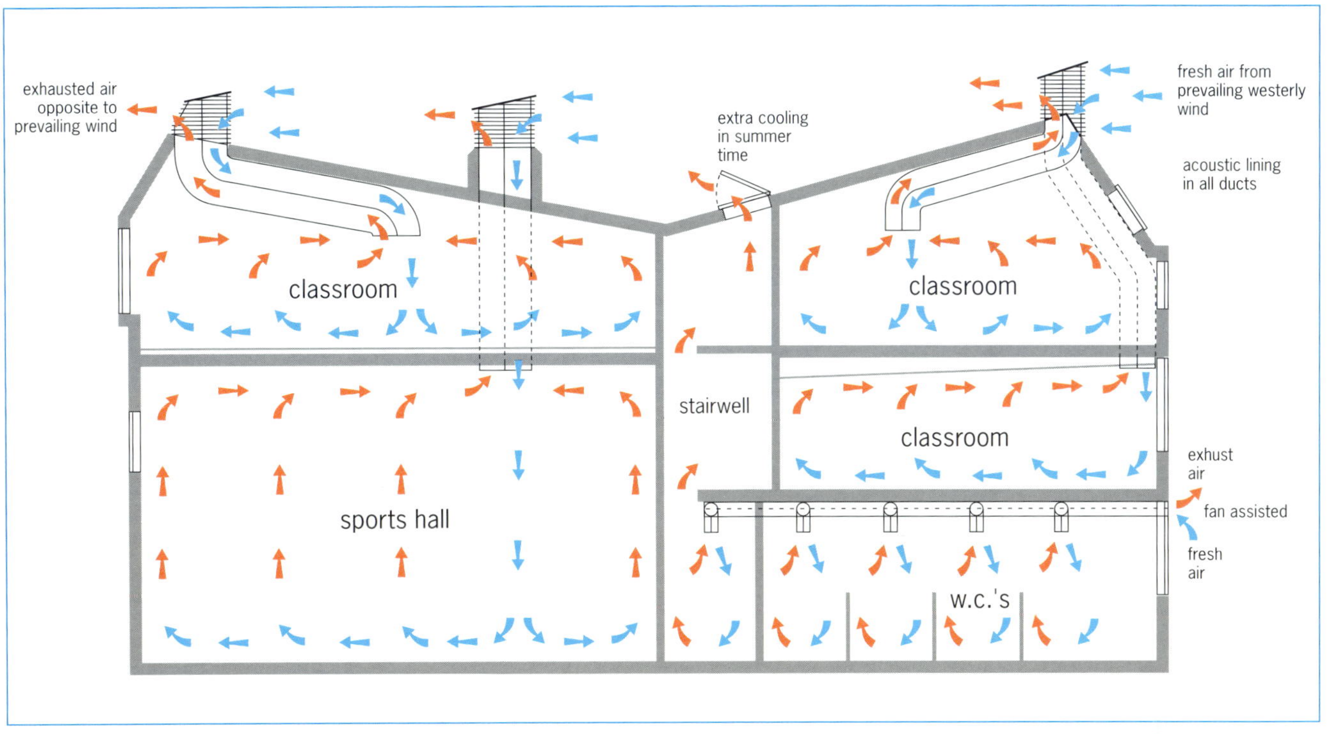

Figure 7.9.1: Section through new extension shows stack ventilation in operation

carried out with the ventilation system open and closed. The results did not display any significant change in level nor was there any significant variation in the sound pressure level around the room.

Ventilation design

The close proximity of the road meant that open windows could not be used for ventilation because road traffic noise would cause problems and airborne pollutants emitted by the heavy road traffic could be carried into the building through low level open windows.

The rooms exposed to traffic noise are therefore ventilated using a wind scoop system with the exception of a manager's office which is provided with a noise-attenuated ventilator unit. This type of unit was originally developed to comply with the requirements of the Noise Insulation Regulations 1975. The unit either comprises a variable speed powered ventilator which is designed to be installed in the building façade and a permanent air vent, or it may be a single unit which combines both. There are normally two speed settings and the Regulations set limits on noise transmission through the units and the self-noise of the fan.

The acoustic consultant suggested that attenuated ventilators should also be fitted in Mathematics classroom 1, the Mathematics office and the staff room as opening the windows in these rooms would result in noise levels exceeding the BB87 guidance of 40 dB $L_{Aeq,1hr}$.

Taking into account the characteristics of the new building and site conditions, adequate ventilation has been achieved as described below.

Teaching areas (Classrooms, ICT Rooms, Science Laboratory and Gymnasium)

All new teaching spaces are naturally ventilated by a wind scoop type system through terminals mounted at roof level. The roof terminals are designed to be omni-directional allowing the intake of fresh air regardless of the prevailing wind direction. Each terminal is divided into equal quadrants; two are positively pressurized by the wind to create a fresh air intake, the remaining two on the leeward side are negatively pressured allowing stale air to be exhausted.

Air is ducted from the terminals either directly into the second floor rooms or down to the ceiling of the first floor classrooms and gymnasium. Each terminal

Fugure 7.9.2: The Roof terminals, viewed from inside during construction

7.9 Case Study: Acoustically attenuated passive stack ventilation of an extension to an inner city secondary school

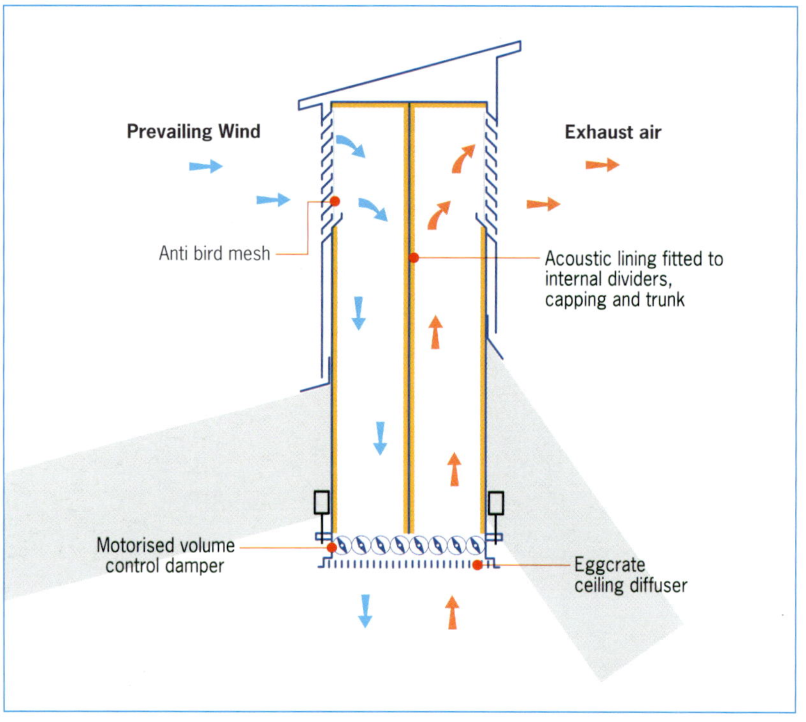

Figure 7.9.3: Section through ventilation stack shows stack operation and acoustic treatment

Figure 7.9.4: Three of the roof teminals, during construction

has been carefully sized according to the volume of the space to be ventilated, the number of people who will normally occupy each space and any potential source of additional heat, for example solar gain or computers. The performance of each terminal has been modelled for a variety of wind speeds to ensure that adequate fresh air can be provided.

Each terminal is individually controlled by dampers set in the ends of the duct to modify the ventilation rate according to the actual conditions in each room. During summer time the control is based on room temperature because higher ventilation rates are required to keep the rooms within acceptable comfort levels. With passive stacks, temperature differences within the room and the length of the ducts will result in improved extract. During the winter the quantity of fresh air needs to be minimized, reducing heat losses via exhaust air, hence there is control by an air quality sensor.

A manual override allows users to have control of the system depending on their experience of room conditions. All windows are openable to allow additional fresh air to be introduced, for example, during changeover of lessons when indoor ambient noise levels are less critical.

The proposed system takes full advantage of the prevailing site conditions. Fresh air is both drawn in and exhausted at the highest level. Wind speeds at 15 m above the ground will be higher than at street level due to fewer obstructions by surrounding buildings. The quality of the air also generally improves with increasing distance from the source of pollution, which in this case is road traffic. Heavy pollutants from vehicle exhausts tend to remain at street level particularly in conditions of high atmospheric pressure.

The ventilation strategy also allows for night time cooling of the building. Studies have shown that air quality by main roads improves at night due to lower traffic flows. At the end of each day stale air left in the building can be fully replaced and then warmed, depending on the season, ready for the next morning.

Rooms for specialist activities (Changing rooms, toilets, shower areas and science laboratories)

Natural ventilation is not suitable for certain parts of the building. Therefore limited mechanical intake and extract are used in areas which require a high rate of ventilation, for example in changing rooms where high levels of water vapour and body odours need to be removed. The mechanical system extracts air at low level, with a simple heat recovery apparatus used to reclaim heat, and replacement air is filtered to remove airborne particles.

The science laboratory requires specific extract ventilation to fume cupboards. This is achieved by mechanical extract ventilation systems that exhaust away from the other roof terminals, with make-up air being naturally induced via a wind scoop and opening window lights.

This case study demonstrates that free-field external noise levels can be reduced by approximately 30 dB inside naturally ventilated classrooms using a sound attenuated passive stack ventilation system.

An investigation was carried out into the acoustic conditions in open-plan learning spaces in a secondary school, construction of which was completed in 1991. Figure 7.10.1 shows the site. The ground and first floor plans can be seen in Figure 7.10.3.

The curriculum model divides the day into 3 hour subject modules. Team teaching is fundamental to the curriculum and to facilitate this, there are several relatively large open-plan learning bases, as shown in Figure 7.10.2, that typically hold around seventy students.

Some of the learning bases are used for teaching particular subjects such as Mathematics or English. All the learning bases are subdivided into smaller areas so that different lessons or activities can take

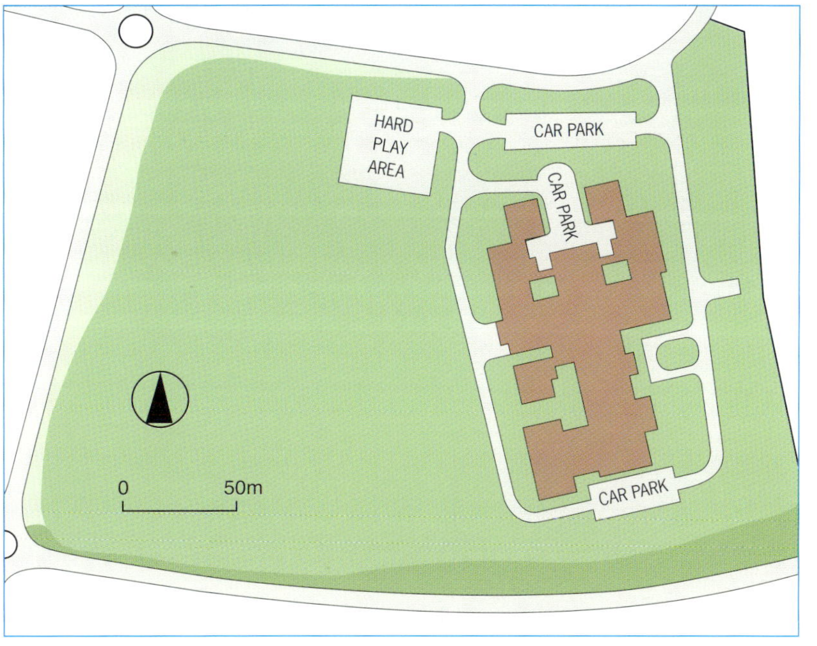

Figure 7.10.1: Site layout

Figure 7.10.2: Open plan learning base 1

7.10 Case Study: An investigation into the acoustic conditions in open-plan learning spaces in a secondary school

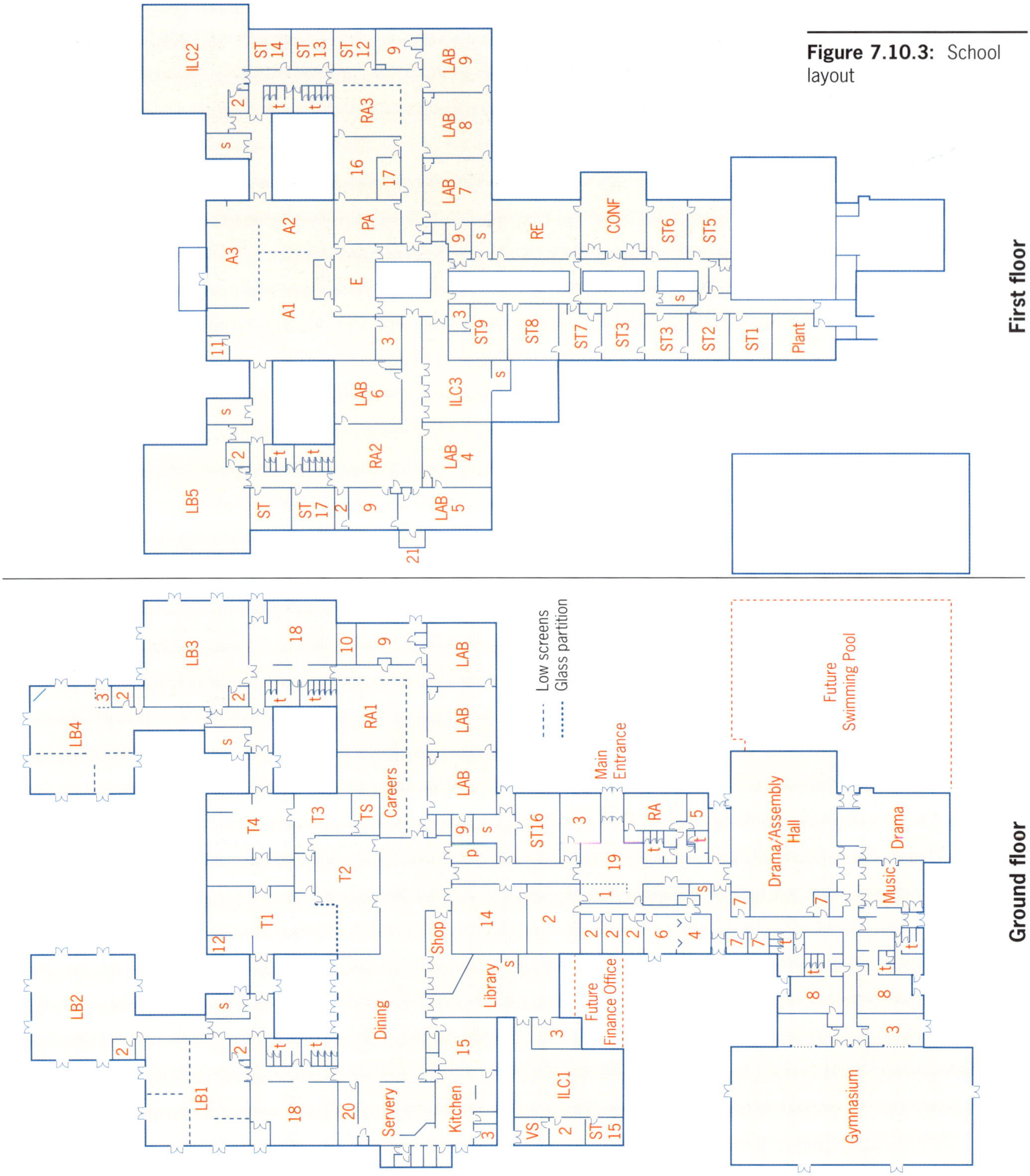

Figure 7.10.3: School layout

Key

1 Reception
2 Office
3 Store
4 Meeting room
5 Medical inspection
6 Principal
7 Music practice
8 Changing room
9 Science prep.
10 Darkroom
11 Kiln
12 Heat bay
13 Technicians base/materials
14 Hospitality suite
15 Training kitchen
16 Sound Laboratory
17 Music Tech.
18 Cloakroom
19 Foyer
20 Wash-up
21 Greenhouse

(LB - Learning Bases)
LB1 English
LB2
LB3 Business Studies
LB4 Humanities
LB5 Mathematics

ILC Independent Learning Centre
TS Technology Store
A1–3 Art
T1–4 Technology
E Electronics
RA Resource Area
RA1 Sixth Form ICT Resource
Conf Conference Room
ST Study Room/Genral Teaching
RE Religious Education
PA Performing Arts
LAB Science Laboratory
VS Video Studio

s Stairs
t Toilets

place at the same time. Typically, moveable screens or lockers are used to separate the different areas within a learning base.

Acoustic measurements

Measurements of sound pressure level, reverberation time, speech intelligibility and airborne sound insulation were made in the school to assess the acoustic environment. These measurements were made in learning base 1 (English learning base), the art area, the workshop and technology areas, and language teaching rooms (study area 1 and study area 5).

Sound pressure levels were measured over 30 minute periods (starting on the hour or half-hour) during the school day to determine $L_{Aeq,30min}$, $L_{A90,30min}$, $L_{A10,30min}$, L_{AFmax} and L_{AFmin}. Observations of classroom activity were noted in order to attribute measured levels to specific activities and events.

In the open-plan area of learning base 1, the Speech Transmission Index (STI) was measured according to BS EN 60268-16 to assess speech intelligibility.

Airborne sound insulation was measured between adjacent language teaching classrooms. These classrooms were enclosed rooms and did not form part of the open-plan teaching space.

In addition to the acoustic measurements, teaching staff completed a questionnaire about the effect of the school layout on their work.

Figure 7.10.4: Students in area A of learning base 1

Figure 7.10.5: Learning base 1 – sound pressure levels in area A

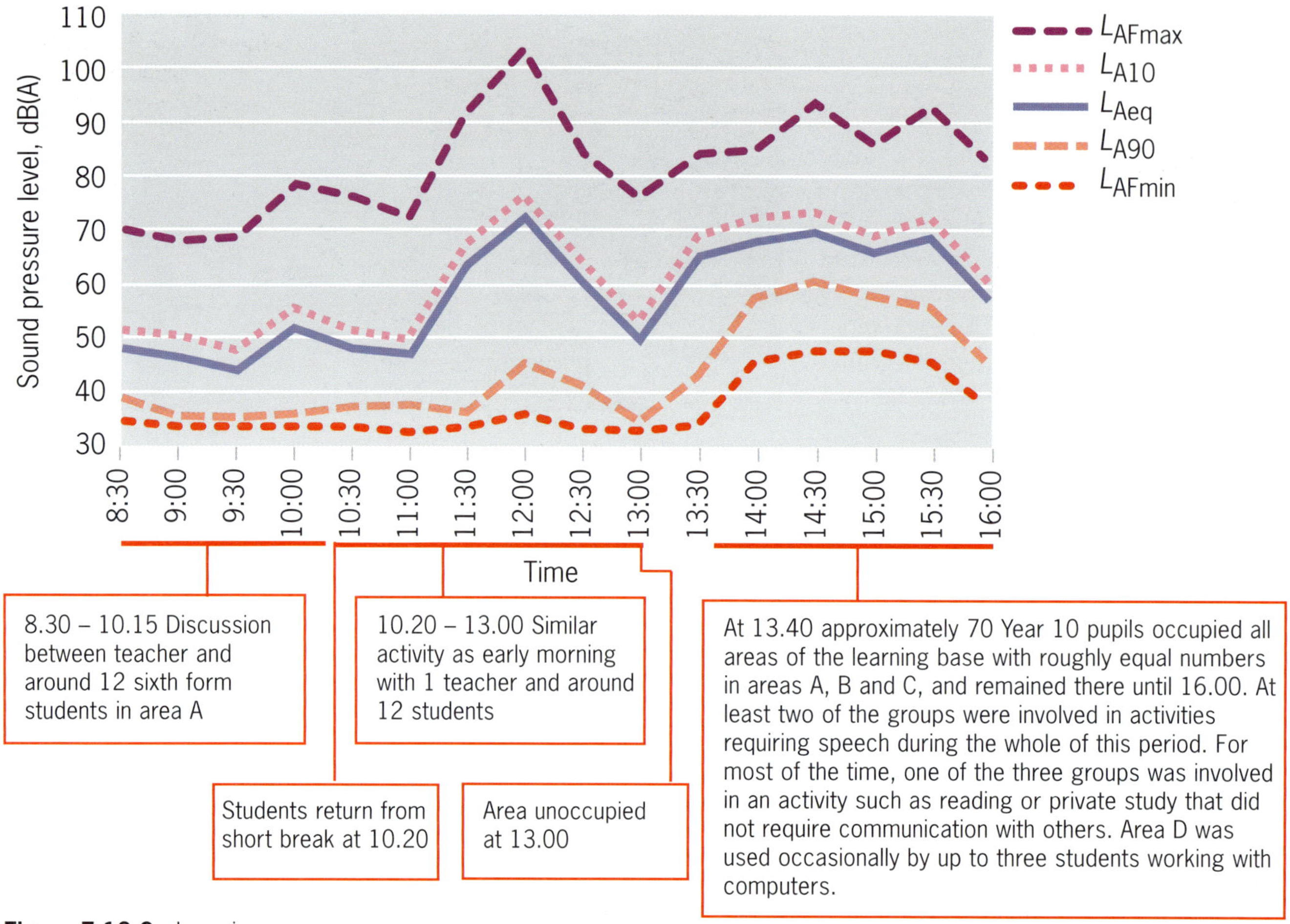

Figure 7.10.6: Learning base 1 – sound pressure levels in area C

Learning base 1

The open-plan layout of learning base 1 is shown in Figure 7.10.2. Figure 7.10.4 shows students working in area A of learning base 1.

Sound pressure levels

Figures 7.10.5 and 7.10.6 show graphs of the continuous sound pressure levels measured in areas A and C. Figure 7.10.7 shows the difference in $L_{Aeq,30min}$ between area A and area C of learning base 1 (area A – area C).

It was intended that observations made during the measurements would allow analysis of individual events that cause disturbance. This aim proved not to be possible. For example, when a telephone rang in learning base 1 there was no observed reaction from the students. The telephone in the room was used to inform teachers that the students could go for lunch and appeared to be viewed as nothing unusual by the students.

Between 08.30 and 11.00 the teacher and students occupied area A. During this time the difference between $L_{Aeq,30min}$ in areas A and C was between 7 and 10 dB (see Figure 7.10.7). The measurements thus demonstrate that there is a maximum of 10 dB attenuation of airborne sound between areas A and C. Therefore, if another class were present in area C carrying out a quiet activity such as private reading, the students in area C would be able to clearly hear the activity noise emanating from area A. These measurements indicate that if the airborne sound insulation were measured between areas A and C then it would not meet the minimum performance standard for airborne sound insulation of 45 dB $D_{nT(0.8s),w}$ required between general teaching areas.

Between 11.00 and 13.00 the level in area C was sometimes higher than in area A. The reason for this is not known because area C was unoccupied, but

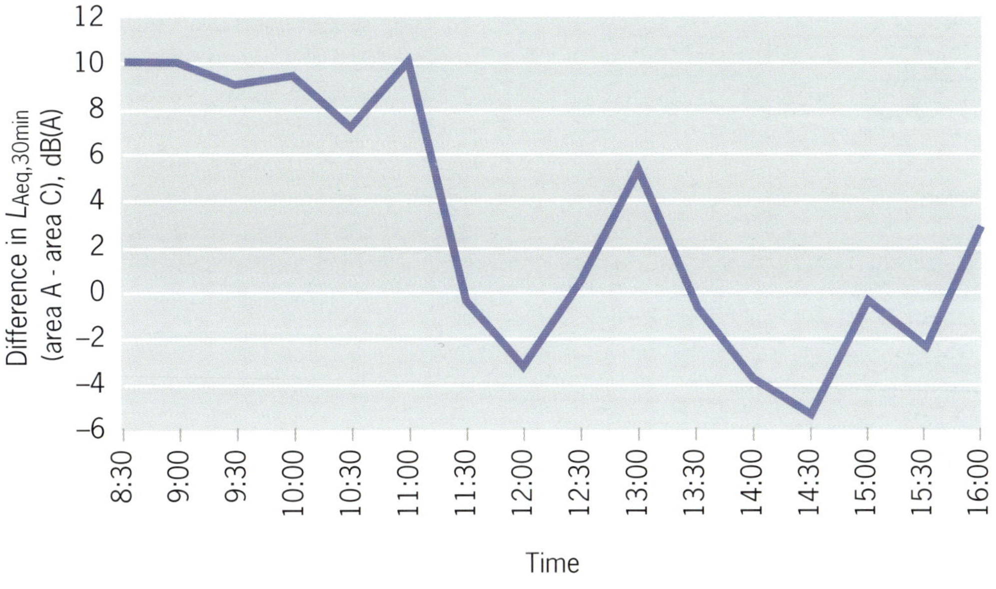

Figure 7.10.7: Learning base 1 – difference in sound pressure levels between area A and area C

could be due to sound generated from the playground outside.

When all areas in the learning base were occupied between 14.00 and 16.00, $L_{Aeq,30min}$ increased to levels between 65 and 70 dB. To gauge the effect of this increase in $L_{Aeq,30min}$ on speech intelligibility, it is instructive to consider the required signal to noise ratio in a classroom which is generally taken to be a minimum of 15 dB, or, ideally, 20 to 25 dB when hearing impaired children are being taught. In these noise levels, a teacher's voice would have to be raised to a level of at least 80 to 85 dB in order to be heard by the students. It is unlikely that a teacher would be able to shout at a sufficiently high level to communicate with hearing impaired students. Thus, based upon measured sound pressure levels, the open-plan space is inadequate in terms of speech intelligibility.

The measurements that indicate inadequate signal to noise ratios were corroborated by the fact that staff reported difficulties in listening to students in the open-plan setting. In addition, some students also reported that it was difficult to hear the teachers when they spoke quietly.

Reverberation time

The mid-frequency reverberation time in learning base 1 was 0.6 seconds, which meets the performance standards in Table 1.5.

Speech intelligibility

For the speech intelligibility measurements, the STI was measured in area C of learning base 1.

STI was measured at two microphone positions with an artificial mouth used to transmit the measurement signal. Six measurements were made at each microphone position. The artificial mouth was sited by the white board in a position that was used by the teacher when addressing the class and referring to information on the white board. The signal level at a point 1 m in front of the artificial mouth was adjusted until a level of 68 dB(A) was measured. The positions of the artificial mouth and the microphones are shown on Figure 7.10.8.

STI measurements were made when the space was empty. Masking sound from a loudspeaker was used to simulate occupied conditions with groups of approximately 12 students in each class base. In the afternoon when all the areas of learning base 1 were occupied, measured levels in the learning base were between 65 and 70 dB $L_{Aeq,30min}$ (see Figures 7.10.5 and 7.10.6). Case Study 7.2 indicates that there is little point in measuring speech intelligibility at such high masking sound levels because the speech intelligibility is likely to be 'Bad', 'Poor' or 'Fair' with low signal to noise ratios where the signal (speech) level is similar to the noise level.

To assess the effect of sound

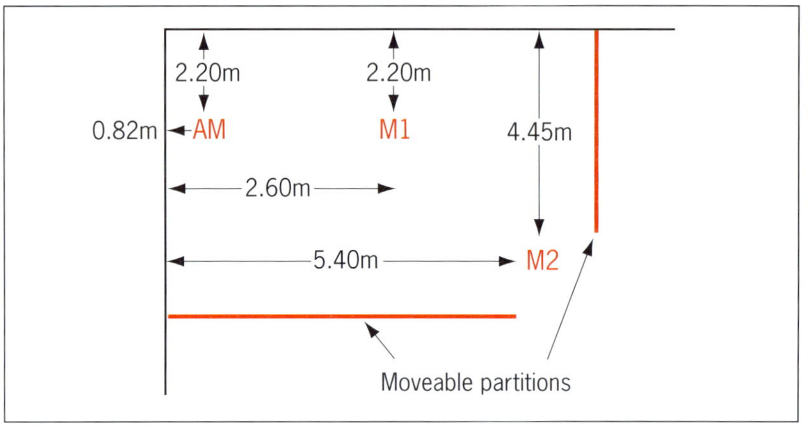

Figure 7.10.8: Learning base 1 – artificial mouth position (AM) and microphone positions M1 and M2 in area C

transmission from adjacent areas on speech intelligibility in area C, measurements were conducted with and without masking sound generated in the learning base. Two masking conditions were used: masking sound in area A; and masking sound simultaneously in areas A and B. The masking sound was produced from a loudspeaker using a white noise signal shaped to the sound spectrum recorded whilst the teacher addressed her students during a tutorial session with the sixth form students. The spectrum was measured at a distance of approximately 5 m from the teacher, the level at that point being 54 dB $L_{Aeq,30min}$. When masking sound was generated in area A only, the masking sound level was adjusted until 54 dB L_{Aeq} was measured in the same position as that used to measure the level of the teacher speaking. When masking sound was generated simultaneously in areas A and B, the same shaped sound signal was fed to the loudspeakers in each area and the level adjusted until 57 dB L_{Aeq} was measured at a position midway between areas A and B.

Measured STI data are shown in Tables 7.10.1 and 7.10.2.

From the tables it can be seen that the speech intelligibility was 'Good' at microphone positions M1 and M2 when there was no masking sound. Hence, when the other areas in the learning base are not occupied, the speech intelligibility is acceptable. When there was masking sound in area A or areas A and B (ie with either one or two of the other teaching areas simulated as being occupied), speech intelligibility was 'Good' at microphone position M1 but 'Fair' at microphone position M2. The reason for the reduction in speech intelligibility at microphone position M2 is because it is closer than position M1 to the other teaching areas, where masking sound was generated. Therefore, when other areas in the learning base are occupied, the speech intelligibility between the teacher and students sitting near microphone position

Table 7.10.1: Learning base 1 – measured STI values at microphone position M1, with and without masking sound

	No masking sound	Masking sound in area A	Masking sound in areas A and B
	STI	STI	STI
Average	0.702	0.666	0.644
Standard deviation	0.041	0.054	0.048
Rating	Good	Good	Good

Table 7.10.2: Learning base 1 – measured STI values at microphone position M2, with and without masking sound

	No masking sound	Masking sound in area A	Masking sound in areas A and B
	STI	STI	STI
Average	0.655	0.569	0.571
Standard deviation	0.037	0.031	0.084
Rating	Good	Fair	Fair

M2 is not acceptable. In the afternoon, when all the areas of learning base 1 were occupied, measured levels in the learning base were between 65 and 70 dB $L_{Aeq,30min}$ (see Figures 7.10.5 and 7.10.6). STI measurements were not made with this masking sound condition as the speech intelligibility would be expected to be 'Bad', 'Poor' or 'Fair' due to the low signal (speech) to noise ratio as in Case Study 2.

The teachers in this school adopted strategies to make the best use of their surroundings, for example, gathering students more closely around them (see Figure 7.10.15) to help overcome problems with speech intelligibility and to reduce disturbance to those involved in other activities within the room. It appeared that co-operation between staff working in the same open-plan area and careful planning of lessons was an important aspect in coping with the speech intelligibility problems in these areas. For example, a teacher in area C notified her colleague in area A that her class would be engaged in quiet reading after they had finished a more noisy activity. This was to reduce disturbance to the reading of a play in area A.

The school has tried to teach languages in the open-plan learning bases, however, it had been decided that such lessons can only be taught effectively in enclosed classrooms. It is not known whether this was due to ambient levels being too high for good speech intelligibility in open-plan areas or whether it was due to disturbance from adjacent areas in a learning base. It is to be expected that conditions for language teaching need to be more closely controlled than for teaching some other subjects. Measuring STI enables speech intelligibility in rooms to be objectively assessed. However it does not enable disturbance to be quantified since this could depend on how distracting the activities are in adjacent areas.

Workshop and technology areas

Sound pressure levels

Figures 7.10.9 and 7.10.10 show graphs of the continuous sound pressure levels recorded in the workshop and technology areas. The levels in these areas are

Figure 7.10.9: Workshop area sound pressure levels

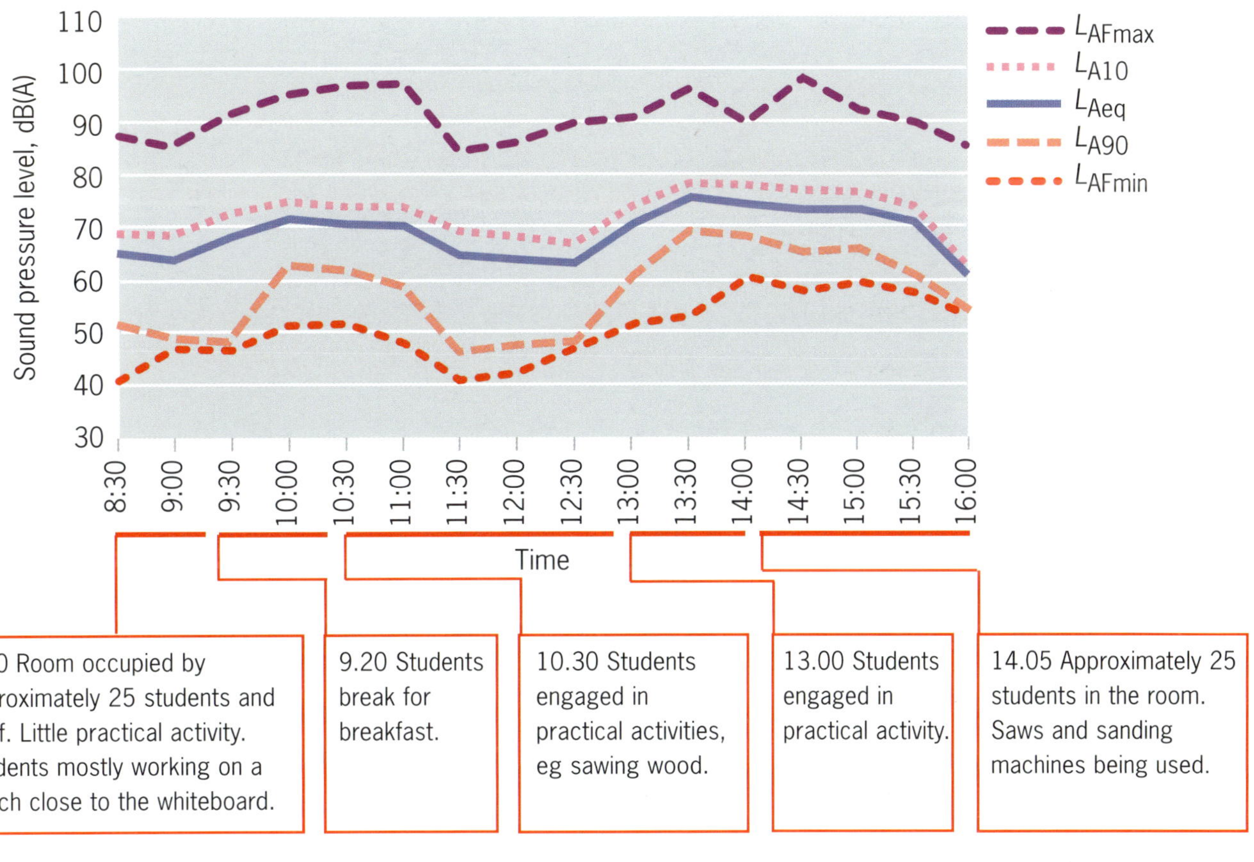

significantly higher than would be expected in classrooms due to the machinery noise. During the period of highest noise, 76 dB $L_{Aeq,30min}$, the signal to noise ratio would result in inadequate speech intelligibility for a teacher talking to a group of students.

Reverberation times

The mid-frequency reverberation time in the workshop was 1.2 seconds, which does not meet the performance standards in Table 1.5.

The mid-frequency reverberation time in the technology area was 0.8 seconds, which does not meet the performance standards in Table 1.5.

A mid-frequency reverberation time of 1.2 seconds in the workshop combined with levels greater than 70 dB $L_{Aeq,30min}$ will provide inadequate speech intelligibility. However, in rooms where students use machine tools such as lathes, good speech intelligibility is essential for safety.

Art area

Sound pressure levels

The art area on the first floor is shown in Figure 7.10.3. During the measurements it was occupied by Year 7 students and teaching staff. The room was divided by partitions into three areas, indicated as A1, A2 and A3.

Measurements were taken in areas A1 and A2, which held between 20 and 25 students. Throughout the day, activities undertaken in the art area did not appear to change significantly. Noise sources included hairdryers which were used to dry items of art work. Figures 7.10.11 and 7.10.12 show graphs of the sound pressure levels recorded in areas A1 and A2 respectively. For most of the day the noise level varies from 65 to 75 dB $L_{Aeq,30min}$. Thus, for a teacher talking to a group of students in the art area, the signal to noise ratio would be inadequate for good speech intelligibility.

Figure 7.10.10: Technology area sound pressure levels

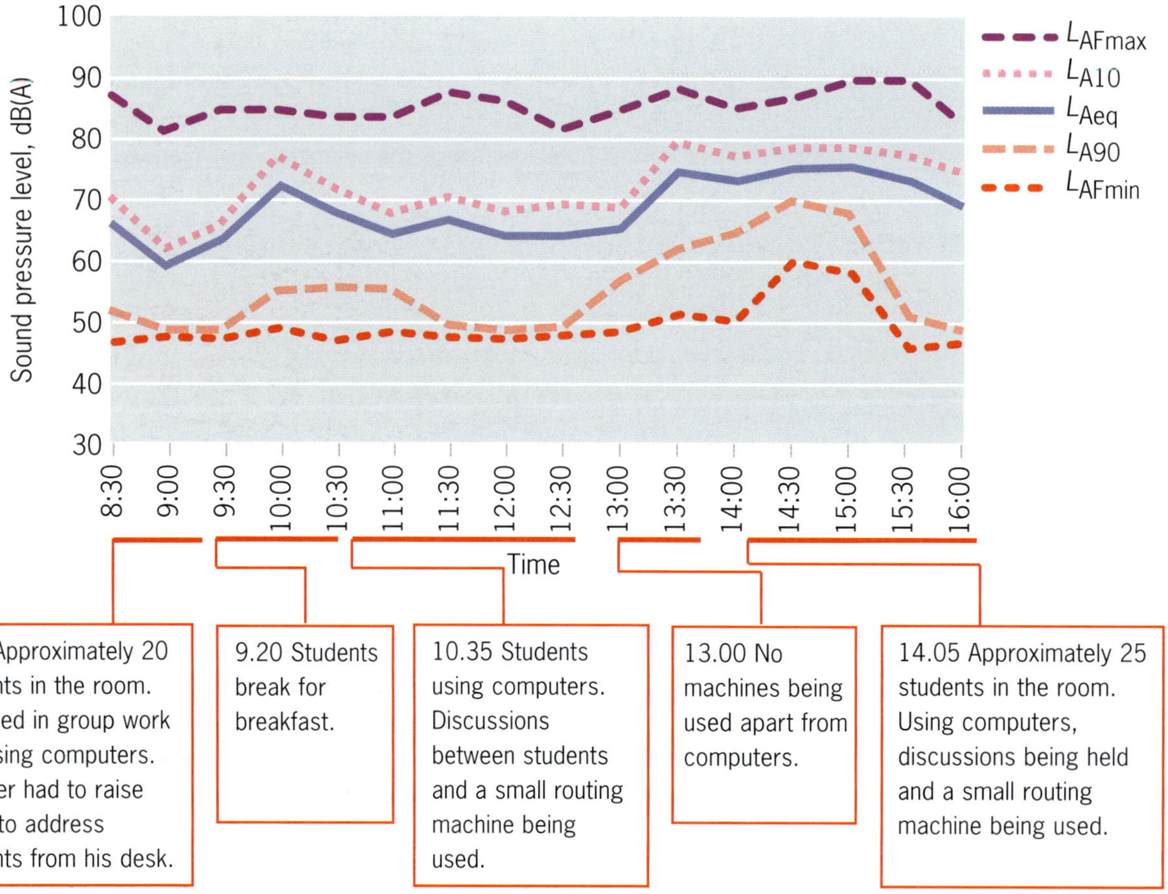

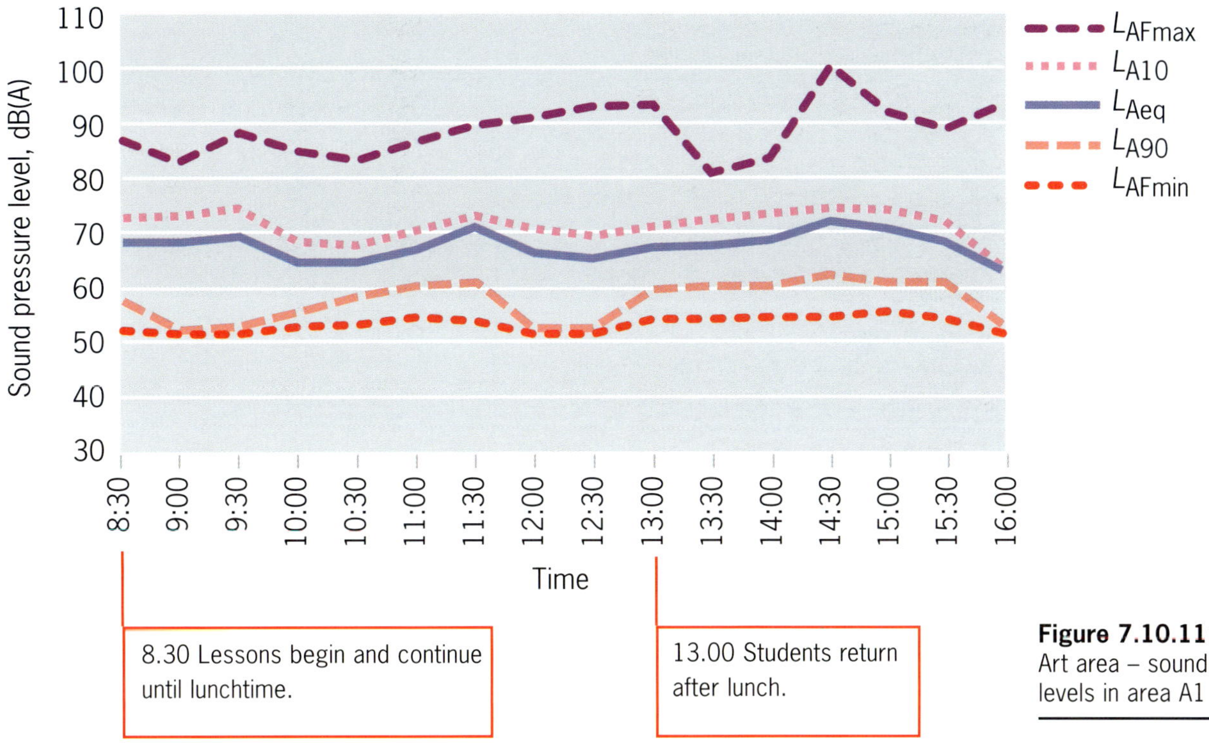

Figure 7.10.11: Art area – sound pressure levels in area A1

Reverberation time

The mid-frequency reverberation time in the art area was 0.9 seconds, which does not meet the performance standards in Table 1.5 for an art room.

Language teaching rooms (study areas 1 and 5)

These rooms were enclosed classrooms that were originally intended to be sixth form study rooms. However, they were subsequently designated as language teaching rooms owing to the difficulties experienced in teaching two different

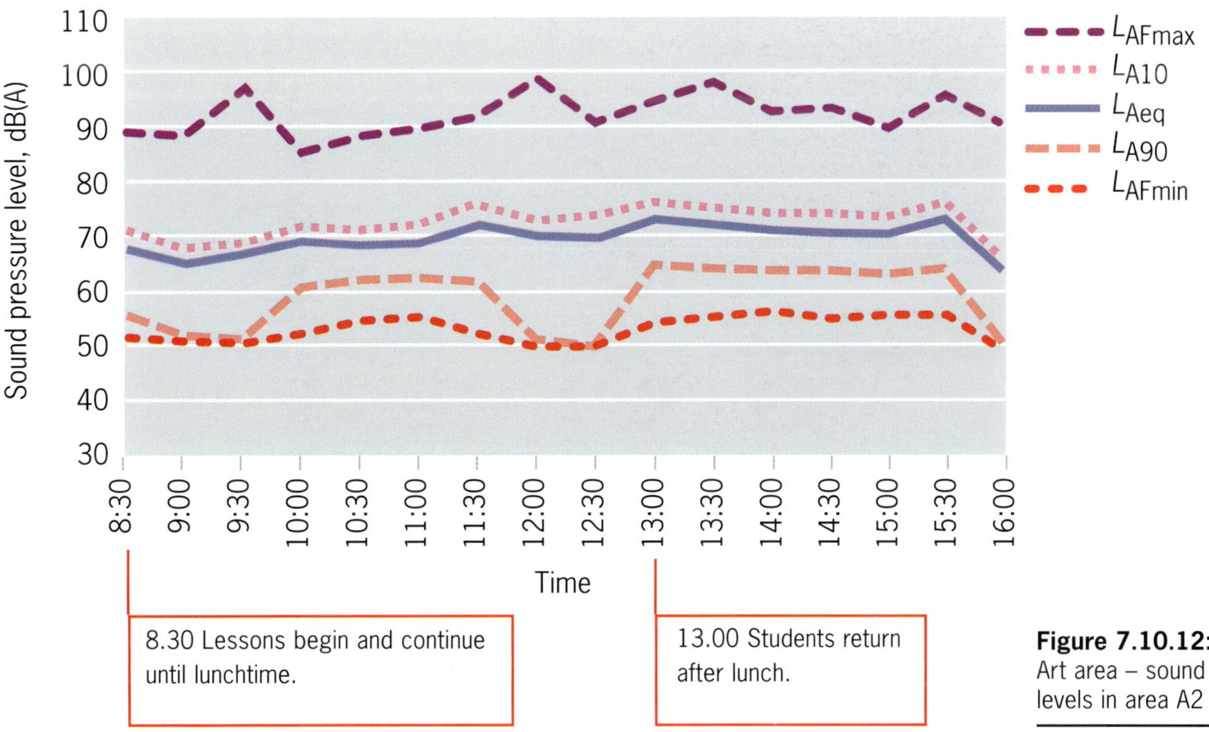

Figure 7.10.12: Art area – sound pressure levels in area A2

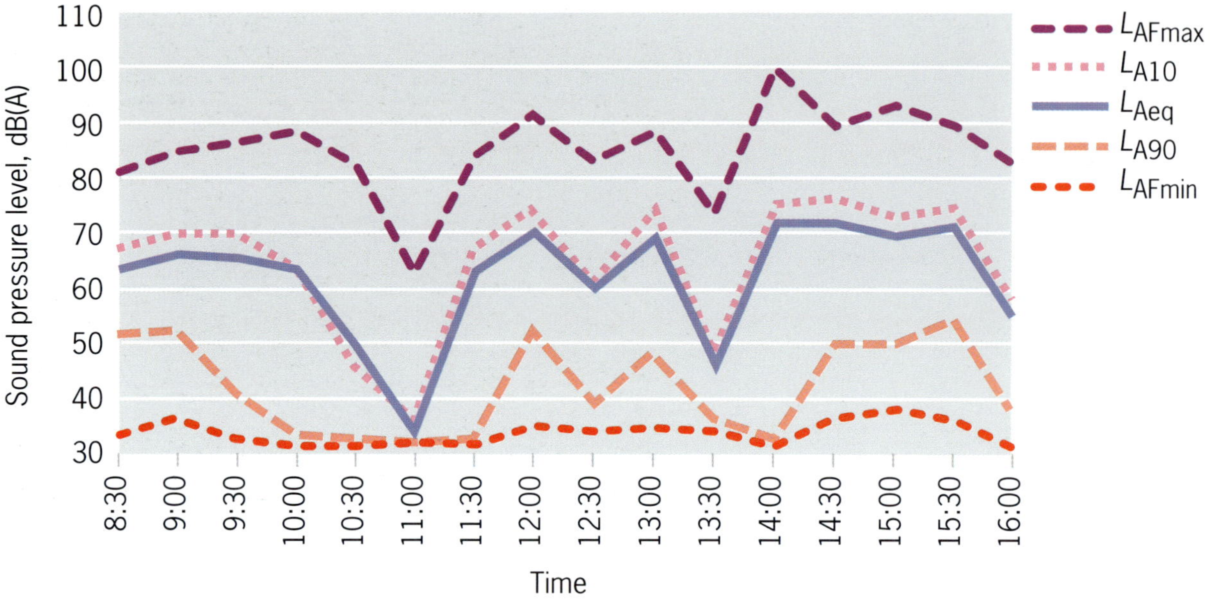

Figure 7.10.13: Language study area 1 sound pressure levels

languages (eg German and French) simultaneously in different areas of the open-plan space.

Sound pressure levels

Figures 7.10.13 and 7.10.14 show the sound pressure levels recorded in study areas 1 and 5 respectively. When the classrooms were unoccupied the measured levels were less than 50 dB $L_{Aeq,30min}$. When there was speech in the room, $L_{Aeq,30min}$ was typically between 65 and 75 dB. In general, $L_{Aeq,30min}$ was between 15 dB and 20 dB higher than $L_{A90,30min}$, indicating that the signal to noise ratio could potentially provide reasonable speech intelligibility. When the spaces were occupied and students and/or staff were speaking, there was a greater difference between $L_{Aeq,30min}$ and $L_{A90,30min}$ in the enclosed classrooms than in the fully occupied open-plan spaces. This indicates that the signal to noise ratios are likely to be higher in the enclosed classrooms than in the open-plan spaces.

Reverberation time

The mid-frequency reverberation time in each study area was 0.5 seconds, which

Figure 7.10.14: Language study area 5 sound pressure levels

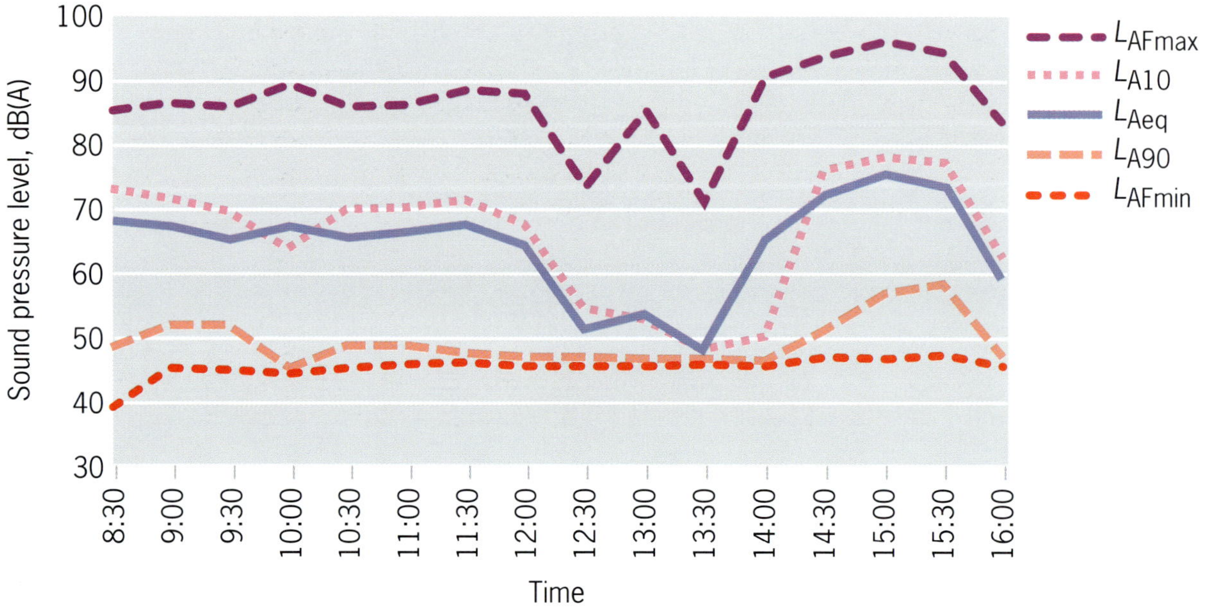

meets the performance standards in Table 1.5.

Airborne sound insulation

The measured airborne sound insulation between study area 1 and study area 2 is 40 dB $D_{nT(0.8s),w}$, which does not meet the performance standards in Table 1.2.

Fugure 7.10.15: Students in learning base 1 gathered around the teacher in area A

Summary

Teaching in an open-plan area in a secondary school requires a different type of working from teaching in traditional enclosed classrooms. This is due in part to the noise levels in open-plan teaching areas. In this school, both students and teachers in the open-plan areas reported being disturbed by noise, whilst in enclosed classrooms very little disturbance was reported. Some of the techniques observed in primary schools in Case Study 9.2 were used when it was important to ensure that students could hear the teacher during noisy periods. For example, students were gathered more closely around their teacher. Also, teaching staff in the area co-operated with each other to minimise disturbance to classes in adjacent areas.

It is concluded that it is difficult to justify the use of open-plan areas in secondary schools in terms of their acoustic environment. This is a similar conclusion to that in Case Study 7.2 for open-plan primary schools. High noise levels in occupied open-plan areas are the primary cause of inadequate speech intelligibility, especially for those students furthest from the teacher. STI measurements demonstrated that for these students, the performance standards in Table 1.6 of Section 1 were not met.

When all areas of the learning base were occupied, measured sound pressure levels were between 65 and 70 dB $L_{Aeq,30min}$. At these levels, the signal to noise ratios are likely to be less than 10 dB and speech intelligibility will be inadequate. When the teaching areas were occupied and students and/or teachers were speaking, there was a greater difference between $L_{Aeq,30min}$ and $L_{A90,30min}$ in the enclosed classrooms than in the open-plan spaces. This suggests that the signal to noise ratios are generally higher in enclosed classrooms than in open-plan areas. Hence, speech intelligibility is likely to be better in enclosed classrooms than in fully occupied open-plan areas.

In many open-plan teaching spaces it is difficult to achieve clear communication of speech between teacher and student, and between students. For this reason, careful consideration should be given as to whether to include open-plan teaching spaces in a secondary school. If open-plan areas are required then rigorous acoustic design is necessary to meet the required performance standards in Section 1.

Introduction to Appendices

The ten appendices provide supporting information for the main sections of Building Bulletin 93, including explanations of acoustic terms, sample calculations and other background information.

There are many technical terms and descriptors used in acoustics, which can not be covered in-depth in these short appendices. However, to help non-acousticians, Appendices 1 to 3 include definitions of those acoustic terms which are used in BB93, and describe the basic principles of the behaviour of sound in buildings. There are many acoustics text books available, some of which are listed in the bibliography These can be referred to for a more complete description of all acoustic terms and descriptors.

Appendix 1: Basic concepts and units

Nature of sound

Sound is usually generated by the vibrations of a surface, which give rise to pressure fluctuations in air or some other elastic medium. Sound is transmitted through the medium as sound waves, and may be described in terms of sound pressure or sound power. Noise is generally defined as unwanted sound.

Decibels

Sound levels are usually measured in decibels (dB) and relate absolute values to a reference value. The decibel scale is logarithmic and it ascribes equal values to proportional changes in sound pressure, which reflects the response of the human ear to sound. For example, an increase in sound pressure from 10 to 20 Pa would sound the same to the human ear as an increase from 1 to 2 Pa. Use of a logarithmic scale has the added advantage that it compresses the very wide range of sound pressures to which the ear may typically be exposed to a more manageable range of numbers.

Sound pressure level

The sound pressure level, L_p, is a measure in decibels of the total instantaneous sound pressure at a point, relative to the pressure of the quietest sound that the human ear can detect. The threshold of hearing for most people corresponds to a sound pressure level of approximately 0 dB and the threshold of aural pain typically occurs at around 140 dB. The sound pressure level is a function of the source and its surroundings. Some typical sound pressure levels are shown in Figure A1.1.

Sound power level

The sound power output of a source in watts can be expressed as the sound power level, L_W, in decibels, relative to 1 pW. The sound power level is a function of the source alone, independent of its surroundings.

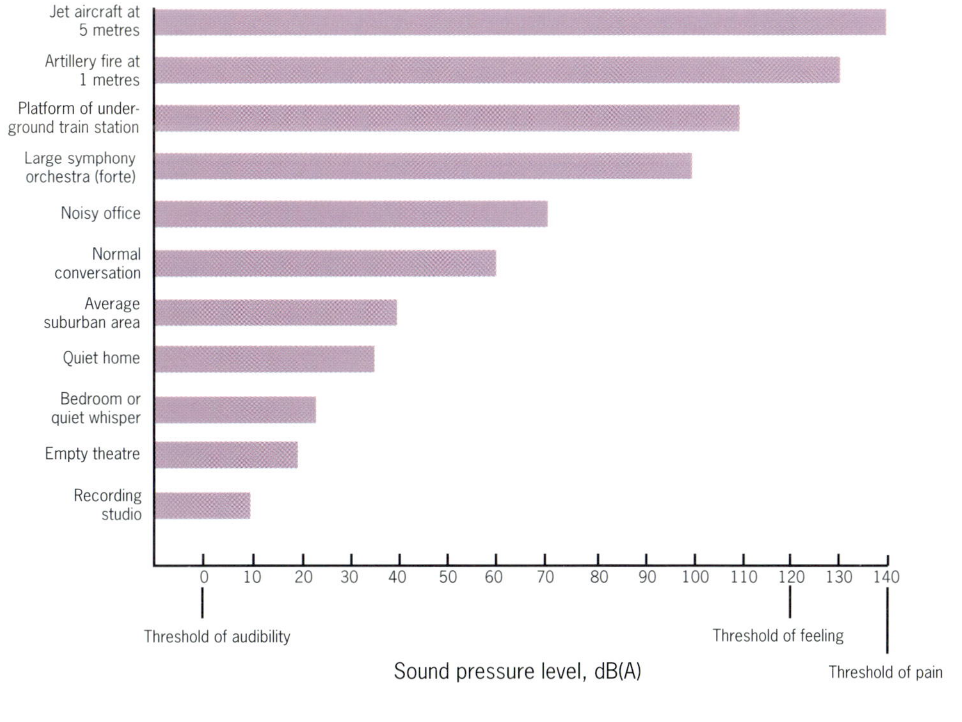

Figure A1.1: Typical sound pressure levels

Addition of sound levels

Because the decibel scale is logarithmic, levels in decibels can not be simply added together. To combine two sound levels, *A* dB and *B* dB, to give the total sound level, *C* dB, the following equation is used:

$$C = 10 \lg (10^{A/10} + 10^{B/10}) \text{ dB} \qquad \text{A1.1}$$

When two identical sounds occur simultaneously, the resulting level is only 3 dB higher than for a single source. By contrast, an increase of 10 dB normally represents a doubling of perceived loudness of the sound. Hence doubling the amount of sound energy results in very much less than a doubling in subjective loudness.

To combine more than two levels, the following equation is used:

$$L = 10 \lg (10^{A/10} + 10^{B/10} + 10^{C/10} + 10^{D/10} + \ldots\ldots) \text{ dB} \qquad \text{A1.2}$$

Frequency of sound

Frequency is analogous to musical pitch. It depends upon the rate of vibration of the air molecules which transmit the sound and is measured as the number of cycles per second or Hertz (Hz). The human ear is sensitive to sound in the range 20 Hz to 20 kHz. Examples of the frequency ranges of musical instruments and the human voice are shown in Figure A1.2. For acoustic engineering purposes, the frequency range is normally divided up into discrete bands. The most commonly used are octave and one-third octave bands.

Octave bands

For an octave band the upper limiting frequency of each band is twice the lower limiting frequency. Octave bands are described by their centre frequency values and bands typically used for building acoustics purposes range from 63 Hz to 4 kHz.

One-third octave bands

Each octave band can be divided into three one-third octave bands. The one-third octave bands are described by their centre frequency values and bands typically used for building acoustics purposes range from 50 Hz to 5 kHz.

Figure A1.2: Frequency range of musical instruments and vocals

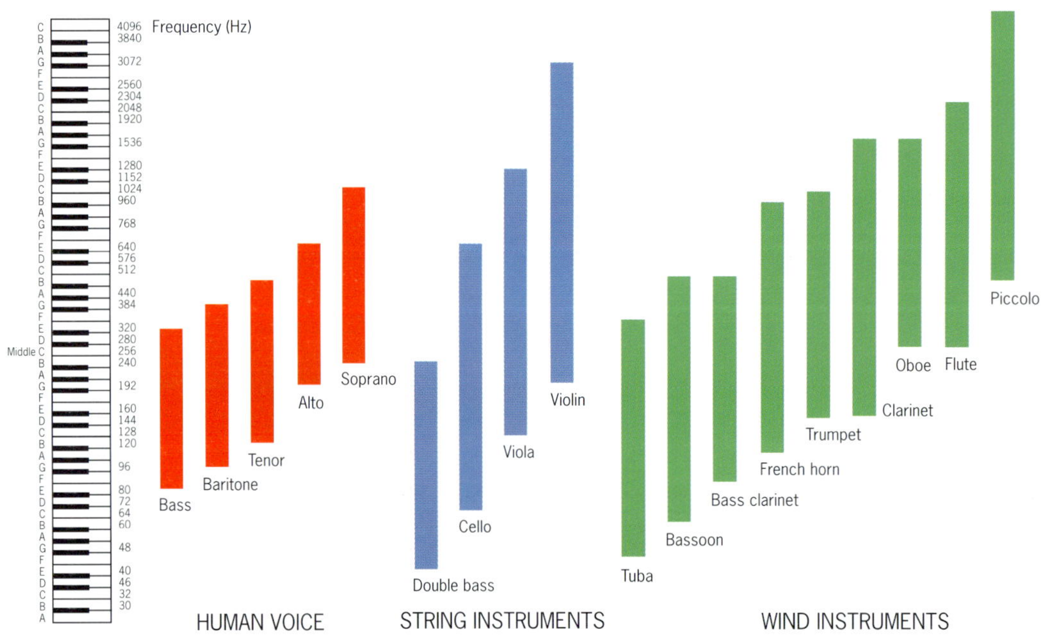

Octave band centre frequency (Hz)	63	125	250	500	1 k	2 k	4 k
A-weighting correction (dB)	–26.2	–16.1	–8.6	–3.2	0	1.2	1.0

Table A.1.1: A-weighting corrections

A-weighted levels

The sensitivity of the ear is frequency dependent. Sound level meters are fitted with a weighting network which approximates to this response and allows sound levels to be expressed as an overall single figure value, in dB(A). For clarity and convenience, the 'A' is often included in the acoustic descriptor, eg L_{Aeq}, rather than in brackets after the units. For example, A-weighted levels can be quoted as 55 dB L_{Aeq}.

The A-weighted level can also be calculated manually from octave band or one-third octave band data. For octave band data, see Table A1.1, corrections are added to the respective sound levels and the resulting values for all octave bands are combined logarithmically (using Equation A1.2).

Measurement of time-varying sounds

Most sounds are not steady and the sound pressure level fluctuates with time. Therefore, it is necessary to express the results of a measurement over a period of time in statistical terms. Some commonly used descriptors are discussed below.

Equivalent continuous sound level

The most widely used unit is the equivalent continuous A-weighted sound pressure level ($L_{Aeq,T}$). It is an energy average and is defined as the level of a notional sound which (over a defined period of time, T) would deliver the same A-weighted sound energy as the actual fluctuating sound.

Percentile level

A percentile level is the highest level exceeded for a certain percentage of a measurement period. The most commonly used percentile levels are:

$L_{A1,T}$ – the A-weighted level exceeded for 1% of the measurement period. This is often used to represent typical maximum levels that occur during the measurement period.

$L_{A10,T}$ – the A-weighted level exceeded for 10% of the measurement period. This is often used to represent the sound level from road traffic.

$L_{A90,T}$ – the A-weighted level exceeded for 90% of the measurement period. This is often used to represent the background level.

Maximum and minimum sound levels

$L_{Amax,T}$ is the maximum sound pressure level measured during the measurement period T. $L_{Amin,T}$ is the minimum sound pressure level measured during the measurement period T.

Sound level meter time constants

To give meaningful results, sound level meters use sound pressure levels averaged over short intervals (within the overall measurement period, T). Time constants for this averaging, defined in international standards, include 'fast' (125 ms) and 'slow' (1 s).

The percentile levels described above are affected by the choice of time constant. By definition, all percentile levels must be measured with the fast time constant.

$L_{Aeq,T}$ is not affected by the sound level meter time constant.

$L_{Amax,T}$ and $L_{Amin,T}$ can be measured with either fast or slow time constants so it is important that the results state which time constant has been used.

Appendix 2: Basic principles of room acoustics

Reflection and absorption of sound

Once emitted from a source, sound waves in a room travel through the air until they reach a boundary surface or other obstacle. When a sound wave reaches a surface it will be partly reflected off the surface back into the room and continue travelling in a new direction, and it will be partly absorbed by the surface with the absorbed energy being dissipated as heat.

Absorption coefficient, α

The amount of sound energy that can be absorbed by a surface is given by its absorption coefficient, α. The absorption coefficient can take values in the range 0 to 1. A surface that absorbs no sound (ie a totally reflective surface) has an absorption coefficient of 0 and a surface that absorbs all sound incident upon it has an absorption coefficient of 1. Thus the higher the value of α, the more sound will be absorbed. In practice, most surfaces have values between 0 and 1. Some typical absorption coefficients are given in Table A6.1 and on the DfES acoustics website.

Absorption classes

The absorption of surfaces varies with frequency. Therefore, absorption coefficients are generally given for each octave band. A surface is categorised as being in a particular absorption class, A to E (according to BS EN ISO 11654:1997), depending on its absorption coefficients across the frequency range. To determine the absorption class the octave band values are plotted on a graph from BS EN ISO 11654:1997 as shown in Figure A2.1. Note that a very reflective surface may be unclassified.

Scattering coefficient, *s*

When sound is reflected from a surface it is partly reflected in a specular direction (ie the angle of incidence equals the angle of reflection) and partly scattered into other directions. The amount of reflected sound energy that will be scattered is given by the surface's scattering coefficient, *s*. This is in the range of 0 to 1 where a perfectly smooth surface giving pure specular reflection has a scattering coefficient of 0 and a very irregular

Figure A2.1: Absorption classes from BS EN ISO 11654: 1997

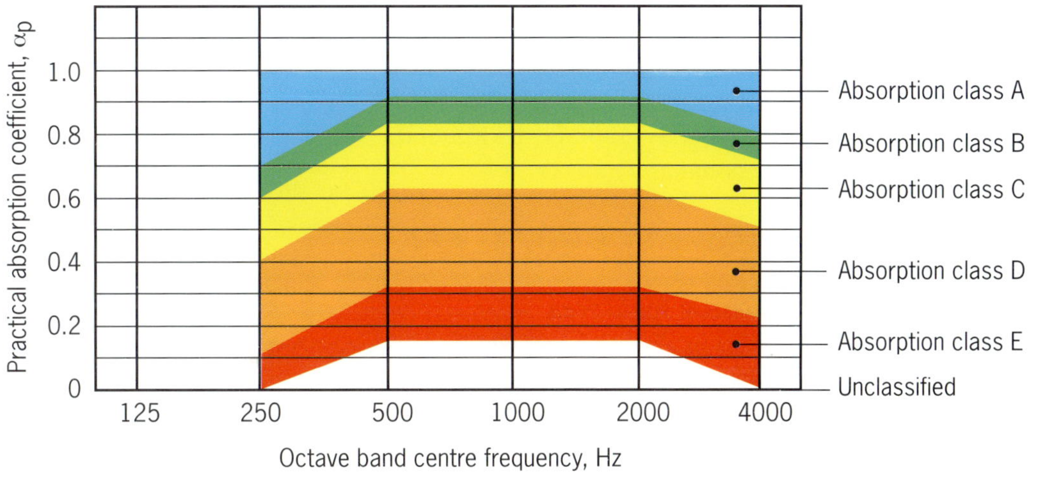

surface scattering all sound away from the specular direction has a scattering coefficient of 1. Scattering coefficients are a relatively new measure in room acoustics so there is little data currently available but they are important in room acoustics computer modelling.

Reverberation time, *T*

After being emitted from a source, sound waves are repeatedly reflected from room surfaces and, as a result of absorption, gradually reduce in strength. The reverberation time, *T*, of a space is a measure of the rate at which the sound decays. It is defined as the time taken for the reverberant sound energy to decay to one millionth of its original intensity (corresponding to a 60 dB reduction in the sound level).

The reverberation time is proportional to the volume of the room and inversely proportional to the quantity of absorption present:

$$T = 0.16\, V / \sum S_i \alpha_i \quad \text{s} \qquad \text{A2.1}$$

where S_i and α_i are respectively the surface area and absorption coefficient of each surface *i* in the room. An example of the application of this equation is given in Appendix 6.

Mid-frequency reverberation time, T_{mf}

The sound absorption of surfaces usually varies with frequency and therefore the reverberation time in a space also varies with frequency. Hence, values of *T* are normally given in frequency bands. In BB93 the reverberation time criteria are set in terms of the average value of the three octave bands, 500 Hz, 1 kHz, and 2 kHz, denoted as T_{mf}

$$T_{\text{mf}} = (T_{500} + T_{1\text{k}} + T_{2\text{k}}) / 3 \quad \text{s} \qquad \text{A2.2}$$

Other acoustic measures

Sound heard in a room generally comprises an extremely complicated combination of many reflected and scattered sound waves. This situation is made manageable by considering only the overall statistics of the sound field such as the reverberation time. Unfortunately, this does not convey all the intricate details of the sound field that determine peoples' subjective responses. There are many other measures used to represent various aspects of subjective response to room acoustics. For school acoustics there is a need to have criteria for subjective speech intelligibility for which the objective measure selected for BB93 is the Speech Transmission Index.

Speech Transmission Index, STI

The intelligibility of speech in a room is a complex function of the location of the speaker, the location of the listener, ambient noise levels, the acoustic characterisics of the space, and the loudness and quality of the speech itself. In addition, if a sound reinforcement system is used, it depends on the design and adjustment of this system. The Speech Transmission Index, STI, is an objective measure defined in BS EN 60268-16:1998, which accounts for all these factors.

To measure the STI, a special sound source is located at the position of the talker (with the normal microphone in place for any sound reinforcement system). The resulting signal is detected at the listening position. Signal processing using the modulation transfer function between transmitted and received signals is carried out to determine the STI.

STI is a value between 0 and 1, the higher the value, the better the speech intelligibility. Speech intelligibility ratings corresponding to STI values are as follows:

STI	Speech Intelligibility
0.1 to 0.3	Bad
0.3 to 0.45	Poor
0.45 to 0.6	Fair
0.6 to 0.75	Good
0.75 to 1	Excellent

Appendix 3: Basic principles of sound insulation

Airborne sound insulation

Speech, AV systems, and musical instruments are all sources of airborne sound in buildings. Sound in a room (the source room) causes the surrounding surfaces, such as walls, ceilings and floors to vibrate. This vibration is transmitted through the building structure and radiated into other rooms (receiving rooms) in the building. Depending upon the building construction, varying amounts of energy are lost during the sound transmission process, resulting in airborne sound insulation between rooms. The greater the airborne sound insulation between two rooms, the lower the resulting sound level in the receiving room.

Measurement of airborne sound insulation

The site measurement procedures for airborne sound insulation are given in BS EN ISO 140-4:1998. Normally, pink noise or white noise is played through an amplifier and loudspeaker in the source room, to provide a high sound level across the frequency range of interest. The sound level in the source room must be high enough to ensure that the levels in the receiving room are above the background noise level.

The resulting sound levels in the source and receiving rooms are measured in one-third octave bands. As the sound levels vary with location, they are averaged either across a number of fixed microphone positions or by using a continuously moving microphone. The resulting time and space averaged sound levels are denoted L_1 in the source room and L_2 in the receiving room.

Level difference, *D*

D is the difference in sound levels in dB between the source room and the receiving room in one-third octave bands:

$$D = L_1 - L_2 \text{ dB} \qquad \text{A3.1}$$

This level difference depends on:

- direct sound transmission through the separating element (ie separating wall or floor)
- flanking sound transmission (see Section 3) through flanking elements (eg flanking walls, suspended ceilings, access floors etc)
- wall and floor dimensions
- reverberation time of the receiving room.

Standardized level difference, D_{nT}

The reverberation time, *T*, measured in a room may be significantly different from the value predicted at the design stage due to a lack of detailed knowledge of finishes, furniture and fittings and their absorption characteristics. This means that the predicted sound level difference, *D*, which depends on *T*, is also subject to change. To avoid problems, a reference reverberation time, T_0, can be used in predictions of *D*. When the building is constructed and *D* is measured, the measured reverberation time, *T*, is referenced to T_0. This gives the standardized level difference, D_{nT}.

$$D_{nT} = D + 10 \lg (T / T_0) \text{ dB} \qquad \text{A3.2}$$

BB93 standardized level difference, $D_{nT(T_{mf,max})}$

D_{nT} is widely used to set sound insulation criteria for dwellings, where T_0 is taken as 0.5 seconds. Although BB93 uses D_{nT} in the sound insulation criteria for schools, a value of $T_0 = 0.5$ seconds would not be appropriate for many school rooms. Hence, T_0 is specified in BB93 as the maximum value of T_{mf} given in Table 1.5 of Section 1. This new descriptor for airborne sound insulation in schools is written as $D_{nT(T_{mf,max})}$ to highlight the alternative value of T_0 that is used.

Sound reduction index, *R*

The sound reduction index, *R*, of an element such as a wall, floor, door or window describes the sound transmitted through that element. It is measured in a laboratory with suppressed flanking transmission. *R* varies with frequency and is expressed as a value for each one-third octave band or octave band.

Apparent sound reduction index, R'

Using field measurements of the level difference, D, it is possible to estimate the value of the sound reduction index, R, for a partition. However, because field measurements include flanking transmission, the resulting quantity is called the apparent sound reduction index, R'.

The apparent sound reduction index, R', of wall or floor constructions in schools (and all other buildings), is usually lower than the laboratory measured value of R. The difference between the results is usually due to flanking transmission and a lower standard of workmanship on site. Guidance on flanking transmission is given in Section 3. Problems due to workmanship can be reduced by close supervision during the construction process.

Weighted sound reduction indices and level differences
R_w, R'_w, D_w, $D_{nT,w}$, $D_{nT(T_{mf,max}),w}$

Most constructions provide higher airborne sound insulation against mid and high frequency sounds (such as speech) than low frequency sounds (such as the bass in music). This typical characteristic is defined in BS EN ISO 717-1:1997 as a rating curve that can be applied to one-third octave band values of R, R', D, D_{nT} or $D_{nT(T_{mf,max})}$ from 100 Hz to 3.15 kHz. The rating curve is used to calculate the following single-number quantities: weighted sound reduction index, R_w; weighted apparent sound reduction index, R'_w; weighted level difference, D_w; weighted standardized level difference, $D_{nT,w}$; weighted BB93 standardized level difference $D_{nT(T_{mf,max}),w}$.

Impact sound insulation

In the case of impact sound, the building construction is caused to vibrate as a result of a physical impact, such as footsteps on floors or stairs. The resulting vibration is radiated into other rooms in the building.

Measurement of impact sound insulation

The site measurement procedures for impact sound insulation are given in BS EN ISO 140-7:1998. Impact sound insulation is measured using an ISO standard tapping machine, which consists of a series of hammers driven by an electric motor so as to produce a continuous series of impacts on the floor under consideration. The resulting sound level in the receiving room is measured in one-third octave bands. The receiving room is usually the space directly below the floor excited by the tapping machine, although the impact sound insulation can also be measured in other neighbouring rooms. As the sound levels will vary with location in the receiving room, they are averaged either across a number of fixed microphone positions or by using a continuously moving microphone.

Impact sound pressure level, L_i

The impact sound pressure level, L_i, is the time and space averaged sound pressure level in the receiving room, while the ISO standard tapping machine excites the floor or stairs above the receiving room.

Standardized impact sound pressure level L'_{nT}

The impact sound pressure level, L_i, depends on the reverberation time, T, of the receiving room. In the same way that D is standardized to give D_{nT} for airborne sound insulation to avoid changes caused by variations of T, an equivalent descriptor is defined for impact sound as the standardized impact sound pressure level, L'_{nT}:

$$L'_{nT} = L_i - 10 \lg (T / T_0) \text{ dB} \qquad \text{A3.3}$$

BB93 standardized impact sound pressure level $L'_{nT(T_{mf,max})}$

L'_{nT} is widely used for dwellings, where T_0 is taken as 0.5 seconds. In a similar manner to airborne sound insulation for schools, a value of T_0 = 0.5 seconds is not appropriate for many school rooms so T_0 is specified in BB93 as the maximum value of T_{mf} given in Table 1.5 of Section 1. This new descriptor for impact sound insulation in schools is written as $L'_{nT(T_{mf,max})}$ to highlight the alternative value of T_o that is used.

Weighted standardized impact sound pressure levels $L'_{nT,w}$ and $L'_{nT(T_{mf,max}),w}$

To reduce the impact sound pressure level data from values in frequency bands to a single-number quantity, BS EN ISO 717-2:1997 contains a rating curve that can be applied to one-third octave band values of L'_{nT} or $L'_{nT(T_{mf,max})}$ from 100 Hz to 3.15 kHz. The rating curve is used to calculate the following single-number quantities: the weighted standardized impact sound pressure level, $L'_{nT,w}$ or the weighted BB93 standardized impact sound pressure level, $L'_{nT(T_{mf,max})}$.

It is important to note that impact sound insulation is measured in terms of an absolute sound level, so that a lower number indicates a better standard of impact sound insulation. This is the opposite of airborne sound insulation, which is based on differences in levels so that a higher number indicates a better standard of airborne sound insulation.

Appendix 4: Classroom sound insulation – sample calculations

Figure A4.1 shows a secondary school classroom adjacent to a science laboratory, plantroom and corridor. In this example, the designer has decided to build the separating walls between these spaces with masonry. The calculations below are used to determine the specification of the masonry wall (eg mass per unit area, thickness, surface finishes) required to meet the performance standards in Section 1.

There are three walls to consider:

- Wall 1 - between classroom and science laboratory
- Wall 2 - between classroom and corridor
- Wall 3 - between classroom and plantroom.

The performance standards for airborne sound insulation are contained in Section 1. For each of the three walls the following apply:

- The performance standard for Wall 1 is in terms of the weighted BB93 standardized level difference, $D_{nT(T_{mf,max}),w}$, in Table 1.2. To determine the blockwork specification, the weighted sound reduction index of the wall is estimated from $D_{nT(T_{mf,max}),w}$.
- The performance standard for Wall 2 is in terms of the weighted sound reduction index in Table 1.3.
- For Wall 3, between the plantroom and the classroom, there are no explicit performance standards in Section 1. Therefore the sound insulation between the plantroom and the classroom needs to ensure that the performance standards are met for the indoor ambient noise level in the adjacent classroom (Table 1.1). The sound insulation is calculated using the noise levels of the actual equipment in the plantroom.

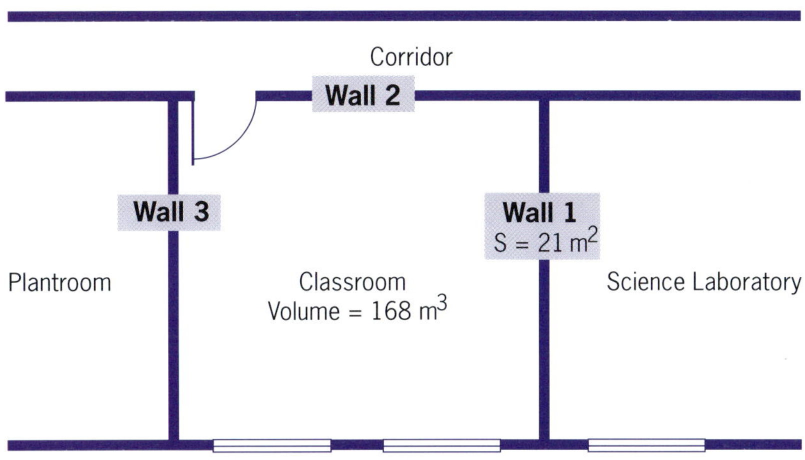

Figure A4.1: Plan of secondary school classroom and adjacent spaces

Wall 1

From Tables 1.1, 1.2 and 1.5 in Section 1 the minimum performance standards for the airborne sound insulation are:

Classroom to Science Laboratory:

40 dB $D_{nT(0.8s),w}$

Science Laboratory to Classroom:

45 dB $D_{nT(0.8s),w}$

As the two room dimensions are similar and the values of $T_{mf,max}$ are both 0.8 s, the specification for the masonry wall is based on the more stringent criterion, 45 dB $D_{nT(0.8s),w}$.

As an initial estimate, the procedure described in Section 3.10 can be used to estimate the weighted sound reduction index, R_w, for the separating wall. The first stage is to calculate $R_{w,est}$.

$$R_{w,est} = D_{nT(Tmf,max),w} + 10 \lg (S \times T_{mf,max} / V) + 8 \text{ dB}$$

$$R_{w,est} = 45 + 10 \lg (21 \times 0.8 / 168) + 8 \text{ dB}$$

$$R_{w,est} = 43 \text{ dB}$$

To obtain R_w the factor X is added to $R_{w,est}$ to account for less favourable mounting conditions and workmanship than in the laboratory test. From Section 3.10, X can be estimated to be 5 dB.

$$R_w = R_{w,est} + X \text{ dB}$$

$$R_w = 43 + 5 \text{ dB}$$

$$R_w = 48 \text{ dB}$$

Therefore, suitable specifications for masonry separating walls and appropriate surface finishes that achieve at least 48 dB R_w can be identified by the designer.

Wall 2

The performance standards for the airborne sound insulation of the corridor wall and door are given in Table 1.3:

Wall between a classroom and a corridor:
40 dB R_w

Door between a classroom and a corridor:
30 dB R_w

In this example there are no ventilators or glazing in the wall. If there were ventilators then they would have to meet the performance standard in Table 1.3. Glazing in the corridor wall does not have a separate performance standard because the performance standard for the wall is for the combined sound insulation of any glazing and the wall. An example of a corridor wall with glazing is included at the end of this appendix.

From Section 3, Figure 3.11, a 44 mm thick timber door with half hour fire rating typically achieves 30 dB R_w if it is "a well fitted solid core doorset where the door is sealed effectively around its perimeter in a substantial frame with an effective stop".

Blockwork specification

The minimum R_w values for the walls are:

Wall 1: 48 dB R_w
Wall 2: 40 dB R_w

Figure 3.9 can be used to draw up an initial specification for the walls along with laboratory test reports from the manufacturer. For Wall 1 and Wall 2 the following specifications might be proposed:

Wall 1: 100 mm medium density blocks (140 kg/m^2) with a 13 mm plaster finish on both sides

Wall 2: 100 mm low density block (70 kg/m^2) with a 13 mm plaster finish on both sides

The R_w specification for Wall 1 is only an estimate of the type of separating wall performance needed to achieve 45 dB $D_{nT(0.8s),w}$ and takes no account of flanking transmission which is usually critical in determining the performance. Therefore, at this stage the designer should seek specialist advice from an acoustic consultant to assess whether the proposed combination of separating and flanking walls is likely to achieve the performance standards.

Wall 3

Wall 3 has to provide sufficient sound insulation to ensure that the indoor ambient noise levels in the classroom do not exceed 35 dB $L_{Aeq,30min}$ (Table 1.1). As there will be other noise sources contributing to the indoor ambient noise level, the level due to noise transmitted through Wall 3 will have to be significantly less than 35 dB $L_{Aeq,30min}$. BB93 does not recommend a standard method for this situation but one approach is to design Wall 3 so that the noise transmitted from the plantroom is at least 10 dB below the indoor ambient noise level in the classroom. Using this method, the noise transmitted through Wall 3 needs to be less than 25 dB $L_{Aeq,30min}$ in the classroom.

The noise transmitted from the plantroom to the classroom depends on the frequency spectrum of the noise in the plantroom and the sound insulation spectrum of the separating wall. For these calculations, plantroom equipment locations and noise emission data are required. Precise equipment details are usually not known until the later stages of a project, therefore generic sound level data are normally used in calculations and the assumptions quoted in the specification.

The calculations are often complex and normally require an acoustic consultant. Guidance for these calculations can be found in the following references:

CIBSE (2002). Guide B5: Noise and Vibration Control for HVCA. CIBSE. ISBN 1903287251.
Fry, A. ed. (1988). Noise control in building services. Oxford: Pergamon. ISBN 0080340679.

Wall 2 with glazing

Walls between classrooms and corridors may contain glazing, therefore this example is a reassessment of Wall 2 with the following areas:

	Area
Masonry	16.6 m^2
Glazing	5.6 m^2
Door	1.8 m^2
Total wall	24 m^2

The door is treated in the same way as in the example above with a value of 30 dB R_w.

The combined R_w criterion for the masonry wall and the glazing also remains at 40 dB R_w.

This combined criterion would be achieved if the masonry wall and the glazing each provide at least 40 dB R_w. A masonry wall specification for this has already been described above and there are three glazing configurations given in Figure 3.10, which also provide 40 dB R_w. However, these glazing configurations can sometimes be relatively expensive due to the use of thick and/or laminated glass and/or wide cavities.

An alternative approach is to improve the masonry wall specification to allow the use of another glazing configuration.

From Figure 3.9 the following masonry wall should give at least 45 dB R_w: 100 mm medium density blocks (140 kg/m^2) with a 13mm plaster finish on each side.

From Figure 3.8, glazing with different R_w values can be assessed to see whether the criterion of 40 dB R_w will be met by the combined value for the wall and glazing. A potential solution would be to use glazing with sound insulation of 35 dB R_w. This is 10 dB lower than the 45 dB R_w sound insulation of the masonry wall. For a glazing area of 25% of the wall area that excludes the door, Figure 3.8 gives a correction factor of approximately 5 dB. The combined R_w is calculated from the R_w for the glazing plus the correction factor, which equals 40 dB R_w. Hence this combination of masonry wall and glazing meets the performance standard. Further reductions in glazing specification could be obtained by reducing the area of glazing or by using a wall with a higher R_w.

Appendix 5: Sound insulation of the building envelope

This appendix describes two methods that can be used to calculate the indoor ambient noise level due to external noise as described in Section 3. The first method calculates the indoor ambient noise level according to the principles of BS EN 12354-3:2000. The second method calculates the indoor ambient noise level using the measured façade sound insulation data from an identical construction at another site.

Principles of the calculation method based on BS EN 12354-3:2000

This section describes the calculation procedure based on BS EN 12354-3:2000.

The chosen frequency range should include all frequency bands that determine the indoor ambient noise level, $L_{Aeq,30min}$. However, for many external noise levels, it is appropriate to calculate the façade insulation using octave bands between 125 Hz and 2 kHz.

Two main equations are used to calculate the internal level in each frequency band.

The first equation gives the internal level due to sound transmission through an element of the building envelope:

$$L_2 = L_{1,in} - R + 10 \lg (S/V) + 11 + 10 \lg T \text{ dB} \qquad \text{A5.1}$$

where

L_2 is the internal level due to the sound transmitted through the element (dB)

$L_{1,in}$ is the external free-field sound level incident on the element (dB)

R is the sound reduction index of the element (dB)

S is the internal surface area of the element (m^2)

V is the room volume (m^3)

T is the room reverberation time (s).

The second equation gives the internal level due to sound transmission through a ventilator installed in the building envelope:

$$L_2 = L_{1,in} - D_{n,e} - 10 \lg V + 21 + 10 \lg T \text{ dB} \qquad \text{A5.2}$$

where

$D_{n,e}$ is the element-normalised sound level difference of the ventilator (dB).

The overall A-weighted internal level is obtained by combining (as in Equation A1.2) the contributions from all elements and ventilators within each frequency band, adding the relevant A-weighting correction to each resultant frequency band level, and then combining all the A-weighted frequency band levels together.

Elements: laboratory sound insulation data

Laboratory testing of building elements should be conducted in accordance with BS EN ISO 140-3:1995 to obtain the required sound reduction index, R.

Ventilators: laboratory sound insulation data

Laboratory testing of ventilator units should be conducted in accordance with BS EN 20140-10:1992 to obtain the required element-normalized level difference, $D_{n,e,w}$.

Some ventilators may have the facility to control the air flow rate, either by being fully opened or fully closed, or by having some form of variable control. In such cases calculations should be based on the performance in the fully open position.

The mounting position of ventilators, for example in the middle or near an edge of a wall, or in a corner, affects the airborne sound insulation. Manufacturers' data on ventilators should give information on the position of the ventilator during the laboratory test according to BS EN 20140-10:1992. This should be used to ensure that the laboratory mounting position is representative of the mounting position in the field.

Some ventilators are available in a variety of sizes but performance data may correspond only to one size. In such cases an estimate can be made for the area correction, A_c, to be added to each frequency band $D_{n,e}$ value. The area correction A_c in dB is given by:

$$A_c = 10 \lg (A_{ref} / A_{actual}) \text{ dB} \qquad \text{A5.3}$$

where
A_{ref} is the area in m^2 of the ventilator from the laboratory sound insulation test
A_{actual} is the area in m^2 of the ventilator to be used.

Excel spreadsheet

An Excel spreadsheet to calculate the sound insulation of building envelopes based on BS EN 12354-3:2000 is available via the DfES acoustics website. This Excel spreadsheet allows users to select from a range of typical building elements and ventilators and enter data obtained from laboratory tests of elements or ventilators.

The spreadsheet provides two options, A and B, to calculate internal levels using octave bands between 125 Hz and 2 kHz.

Option A allows the user to enter measured free-field octave band data as a 'user-defined spectrum'. Option A is the preferred option. However, if measured data are not available then option B can be used to give an estimate based upon an A-weighted value and an assumed spectrum shape, either C or C_{tr} from BS EN ISO 717-1:1997. Estimations using option B can be useful with some prediction methods such as CRTN for road traffic noise and CRN for railway noise, which generate only A-weighted levels. The user can enter external free-field L_{Aeq} levels and choose from the generic spectra described in BS EN ISO 717-1:1997. The C spectrum (Spectrum No. 1) is used to represent railway traffic at medium and high speed, and road traffic travelling at greater than 80 km/h. The C_{tr} spectrum (Spectrum No. 2) is used to represent noise from urban road traffic, jet aircraft at large distances, propeller driven aircraft, and railway traffic at low speeds.

Spectra of aircraft noise at relatively close distances are likely to depend greatly on aircraft type and operation, and specialist advice should be sought.

Principles of the calculation method based on field test data

This section describes the calculation of the indoor ambient noise level using the measured façade sound insulation data from an identical construction at another site.

The chosen frequency range should include all frequency bands that determine the indoor ambient noise level, $L_{Aeq,30min}$. However, for many external noise levels, it is appropriate to calculate the façade insulation using octave bands between 125 Hz and 2 kHz.

Field tests of building envelope insulation according to BS EN ISO 140-5:1998 give results in frequency bands expressed as the standardized level difference, $D_{2m,nT}$.

The internal level in each frequency band may be calculated according to the following equation:

$$L_2 = L_{1,in} - D_{2m,nT} + 6 + 10 \lg T \text{ dB} \qquad \text{A5.4}$$

where
L_2 is the internal level due to the sound transmitted through the facade (dB)
$L_{1,in}$ is the external free-field sound level incident on the facade (dB)
$D_{2m,nT}$ is the standardized level difference (dB)
T is the room reverberation time (s).

The overall A-weighted internal level is obtained by adding the relevant A-weighting to each frequency band level, and then combining all the A-weighted frequency band levels, according to Equation A1.2.

Appendix 6: Calculation of room reverberation times

For empty rooms with volumes less than 200 m^3, simple room geometry and a reasonable distribution of sound absorption, the reverberation time, T, can be calculated using Sabine's formula and absorption coefficients appropriate to the room surfaces as shown below.

$$T = \frac{0.16V}{A} \text{ seconds}$$

where V is the volume of the room in m^3 and A is the absorption area in the room in m^2.

Table 1.5 gives the recommended mid-frequency reverberation times for rooms. The mid frequency reverberation time, T_{mf}, is the arithmetic average of the reverberation times in the 500 Hz, 1000 Hz and 2000 Hz octave bands.

$$T_{mf} = \frac{T_{500\text{ Hz}} + T_{1000\text{ Hz}} + T_{2000\text{ Hz}}}{3} \text{ s}$$

For n surfaces in a space, the total absorption area, A, can be found using the following equation:

$$A = \alpha_1 S_1 + \alpha_2 S_2 + \ldots + \alpha_n S_n$$

where $\alpha_1,\ldots,\alpha_n$ are the absorption coefficients of the different surfaces in the room and $S_1,\ldots,S_n$ are the areas of the surfaces having absorption coefficients $\alpha_1\ldots\alpha_n$.

Absorption coefficients can be obtained from the spreadsheet on the DfES acoustics website and/or from manufacturers' data. Values for some common materials, used in the worked example which follows, are given in Table A6.1.

Two decimal places should be used for the absorption coefficient values for calculations.

In empty teaching rooms with volumes less than 200 m^3 and simple room geometry, the absorption area, A, needed to give the required reverberation time, T, can be obtained by rearranging Sabine's formula as follows:

$$A = \frac{0.16V}{T} \text{ m}^2$$

For such rooms the formula can also be used to estimate the amount of additional

Note: Reverberation time calculations using Sabine's formula in all octave bands can be carried out using the method illustrated in the worked example that follows. However, neglecting air absorption slightly underestimates the equivalent sound absorption area in a room.

Table A6.1: Absorption coefficient data for some common materials

Material	Sound absorption coefficient α 250 Hz	500 Hz	1000 Hz	2000 Hz	4000 Hz
Fair-faced concrete or plastered masonry	0.01	0.01	0.02	0.02	0.03
Fair-faced brick	0.02	0.03	0.04	0.05	0.07
Painted concrete block	0.05	0.06	0.07	0.09	0.08
Windows, glass façade	0.08	0.05	0.04	0.03	0.02
Doors (timber)	0.10	0.08	0.08	0.08	0.08
Glazed tile/marble	0.01	0.01	0.01	0.02	0.02
Hard floor coverings (eg linoleum, parquet) on concrete floor	0.03	0.04	0.05	0.05	0.06
Soft floor coverings (eg carpet) on concrete floor	0.03	0.06	0.15	0.30	0.40
Suspended plaster or plasterboard ceiling (with large airspace behind)	0.15	0.10	0.05	0.05	0.05

absorption area required to give a desired mid-frequency reverberation time. (Note that the absorption due to any surface that is to be covered with additional absorption must be discounted.)

The procedure is illustrated in the worked example shown below.

Specialist advice may be needed for large (>200 m^3) rooms or rooms where music is to be performed

Worked example

A school laboratory is required to have a mid-frequency reverberation time of less than 0.8 seconds. The room is rectangular in plan, is 7 m wide, 9 m deep and has a ceiling height of 3 m. There is one door and the glazing is located in one of the 7 m x 3 m walls. The room volume is 7 m x 9 m x 3 m = 189 m^3. The glazing has an area of 6 m^2 and the door has an area of 2 m^2.

Step 1 Calculate the surface area related to each material in the room (ie floor, walls, doors, ceiling and windows).

Surface	Surface finish	Area (m^2)
Floor	Hard floor covering	63
Door	Timber	2
Walls (excluding door and glazing areas)	Painted concrete block	88
Ceiling	Suspended plaster	63
Windows	Glass	6

Step 2 Obtain values of absorption coefficients for the room surfaces. In this case, the values are taken from Table A6.1.

		Absorption coefficient α		
Surface	**Area (m^2)**	**500 Hz**	**1000 Hz**	**2000 Hz**
Floor	63	0.04	0.05	0.05
Door	2	0.08	0.08	0.08
Walls	88	0.06	0.07	0.09
Ceiling	63	0.10	0.05	0.05
Windows	6	0.05	0.04	0.03

Step 3 Calculate the absorption area (m^2) related to each surface in octave frequency bands (Absorption area = surface area × absorption coefficient).

		Absorption area (m^2)		
Surface	**Area (m^2)**	**500 Hz**	**1000 Hz**	**2000 Hz**
Floor	63	63 × 0.04 = 2.52	63 × 0.05 = 3.15	63 × 0.05 = 3.15
Door	2	0.16	0.16	0.16
Walls	88	5.28	6.16	7.92
Ceiling	63	6.30	3.15	3.15
Windows	6	0.30	0.24	0.18

Step 4 Calculate the sum of the absorption areas (m^2) obtained in Step 3.

	500 Hz	1000 Hz	2000 Hz
Total absorption area (m^2)	14.56	12.86	14.56

Step 5 Calculate the reverberation time for the room using Sabine's formula.

	500 Hz	1000 Hz	2000 Hz
$T = \frac{0.16V}{A}$ seconds	2.08	2.35	2.65

Step 6 Calculate the mid-frequency reverberation time (T_{mf}) from the reverberation times in the 500 Hz, 1000 Hz and 2000 Hz octave bands.

$$T_{mf} = \frac{2.08 + 2.35 + 2.65}{3} = 2.36 \text{ seconds}$$

This reverberation time exceeds the required value.

Step 7 Identify a sound absorbing material that is suitable for use in a school laboratory and determine the best position for the material.

A manufacturer produces a non-flammable sound absorbing material that can be cleaned relatively easily. The following absorption coefficient data is provided for the material.

	250 Hz	500 Hz	1000 Hz	2000 Hz	4000 Hz
α	0.20	0.45	0.85	1.00	1.32

Because the room is used as a laboratory, it is decided that the most appropriate place for the sound absorbing material is on the ceiling or high on the walls.

Step 8 Estimate the required area of the sound absorbing material and calculate the mid-frequency reverberation time when it is in place.

As a first estimate, it is decided to cover the entire ceiling with the sound absorbing material. The total absorption areas in the 500 Hz, 1000 Hz and 2000 Hz octave frequency bands are then calculated.

		Absorption area (m^2)		
Surface	**Area (m^2)**	**500 Hz**	**1000 Hz**	**2000 Hz**
Floor	63	2.52	3.15	3.15
Door	2	0.16	0.16	0.16
Walls	88	5.28	6.16	7.92
Ceiling	63	63.0 x 0.45 = 28.35	63 x 0.85 = 53.55	63 x 1.00 = 63.00
Windows	6	0.30	0.24	0.18
Total absorption area		36.61	63.26	74.41
		Reverberation time (s)		
$T = \frac{0.16V}{A}$ seconds		0.83	0.48	0.41

Note: Because the mid-frequency reverberation time is required, calculations need only be conducted in the 500 Hz, 1000 Hz and 2000 Hz octave bands. However, should reverberation times need to be calculated for all octave bands, the calculation method is the same as that illustrated for each octave band.

Step 9 Calculate the new mid-frequency reverberation time.

$$T_{mf} = \frac{0.83 + 0.48 + 0.41}{3} = 0.57 \text{ seconds}$$

This reverberation time meets the reverberation time requirements in Section 1.1 for the school laboratory.

Appendix 7: Calculation of sound absorption required in corridors, entrance halls and stairwells

Approved Document E contains guidance on the addition of sound absorption to common areas in buildings containing dwellings. Where the addition of sound absorption to common areas in schools, such as corridors, entrance halls or stairwells, is required, it is advised that the approach described in Approved Document E be used.

Approved Document E describes two methods, A and B, for controlling reverberation in common internal parts of buildings. These methods are reproduced below from Approved Document E.

Method A

For entrance halls, corridors or hallways, cover an area equal to or greater than the floor area, with a Class C absorber or better. It will normally be convenient to cover the ceiling area with the additional absorption.

For stairwells or a stair enclosure, calculate the combined area of the stair treads, the upper surface of the intermediate landings, the upper surface of the landings (excluding ground floor) and the ceiling area on the top floor.

Either, cover at least an area equal to this calculated area with a Class D absorber, or cover an area equal to at least 50% of this calculated area with a Class C absorber or better. The absorptive material should be equally distributed between all floor levels. It will normally be convenient to cover the underside of intermediate landings, the underside of the other landings, and the ceiling area on the top floor.

Method A can generally be satisfied by the use of proprietary acoustic ceilings. However, the absorptive material can be applied to any surface that faces into the space.

Method B

In comparison with Method A, Method B takes account of the existing absorption provided by all surfaces. In some cases, Method B should allow greater flexibility and require less additional absorption than Method A.

For an absorptive material of surface area S in m^2, and sound absorption coefficient α, the absorption area A is equal to the product of S and α. The total absorption area, A_T, in square metres is defined as the hypothetical area of a totally absorbing surface, which if it were the only absorbing element in the space would give the same reverberation time as the space under consideration.

For n surfaces in a space, the total absorption area, A_T, can be found using the following equation.

$$A_T = \alpha_1 S_1 + \alpha_2 S_2 + \ldots + \alpha_n S_n$$

For entrance halls, provide a minimum of 0.20 m^2 total absorption area per cubic metre of the volume. The additional absorptive material should be distributed over one or more of the surfaces.

For corridors or hallways, provide a minimum of 0.25 m^2 total absorption area per cubic metre of the volume. The additional absorptive material should be distributed over one or more of the surfaces.

Absorption areas should be calculated for each octave band between 250 Hz and 4 kHz inclusively.

Absorption coefficient data (to two decimal places) should be determined as follows:

For specific products, use laboratory measurements of absorption coefficient data determined using BS EN 20354:1993 Acoustics – Measurement of sound absorption in a reverberation room. The measured third-octave band data should be converted to practical sound absorption coefficient data, α_p, in octave bands, according to BS EN ISO 11654:1997 Acoustics – Sound absorbers for use in buildings – Rating of sound absorption.

For generic materials, use the octave band data in Table 7.1 of Approved Document E or the more comprehensive data on the DfES acoustics website. These contain typical absorption coefficient data for common materials used in buildings and may be supplemented by other published data.

Appendix 7: Calculation of sound absorption required in corridors, entrance halls and stairwells

Worked example

A school has an entrance hall that has parquet on a concrete floor (7 m x 5 m) with a ceiling height of 3.2 m. The ceiling area is equal to that of the floor. The 5 m x 3.2 m façade is completely glazed and incorporates a glass door. Wooden doors having total surface area of 2.4 m x 2.7 m lead to the corridor described below. The walls are of fair-faced brick.

The corridor has equal floor and ceiling areas of 20 m x 2.4 m and a ceiling height of 2.7 m. The 20 m x 2.7 m external wall of the corridor has half its surface area (27 m^2) glazed. The corridor has parquet on a concrete floor and its unglazed walls are of painted concrete blocks.

Each end of the corridor consists of wooden doors, of surface area 2.4 m x 2.7 m. In addition there are two wooden doors, each of area 2 m x 1.8 m, leading to classrooms off the corridoor.

The sound absorption coefficients necessary to control reverberation in the two spaces using Method B are calculated as shown below.

Example calculation for an entrance hall (Method B)

Step 1 Calculate the surface area related to each absorptive material (ie for the floor, walls, doors and ceiling).

Surface	Surface finish	Area (m^2)
Floor	Parquet on concrete	35.00
Doors (wooden)	Timber	6.48
Walls (excluding door area)	Fair-faced brick	54.32
Façade (and door)	Glazing	16.00
Ceiling	*To be determined*	35.00

Step 2 Obtain values of absorption coefficients for the floor, walls, glazing and doors. (The values below are taken from Table 7.1 of Approved Document E.)

Surface	Area (m^2)	Absorption coefficient (α) in octave frequency bands				
		250 Hz	500 Hz	1000 Hz	2000 Hz	4000 Hz
Floor	35.00	0.03	0.04	0.05	0.05	0.06
Doors (wooden)	6.48	0.10	0.08	0.08	0.08	0.08
Walls	54.32	0.02	0.03	0.04	0.05	0.07
Glazing (façade & door)	16.00	0.08	0.05	0.04	0.03	0.02
Ceiling	35.00	*To be determined*				

Step 3 Calculate the absorption area (m^2) related to each surface in octave frequency bands.
(Absorption area = surface area x absorption coefficient)

Surface	Absorption area (m^2)				
	250 Hz	500 Hz	1000 Hz	2000 Hz	4000 Hz
Floor	1.05	1.40	1.75	1.75	2.10
Doors (wooden)	0.65	0.52	0.52	0.52	0.52
Walls	1.09	1.63	2.17	2.72	3.80
Glazing (façade and door)	1.28	0.80	0.64	0.48	0.32

Step 4 Calculate the sum of the absorption areas (m^2) obtained in Step 3

	250 Hz	500 Hz	1000 Hz	2000 Hz	4000 Hz
Total absorption area (m^2)	4.07	4.35	5.08	5.47	6.74

Step 5 Calculate the total absorption area (A_T) required for the entrance hall.
The volume is 112 m^3 and therefore $A_T = 0.2 \times 112.0 = 22.4\ m^2$.

Step 6 Calculate additional absorption area (m^2) to be provided by the ceiling. If values are negative in any octave band then there is sufficient absorption from the other surfaces to meet the requirement without any additional absorption in this band.
(Additional absorption = A_T –total absorption area (from Step 4))

	250 Hz	500 Hz	1000 Hz	2000 Hz	4000 Hz
Additional absorption area (m^2)	18.33	18.05	17.32	16.93	15.66

Step 7 Calculate required absorption coefficient (α) to be provided by ceiling
(α = Additional absorption area / area of ceiling)

	250 Hz	500 Hz	1000 Hz	2000 Hz	4000 Hz
Required absorption coefficient	0.52	0.52	0.49	0.48	0.45

Step 8 Identify a ceiling product from manufacturers' laboratory measurement data that provides absorption coefficients that exceed the values calculated in Step 7.

Example calculation for a corridor (Method B)

Step 1 Calculate the surface area related to each absorptive material (ie for the floor, walls, doors and ceiling).

Surface	Surface finish	Area (m^2)
Floor	Parquet on concrete base	48.00
Glazing		27.00
Doors	Timber	20.16
Wall (excluding door area and glazing)	Painted concrete block	73.80
Ceiling	*To be determined*	48.00

Step 2 Obtain values of absorption coefficients for the floor, walls, glazing and doors. (The values below are taken from Table 7.1 of Approved Document E.)

		Absorption coefficient (α) in octave frequency bands				
Surface	**Area**	**250 Hz**	**500 Hz**	**1000 Hz**	**2000 Hz**	**4000 Hz**
Floor	48.00	0.03	0.04	0.05	0.05	0.06
Glazing	27.00	0.08	0.05	0.04	0.03	0.02
Doors	20.16	0.10	0.08	0.08	0.08	0.08
Wall (excluding door area and glazing)	73.80	0.05	0.06	0.07	0.09	0.08
Ceiling	48.00	*To be determined*				

Step 3 Calculate the absorption area (m^2) related to each surface in octave bands.
(Absorption area = surface area x absorption coefficient)

	Absorption area (m^2)				
	250 Hz	**500 Hz**	**1000 Hz**	**2000 Hz**	**4000 Hz**
Floor	1.44	1.92	2.40	2.40	2.88
Glazing	2.16	1.35	1.08	0.81	0.54
Doors	2.02	1.61	1.61	1.61	1.61
Wall (excluding door area and glazing)	3.69	4.43	5.17	6.64	5.90

Step 4 Calculate the sum of the absorption areas (m^2) obtained in Step 3.

	250 Hz	500 Hz	1000 Hz	2000 Hz	4000 Hz
Total absorption area (m^2)	9.31	9.31	10.26	11.46	10.93

Step 5 Calculate the total absorption area (A_T) required for the corridor.
The volume is 129.6 m^3 and therefore A_T = 0.25 x 129.6 = m^2.

Step 6 Calculate additional absorption area (m^2) to be provided by ceiling. If values are negative in any octave band then there is sufficient absorption from the other surfaces to meet the requirement without any additional absorption in this band.
(Additional absorption = A_T – total absorption area (from Step 4))

	250 Hz	500 Hz	1000 Hz	2000 Hz	4000 Hz
Additional absorption area (m^2)	23.09	23.09	22.14	20.94	21.47

Step 7 Calculate required absorption coefficient (α) to be provided by ceiling
(α = Additional absorption area / area of ceiling)

	250 Hz	500 Hz	1000 Hz	2000 Hz	4000 Hz
Required absorption coefficient	0.48	0.48	0.46	0.44	0.45

Step 8 Identify a ceiling product from manufacturers' laboratory measurement data that provides absorption coefficients that exceed the values calculated in Step 7.

Appendix 8: Equipment specifications for sound field systems in schools

Standard loudspeakers

A standard specification for loudspeakers is difficult, since there are circumstances when specialised solutions are required. The specification provided below is a general recommendation for typical loudspeakers used in a set of four to six in a classroom within the normal range of sizes.

Specification descriptor	Standard symbol	Recommended value or range	Comments
Characteristic sensitivity	L_p	>85 dB @ 1 W	
RMS power		>10 W continuous band limited pink noise 150 Hz to 8 kHz	
Frequency response		+/– 3 dB over range 150 Hz to 8 kHz	
Coverage angle		Min 90° H x 60° V Min 90° H x 90° V	Wall mounting type Ceiling mounting type
Loudspeaker type		2-way, 100 V preferred	Must be matched to amplifier and system suitably wired.
Enclosure			Flame retarding for wall mounted applications. Fire rated back enclosure for all ceiling loudspeakers that penetrate a fire rated ceiling. Acoustically rated back enclosure for all ceiling loudspeakers that penetrate a ceiling separating classroom floors.
Brackets			Brackets for wall mounted enclosures should provide lockable adjustment vertically and horizontally. Fixing to wall and loudspeaker to be a minimum of two secure screws or bolts each. Secondary safety bond to be provided between loudspeaker and mounting surface.

NXT or other distributed mode loudspeakers

These loudspeakers use new and emerging technology. An 'exciter' causes a panel to vibrate and the panel emits sound. The characteristics of the exciter, its location on the panel and the panel material are all important for correct operation. Products include ceiling tiles, wall mounting posters, projection screens and whiteboards that serve also as loudspeakers. There are advantages in the use of these loudspeaker types, as they provide a better average sound coverage in a room and provide a better average speech intelligibility under some conditions. The reason for this is not fully understood. Fewer NXT type loudspeakers are required, and it may be found that one wall mounted whiteboard model will suffice for the smaller classroom.

Specification descriptor	Standard symbol	Recommended value or range	Comments
Characteristic sensitivity	L_p	>85 dB @ 1 W	
RMS power		>10 W continuous band limited pink noise 150 Hz to 8 kHz	
Frequency response		+/– 3 dB over range 150 Hz to 8 kHz	
Coverage angle		Average 120° H x 120° V	
Loudspeaker type		100 V preferred for ceiling type as several will be connected in parallel.	Must be matched to amplifier and system suitably wired.
Enclosure		Most require no enclosure, but usually have to be spaced away from the wall on a mounting frame. Units for classroom use should be of Class 1 or better for spread of flame. Where used in an acoustically separating ceiling, provision must be made to maintain the sound insulation behind the loudspeakers.	
Brackets		Adjustment of aiming angle is not necessary. Brackets must provide a minimum of two fixing points.	

Mixer amplifier

Specification descriptor	Standard symbol	Recommended value or range	Comments
Inputs	1 mic/line 1 stereo line 1 mic/line	Mic (-50 dBu sensitivity) line (-10 dBu sensitivity) switchable Stereo phono or 3.5 mm jack (-10 dBu sensitivity) Min/line switchable as above	• 1 mono, compatible with teacher radio microphone receiver • 1 stereo (mixed to mono), to enable music playback, connection to computer audio output. Alternatively built-in cassette player • Prefer minimum of 1 additional mono input to enable second microphone for class discussion use when child using personal FM system is present.
Equivalent input noise	Mic input	<-110 dB	
Frequency response		+/– 3 dB over range 80 Hz to 15 kHz	Mic or line input
Outputs	1 line level 1 speaker	-10 dBu, unbalanced, phono or mini-jack continuous band limited pink noise 150 Hz to 8 kHz 100 V, 40 W continuous band limited pink noise 150 Hz to 8 kHz	For connection of personal FM or induction loop amplifier. Can be formed by resistively attenuating the speaker output. For connection of 100 V type loudspeakers. Amplifiers are available with both low impedance and 100 V outputs. Usually only one should be used at a time. Amplifier MUST match with type of loudspeaker used.
Dynamic range		>75 dBA from amplifier noise floor to clipping point	Allows for usable listening range and scope for adjustment of controls.
Distortion	THD+N	< 2% from 150 Hz to 8 kHz	
Equalisers	Bass Treble	min +/– 6 dB variation @ approx 100 Hz min +/– 6 dB variation @ approx 10 kHz	Minimum 2 band equaliser operating on the mixed output signal. Preferred minimum 2 band equaliser operating on each input.
Hum and noise		>85 dB below maximum output level	Under normal range of control settings.

Radio microphone system

A diversity receiver is preferred. See sidebar for further details.

Specification descriptor	Standard symbol	Recommended value or range	Comments
System main parameters		Wideband FM radio microphone system operating in the VHF high band channels allocated for use in personal FM systems.	These channels are provided for service to the hearing impaired without requirement for a licence.
		Must conform to IR 2030, published by the Radiocommunications Agency under the category Short Range Devices. See www.radio.gov.uk for latest standards.	
		If necessary to accommodate a large number of channels within a single school or site, licensed radio microphone units operating in the UHF band can be used.	These channels require a licence, with an associated annual fee.
Channel selection		It is preferred that the system has a user programmable channel selection.	This enables a spare unit to support all units within a school or group of schools. Also enables channels to be easily changed in the event of interference or the desire to tune the system to match a compatible personal FM receiver brought in by a student.
Microphone input		Compatible with plug-in dynamic and electret microphones. Robust connector with locking mechanism and high quality cable retention is required. A permanently wired microphone is not acceptable.	
Transmitter antenna connection		Can be used with a ¼ wave cable antenna. Robust connector with locking mechanism and high quality cable retention is required. A permanently wired antenna is not acceptable.	
Transmitter controls and indicators	Volume or gain	Transmitter should be provided with a means to adjust the level of the signal. This should be recessed or screwdriver controlled to minimise the risk of accidental adjustment.	Some cheaper transmitters provide no gain adjustment. This limits use with other microphones and some users. This actually controls the modulation of the radio section of the transmitter.
	On/off Switch	A switch should be provided to enable the transmitter to be switched off to preserve battery life. This should be recessed to prevent accidental operation.	An on/off switch should not be used unless the receiver is also turned off. If the TX is off, the receiver may pick up an alternative source on the same channel.

continued	Mute switch	A switch should be provided to enable the audio signal output from the transmitter to be muted without turning off the transmitter.	This allows the audio to be turned off to allow private conversation, etc.
	Battery switch	A means of indicating the battery transmitter to be switched off to preserve battery life. This should be recessed to prevent accidental operation.	An alternative means of testing batteries can be provided.
	Transmitter level indicator	It is preferred that there is a means of checking the operating level of the transmitter, either on the transmitter unit, or on the receiver.	Transmitter level is actually easy to measure at either end.
	Channel selector	Channel selection should be available by means of an easy to understand control that is protected against accidental operation.	
Receiver antenna		Diversity receiver with dual antennas. Built-in or detachable telescopic or helical antennae.	Diversity provides protection against signal loss due to reflections in the room. Reduces signal 'drop-out'.
Receiver controls and indicators	Volume control	An output volume control aids in setting up a system.	Alternatively, given compatibility with the amplifier, the gain may be adjusted there instead.
	On/off switch	A front panel operable on/off switch is required.	
System frequency response		100 Hz to 10 Hz ± 1 dB	Performance of whole transmitter to receiver system without microphone connected.
Dynamic range		≥ 85 dB	This sets the maximum signal to noise ratio available from the equipment. It is the performance of the whole transmitter to receiver system without the microphone connected.
Distortion	THD+N	< 0.5% from 150 Hz to 8 kHz at any signal level	
Transmitter battery	Life	≥ 6 hours from a rechargeable nicad battery under continuous transmission conditions Battery compartment should be robust, enabling battery to slide in. A loose, plug-on battery connection is not acceptable.	Battery life should be measured under real operating conditions. Many published figures are not trustworthy as they are actually for a standby condition.

Headworn microphone

Specification Descriptor	Standard Symbol	Recommended value or range	Comments
Microphone type		Omni-directional headworn microphone. Robust cable and connector with locking mechanism and good cable retention. Condenser microphone types must be compatible with radio microphone transmitter powering system or contain easily changed battery with long service life.	
Frequency response		100 Hz to 12 kHz ± 3 dB when used in recommended operating position	Microphone response is partly dependent upon surrounding surfaces. Microphone response should be considered when used as intended, not in an anechoic measurement.
Sensitivity		Microphone sensitivity should match the gain range of the transmitter enabling full transmitter modulation to be achieved when worn as recommended and used with a raised voice level.	
Dynamic range		> 65 dBA	
Sensitivity		> -46 dBV re 1 V/Pa Suitable for use in close proximity to the mouth	

Appendix 9: Noise at Work Regulations relating to teachers

There is growing concern about the possibility of long-term hearing damage to those teachers who generally work in the noisier school environments, for example PE teachers, CDT teachers and music teachers. The Health and Safety Executive has recently carried out a study of noise exposure among these teachers due to their potential for exposure to high noise levels. It is known that orchestral musicians are at risk of noise induced hearing damage[1,2] and therefore peripatetic music teachers may experience additional risk. It is necessary under the Noise at Work Regulations[3] to ensure that teachers of 'noisy' subjects are not exposed to levels of noise likely to cause, or increase, risk of hearing damage. The noise levels to which teachers are exposed can be reduced to some extent by good acoustic design of schools, for example by ensuring that the reverberation time is short so as not to increase the noise level in a room. However, there will inevitably be some occasions when the noise associated with a particular teaching activity approaches, or is above, the levels known to pose a risk to hearing.

The risk of noise induced hearing damage is a function of both noise level and the duration of exposure to the noise. The noise levels to which employees may be exposed, and the wearing of hearing protection, are currently subject to the Noise at Work Regulations 1989[3]. However, in February 2003 the European Union published a new directive, the Physical Agents (Noise) Directive[4], relating to noise at work which will result in the UK legislation being changed in 2006.

The main points of the current regulations, the European Directive and the likely changes to UK legislation in 2006 are summarised below.

Action levels

The current regulations are expressed in terms of action levels, that is levels of noise exposure at which certain actions are required by employers and/or employees (together with manufacturers of equipment).

Action levels are defined in terms of daily personal noise exposure $L_{EP,d}$ which takes account of both level and exposure time. $L_{EP,d}$ is similar to $L_{Aeq,T}$ (see Appendix 1) but is always normalised to an exposure time of 8 hours. For example, a person exposed to a continuous noise level of 85 dB(A) for 8 hours per day experiences a daily personal noise exposure of 85 dB(A) $L_{EP,d}$. For each halving of the daily exposure time, the $L_{EP,d}$ reduces by 3 dB(A), so that a daily exposure of 4 hours to a noise level of 85 dB(A) is equivalent to 82 dB(A) $L_{EP,d}$, and a daily exposure of 2 hours to a noise level of 85 dB(A) is equivalent to 79 dB(A) $L_{EP,d}$. Similarly, a doubling of exposure time increases $L_{EP,d}$ by 3 dB(A).

Current Regulations

The Regulations apply to employers, employees and self-employed people (who have the duties of both employers and employees). Peripatetic music teachers, for example, may fall into this last category.

Whatever the level of noise, employers have a duty under the Regulations to reduce the noise level to the lowest level reasonably practicable.

The first action level is 85 dB(A) $L_{EP,d}$. If any employee is likely to be exposed to this level or above, employers' duties under the Regulations include the following :

- to ensure that a competent person makes a noise assessment which is adequate to identify which employees are exposed to this level or above
- to inform employees of the risks of noise and ways in which the risk may be reduced
- to provide, and maintain, hearing protection for those who request it.

Employees also have a duty at this action level to maintain any equipment that is provided by the employer to reduce the risk of hearing damage, and to report any defects in the equipment.

The second action level is 90 dB(A) $L_{EP,d}$. If any employee is likely to be exposed to this level or above, employers' additional duties under the Regulations include the following :

• to reduce noise exposure of employees through noise control measures other than hearing protection

• to mark hearing protection zones where noise reaches the second action level with recognised signs

• to provide hearing protection to all employees and to ensure that it is worn.

At this action level employees must again maintain any equipment provided, and must also wear the hearing protection provided.

The regulations also specify a peak action level of 200 Pascals (equivalent to an unweighted sound level of 140 dB). This represents an instantaneous sound level, caused for example by a loud bang. Where this level is exceeded, employers and employees have the same duties as at the second action level. Exposure to the peak action level is normally linked with the use of cartridge operated tools, guns or similar loud explosive noises, but can occur during the loud playing of a musical instrument[1].

Changes to the Regulations

New legislation will come into force in February 2006 to comply with the Physical Agents (Noise) Directive. For musicians, who may include music teachers, the new legislation will not be enforced until 2008.

The main changes to the legislation are that, in effect, the action levels will be lowered by 5 dB(A). In general, the actions currently required at 85 and 90 dB(A) $L_{EP,d}$ will be mandatory at 80 and 85 dB(A) $L_{EP,d}$ respectively. In addition an overall personal exposure level of 87 dB(A) $L_{EP,d}$ is to be introduced; this is the limit of exposure at the ear which means that the level at the ear (with or without hearing protection) must never exceed 87 dB(A) $L_{EP,d}$.

Two peak exposure limit values are included in the directive: 112 Pa at which the duties required at 80 dB(A) $L_{EP,d}$ apply, and 140 Pa at which the duties at 85 dB(A) $L_{EP,d}$ are required. 200 Pa remains as an overall peak exposure limit.

Another feature of the new directive which may be relevant to teachers is that it will be possible to assess noise exposure on a weekly, rather than a daily basis, if exposure varies significantly from day to day.

Further information

The above summary of some aspects of the Noise at Work Regulations is included for information, but does not purport to be a complete statement of the Regulations. Employers and employees who believe that they may have duties under the Regulations should obtain a copy of the Regulations and should be familiar with the requirements. For a full version of the regulations see the HMSO web site[5]. For information on the effects of noise, the current regulations, the new Directive and its implications for the UK see the Health and Safety Executive website[6]. The text of the Directive may be found on the website of the Official Journal of the Euopean Union. For a discussion of the implications for music teachers see www.musiced.co.uk.

References

[1] R Canham and B Shield. Noise surveys of orchestral musicians at the Barbican concert hall. Proc. Institute of Acoustics 21(6), 217-225, 1999.

[2] A Wright Reid. A sound ear. Association of British Orchestras 2001.

[3] Noise at Work Regulations 1989.

[4] Directive 2003/10/EC of the European Parliament and of the Council of 6 February on the minimum health and safety requirements regarding the exposure of workers to the risks arising from physical agents (noise). Official Journal of the European Union L042, 15 February 2003, 38-44.

[5] www.hmso.gov.uk

[6] www.hse.gov.uk

Appendix 10: Example submission to Building Control Body

All submissions to the Building Control Body (BCB) should clearly identify the relevant performance standards from Section 1, how they will be met, and the performance that the design is expected to achieve. Calculations, test reports etc should preferably be included in appendices to the submission, rather than in the main body of the submission. The extent of acoustic information required to satisfy the BCB may vary between Authorities and individuals. This example provides an indication of the minimum level of information that should be provided. The right hand column contains a commentary on the submission.

A set of symbols has been created for use on plans in submissions to allow a quick visual inspection of the BB93 performance standards for each acoustic criterion and the performance that the design is expected to achieve. The symbols can be downloaded from the DfES acoustics website. Hand-produced drawings would also be acceptable.

Example submission

The following items are provided in support of the submission to the Building Control Body to demonstrate compliance with the acoustic requirements of Part E of the Building Regulations.

The ground floor plan of the rooms and the acoustic performance standards are shown in Figure A10.1. On the first floor there are classrooms above the ground floor classrooms and music classrooms.

This example submission focuses on only a few of the rooms, although the same level of detail would be required for all relevant rooms in the school.

A 10.1 Indoor ambient noise levels in unoccupied spaces

The performance standards in Table A10.1 for indoor ambient noise levels

Table A10.1: BB93 performance standards – indoor ambient noise level

Room	BB93 performance standard $L_{Aeq,30min}$ (dB)
Classroom	≤ 35

Figure A10.1: The plan of the rooms and the acoustic performance standards

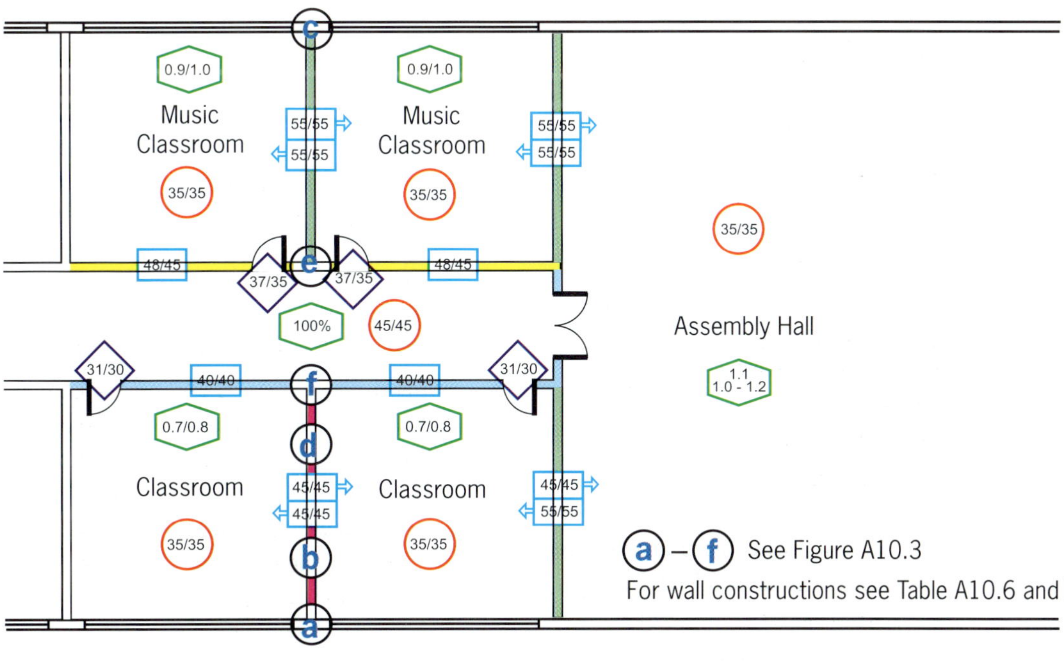

Ⓐ–Ⓕ See Figure A10.3

For wall constructions see Table A10.6 and Figure A10.2

Details of the noise survey should be provided in an appendix. Sufficient information should be provided to allow the BCB to confirm that the measurement times and positions are appropriate and representative for the proposed school.
The external noise spectrum is required to calculate the indoor ambient noise level due to sound transmission through the façade.
In this example, only a single noise measurement has been taken at the proposed façade position for the few classrooms under consideration. Normally, noise measurements would be taken at the positions of all the proposed school façades. In some cases these measurements would be adjusted to take account of some facades being shielded from the noise by the proposed building.

On sites where there is a greater incidence of individual noisy events such as from aircraft overflights or near a railway a more detailed noise survey would normally be expected.

Depending upon the site, each elevation could be exposed to a different level of noise. Therefore a different ventilation strategy could be used on each elevation. This would require separate calculations.

have been taken from Table 1.1, Section 1.

A noise survey was carried out at the site on XX.YY.ZZZZ to establish the noise climate. Free-field external noise levels in terms of $L_{Aeq,30min}$ and $L_{A1,30min}$ were measured at a position corresponding to the proposed school facade closest to the dominant external noise source, the nearby road. The measured data are shown in Table A10.2.

Table A10.2: Noise survey data – $L_{Aeq,30min}$ and $L_{A1,30min}$ (external noise)

Time	$L_{Aeq,30min}$ (dB)	$L_{A1,30min}$ (dB)
08:00 – 08:30	57.5	66.1
08:30 – 09:00	56.8	64.8
09:00 – 09:30	58.0	64.8
09:30 – 10:00	57.7	66.0
10:00 – 10:30	57.4	66.9
10:30 – 11:00	56.9	64.6
11:00 – 11:30	56.1	63.9
11:30 – 12:00	59.1	66.3
12:00 – 12:30	59.4	67.1
12:30 – 13:00	59.8	64.9
13:00 – 13:30	58.6	66.4
13:30 – 14:00	58.5	66.2
14:00 – 14:30	57.4	65.3
14:30 – 15:00	57.9	66.8
15:00 – 15:30	58.2	65.6
15:30 – 16:00	57.4	66.0
16:00 – 16:30	59.9	71.3
16:30 – 17:00	56.6	65.0

The 30 minute time period with the highest external noise level during the school day, 59.9 dB $L_{Aeq,30min}$, is highlighted in the table.

The $L_{eq,30min}$ noise spectrum corresponding to the 30 minute time period with the highest external noise level, 59.9 dB $L_{Aeq,30min}$, is shown in Table A10.3.

The construction of the external envelope of the school will be a cavity brick/block wall with 6/12/6 glazing. The sound transmitted through the façade has been calculated for the classroom to determine the indoor ambient noise level. The upper limit for the reverberation time of the classroom from Table 1.5, Section 1 has been used in the calculation.

Ventilation will be provided by an acoustic ventilator and a passive stack roof ventilator with acoustic attenuation treatment. The ventilation requirement has been calculated based on 3 litre/s per person.

The calculations have been carried out using the Excel spreadsheet based on BS EN 12354-3:2000 from the DfES acoustics website The results are shown in Table A10.4.

The indoor ambient noise level is calculated to be 34.7 dB $L_{Aeq,30min}$ which is just below the upper limit for classrooms, 35 dB $L_{Aeq,30min}$, and therefore satisfies the performance standards in Table 1.1, Section 1.

Note 1 of Table 1.1, Section 1 gives guidance on indoor levels from individual external noisy events. The facade will offer a similar reduction in performance for $L_{A1,30min}$ as for $L_{Aeq,30min}$, hence the indoor level should not regularly exceed 55 dB $L_{A1,30min}$.

For these classrooms there are no noise sources due to building services that require consideration.

Table A10.3: Noise survey data – $L_{eq,30min}$ external noise spectrum

	$L_{eq,30min}$ (dB)				
Octave band centre frequency (Hz)	125	250	500	1 k	2 k
Time 16.00 – 16.30	62.0	58.5	56.1	55.9	52.7

Table A10.4: Calculation of indoor ambient noise level in the classroom due to external noise transmitted through the façade

		Octave band centre frequency (Hz)				
		125	250	500	1 k	2 k
External $L_{eq,30min}$ (dB)		62.0	58.5	56.1	55.9	52.7
	Area (m^2)	Façade element sound insulation R (dB)				
Double glazing 6/12/6	14	20.0	19.0	29.0	38.0	34.0
Brick/block cavity wall	7	41.0	45.0	45.0	54.0	58.0
	Number of ventilators	Ventilator sound insulation $D_{n,e}$ (dB)				
Acoustic vent	1	20.1	23.4	28.7	40.3	52.6
Passive stack roof ventilator	1	20.0	26.0	31.0	36.0	44.0
Reverberation time (s)		0.8				
Room volume (m^3)		168				
Internal $L_{Aeq,30min}$ (dB)		34.7				

The submission should reference the calculation procedure (eg BS EN 12354-3:2000, BS 8233:1999 etc). The submission should also contain references to the source(s) of all sound insulation data used in the calculations. The following options are suitable: copies of laboratory sound insulation test certificates in the appendices of the submission, reference to laboratory accreditation number and test report number, reference to books or papers.

In this example there are no significant internal noise sources. Hence, only external noise sources have been considered. When there are significant internal noise sources, Section 1.1.1 describes which internal sources should be considered in the calculation of the indoor ambient noise level.

A 10.2 Airborne sound insulation between spaces

The performance standards for the airborne sound insulation between spaces have been taken from Tables 1.1 and 1.2, Section 1.

The performance standard for airborne sound insulation between rooms has been assessed in both directions but only the highest $D_{nT(Tmf,max),w}$ values are shown in Table A10.5.

The minimum weighted sound

Table A10.5: Airborne sound insulation between spaces

Source room	Receiving room	BB93 performance standard $D_{nT(Tmf,max),w}$ (dB)
Classroom	Classroom	45
Classroom	Assembly hall	45
Assembly hall	Classroom	55
Music classroom	Music classroom	55
Assembly hall	Music classroom	55
Music classroom	Assembly hall	55

Table A10.6: Separating walls – airborne sound insulation between spaces

Source room	Receiving room	Separating wall construction *(Refer to Figure A10.2 for details)*	Minimum laboratory sound insulation for separating wall* R_w (dB)	Separating wall laboratory performance R_w (dB)	Laboratory test report
Classroom	Classroom	(3)	49	53	Test report No. XXXXXX, Laboratory accreditation No. ZZZZ
Classroom	Assembly hall	(2)	42	45	Test report No. ZZZZZZ, Laboratory accreditation No. ZZZZ
Assembly hall	Classroom	(3)	49	53	Test report No. XXXXXX, Laboratory accreditation No. ZZZZ
Music classroom	Music classroom	(4)	60	63	Test report No. YYYYYY, Laboratory accreditation No. ZZZZ
Assembly hall	Music classroom	(3)	50	53	Test report No. XXXXXX, Laboratory accreditation No. ZZZZ
Music classroom	Assembly hall	(3)	52	53	Test report No. XXXXXX, Laboratory accreditation No. ZZZZ

*Estimated using the approach described in Section 3.10.

The submission should reference the calculation procedure (e.g BS EN 12354-1:2000) that has been used to estimate the airborne sound insulation of the separating element and associated flanking elements. Details and assumptions made in the calculations should be contained in the appendices of the submission.

The submission should include all relevant flanking details and reference the calculation tools or software used to estimate the sound insulation due to the combination of direct and flanking transmission.

The submission should reference the source(s) of all sound insulation data. Copies of laboratory sound insulation test certificates can be included in the appendices of the submission, or reference can be made to the test report number and the laboratory accreditation number.

reduction indices (R_w) required to achieve the $D_{nT(T_{mf,max}),w}$ performance standards for the separating walls have been estimated using the approach described in Section 3.10.

The wall constructions to be used in the school are shown in Figure A10.2.

Table A10.6 contains information on the separating walls to be used between different rooms.

The proposed flanking details are shown in Figure A10.3. Each detail in Figure A10.3 corresponds to the detail identified on Figure A10.1. These flanking details have been used to estimate the field airborne sound insulation according to BS EN 12354-1:2000, to confirm that the performance standards in terms of $D_{nT(T_{mf,max}),w}$ can be achieved.

The floor construction to be used to satisfy the performance standards is shown in Figure A10.4. Test report No. XYZXYZ, laboratory accreditation No. ZZZZ contains the airborne sound insulation data for this floor.

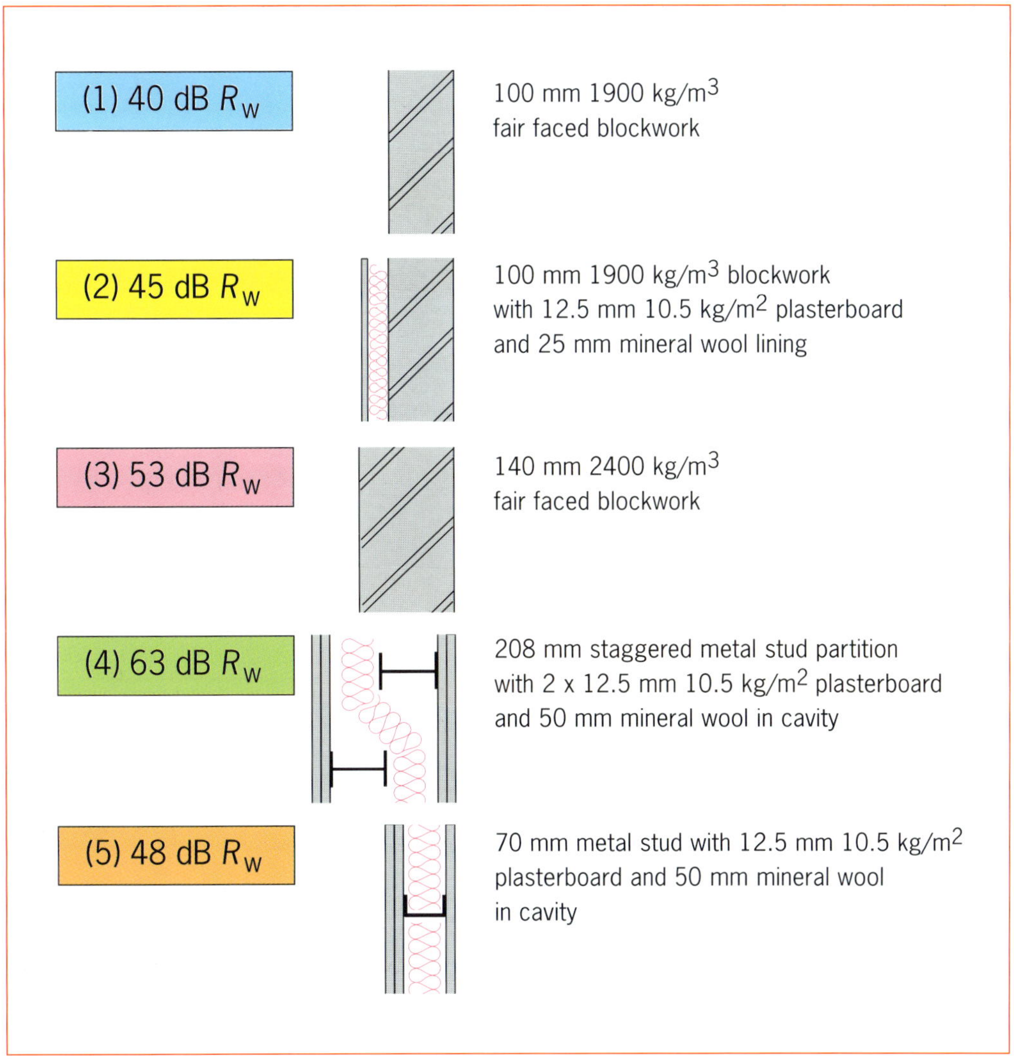

Figure A10.2: Wall constructions

A10.3 Airborne sound insulation between circulation spaces and other spaces used by students

The performance standards in Table A10.7 for airborne sound insulation between circulation spaces and other spaces used by students have been taken from Table 1.3, Section 1.

The walls and doorsets to be used in the school are referenced in Table A10.8 along with references to the laboratory sound insulation test certificates. Doorsets will have a vision panel and neoprene blade seals to the jambs and head, and drop threshold seals.

The submission should reference the source(s) of all laboratory sound insulation data for the walls, doorsets and ventilators. Copies of laboratory sound insulation test certificates can be included in the appendices of the submission, or reference can be made to the test report number and the laboratory accreditation number.

Table A10.7: BB93 performance standards - airborne sound insulation between circulation spaces and other spaces used by students

Space used by students	Circulation space	Separating element	BB93 performance standard R_w (dB)
Classroom	Corridor	Wall Doorset	≥ 40 ≥ 30
Music classroom	Corridor	Wall Doorset	≥ 45 ≥ 35

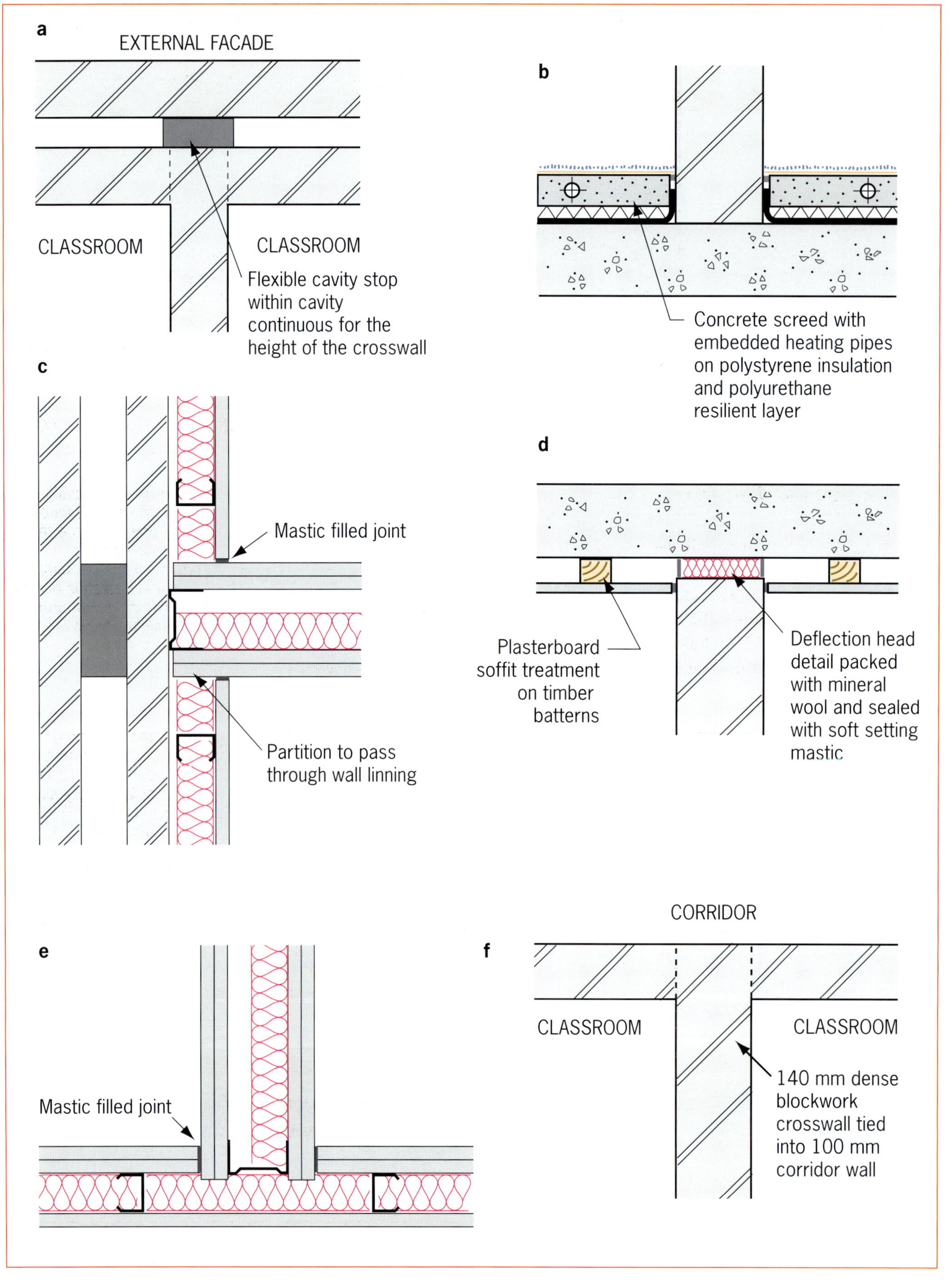

Figure A10.3: Flanking details

Table A10.8: Separating elements - airborne sound insulation between circulation spaces and other spaces used by students

Separating element	Separating element *(Refer to Figure A10.2 for wall details)*	BB93 performance standard R_w (dB)	Separating element laboratory performance R_w (dB)	Laboratory test report
Wall	(1)	≥ 40	40	Test report No. XXXXXX, Laboratory accreditation No. XXXX
Doorset	Product X, Manufacturer X	≥ 30	31	Test report No. YYYYYY, Laboratory accreditation No. XXXX
Wall	(2)	≥ 45	45	Test report No. ZZZZZZ, Laboratory accreditation No. XXXX
Doorset	Product Y, Manufacturer X	≥ 35	37	Test report No. YYYYYY, Laboratory accreditation No. XXXX

A10.4 Impact sound insulation of floors

The performance standards in Table A10.9 for impact sound insulation have been taken from Table 1.4, Section 1.

The floor construction to be used in the school is in Table A10.10 along with references to the estimation method and laboratory sound insulation test certificates.

The separating floor construction shown in Figure A10.4 with a permanent carpet (ie glued to the floor) will be used in the classroom and the music classroom to achieve the performance standards. The first floor science laboratories will have vinyl flooring instead of carpet.

The submission should include all relevant flanking details and reference the calculation tools or software used to estimate the sound insulation due to the combination of direct and flanking transmission.

Table A10.9: BB93 performance standards - impact sound insulation

Ground floor room	First floor room	BB93 performance standard $L'_{nT(T_{mf,max}),w}$ (dB)
Classroom	Classroom	≤ 60
Music classroom	Classroom	≤ 55
Science laboratory	Classroom	≤ 65

The submission should reference the source(s) of all sound insulation data. Copies of laboratory sound insulation test certificates can be included in the appendices of the submission, or reference can be made to the test report number and the laboratory accreditation number.

Table A10.10: Separating floor - impact sound insulation

Ground floor room	First floor room	Floor construction	Estimation method	Estimated $L'_{nT(Tmf,max),w}$ (dB)
Classroom	Classroom	As Figure A10.4 with 5 mm permanent needlefelt carpet	Predictions using BS EN 12354-2:2000 and test report No. YYYYYY, laboratory accreditation No. XXXX	53
Music classroom	Classroom			53
Science laboratory	Classroom	As Figure A10.4 with permanent vinyl flooring instead of carpet	Predictions using BS EN 12354-2:2000 and test report No. YYYYYY, laboratory accreditation No. XXXX	62

Figure A10.4: Separating floor construction

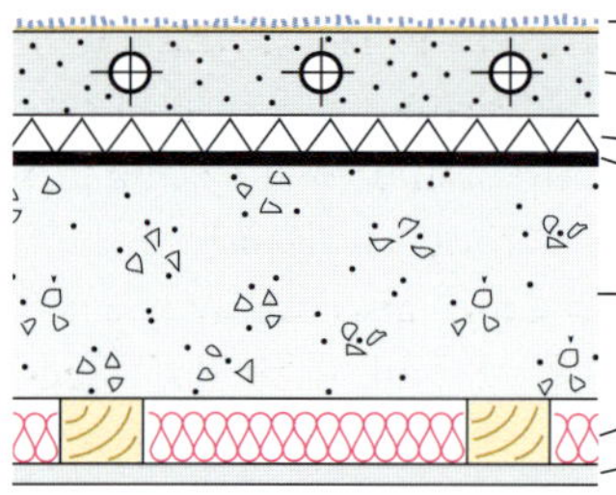

Room data sheets may be a convenient way to present information about the various room finishes to the BCB as they may already have been generated for the school. These data sheets may also be an appropriate place to identify the other acoustic standards proposed for the school.

A10.5 Reverberation in teaching and study spaces

The performance standards in Table A10.11 for the reverberation times have been taken from Table 1.5, Section 1.

The reverberation times for the classrooms and the assembly hall have been calculated as described in Appendix 6 of BB93.

Table A10.11: BB93 performance standards - reverberation times

Room	BB93 performance standard T_{mf} (s)
Classroom	<0.8
Assembly hall	0.8 – 1.2

Table A10.12: Reverberation times for the classroom

		Octave band centre frequency (Hz)				
		125	250	500	1 k	2 k
	Area (m^2)					
		Absorption coefficients				
Painted blockwork	60	0.10	0.05	0.06	0.07	0.09
Carpet on concrete floor	56	0.05	0.05	0.1	0.2	0.45
Glazing	14	0.1	0.07	0.05	0.03	0.02
Suspended plasterboard ceiling	30	0.2	0.15	0.10	0.05	0.05
Absorbent ceiling - Product X Manufacturer Y	26	0.45	0.50	0.55	0.30	0.15
Absorbent tile for back wall - Product X Manufacturer Y	16	0.17	0.38	0.48	0.68	0.88
Volume (m^3)		168				
	Reverberation times					
T (s)		0.88	0.89	0.77	0.75	0.54
T_{mf} (s)		0.7				

Classroom

The absorption coefficients and the predicted reverberation times for the classroom are shown in Table A10.12. The predicted T_{mf} is 0.7 seconds which satisfies the performance standard of <0.8 seconds in Table 1.5, Section 1.

The predicted octave band reverberation times also satisfy the guidance in Note 1 of Table 1.5, Section 1 that the reverberation times in the 125 Hz and 250 Hz octave bands increase up to a value not more than 30% above T_{mf}.

The central section of the ceiling will be a suspended plasterboard ceiling to provide a reflective central section with the absorbent ceiling around the ceiling perimeter as shown in Figure 4.2(a), Section 4.

Assembly hall

The absorption coefficients and the predicted reverberation times for the assembly hall are shown in Table A10.13. The predicted T_{mf} is 1.1 seconds which satisfies the performance standard of 0.8 – 1.2 seconds in Table 1.5, Section 1.

The predicted octave band reverberation times also satisfy the guidance in Note 2 of Table 1.5, Section 1 that the reverberation times in the 125 Hz and 250 Hz octave bands increase up to a value not more than 50% above T_{mf}.

The central section of the ceiling will be a suspended plasterboard ceiling to provide a reflective central section with the absorbent ceiling around the ceiling perimeter as shown in Figure 4.2(a), Section 4.

The submission should reference the source(s) of all absorption data used in the reverberation time calculations. The following options are suitable: copies of laboratory sound insulation test certificates in the appendices of the submission, reference to test report number and laboratory accreditation number, or reference to books or papers.

Table A10.13: Reverberation times for the assembly hall

		Octave band centre frequency (Hz)				
		125	250	500	1 k	2 k
	Area (m^2)					
		Absorption coefficients				
Fair-faced blockwork	170	0.05	0.05	0.05	0.08	0.14
Parquet on concrete floor	192	0.04	0.04	0.07	0.06	0.06
Glazing	76	0.15	0.05	0.03	0.03	0.02
Suspended plasterboard ceiling	140	0.2	0.15	0.10	0.05	0.05
Absorbent ceiling - Product X Manufacturer Y	52	0.51	0.68	0.85	0.92	1.00
Absorbent tile for back wall and top of side walls - Product X Manufacturer Y	90	0.46	0.67	0.83	0.93	0.95
Volume (m^3)		1152				
		Reverberation times				
T (s)		1.50	1.35	1.18	1.12	1.02
T_{mf} (s)		1.11				

The submission should reference the source(s) of all absorption data for products (eg ceiling tiles) used in corridors, entrance halls and stairwells. The following options are suitable: copies of laboratory sound insulation test certificates in the appendices of the submission, reference to test report number and laboratory accreditation number, or reference to books or papers.

A10.6 Sound absorption in corridors, entrance halls and stairwells

The absorption treatment in the corridors has been assessed in accordance with Method A, as described in Appendix 7. It is proposed to install acoustically absorbing material in the form of a suspended ceiling, to provide an area equal to the total floor area. The ceiling system has an absorption class equal to, or better than, Absorption Class C according to manufacturers' literature XYZ.

The submission should reference the source(s) of rain noise measurement or prediction data.

A10.7 Rain Noise

Laboratory measurement data for rain noise on the proposed lightweight metal roof construction is currently unavailable. However, the same roof construction has been used in another school, School X, in which field measurements were carried out. These tests are described in field test report XYZXYZ. These measurements demonstrate that the rain noise is not audible when the roof construction comprises a suspended plasterboard ceiling (10 kg/m^2) with a 150 mm void. Hence, the same roof construction with the suspended ceiling will be used.

Government Documents

Statutory Instruments: 1999: 2. *The Education (School Premises) Regulations 1999: Education, England & Wales.* London: The Stationery Office. ISBN 0110803310. www.legislation.hmso.gov.uk/si/si1999/19990002.htm

Office of the Deputy Prime Minister (2003). *The Building Regulations, Approved Document E, Resistance to the passage of sound (2003 Edition).* London: The Stationery Office. ISBN 0117536423. www.safety.odpm.gov.uk/bregs/approvede

Approved Document M, Access and facilities for disabled people, in support of the Building Regulations, The Stationery Office, 1999
ISBN 0 11 753469. To be replaced shortly by Approved Document M, Access to and use of buildings.
www.odpm.gov.uk

Department of the Environment (1994). *Policy Planning Guide: Planning and Noise, PPG24.* London: The Stationery Office. ISBN 0117529249. www.planning.odpm.gov.uk/ppg/ppg24

Department of Transport and Welsh Office (1988). *Calculation of Road Traffic Noise.* London: The Stationery Office. ISBN 0115508473.

Department of Transport (1995). *Calculation of Railway Noise.* London: The Stationery Office. ISBN 0115517545.

DfES regulations and general guidance

A full list of all relevant DfES publications is on www.teachernet.gov.uk/sbdupublications. Many of the publications are available free either on the DfES website or from DfES Publications, P O Box 5050, Nottinghamshire, NG15 0DJ, Tel: 0845 60222 60.

Department for Education and Employment (1999). *Science accommodation in secondary schools, Building Bulletin 80.* London: The Stationery Office. ISBN 0112710395.

Department for Education and Employment (1996). *Design and technology accommodation in secondary schools, Building Bulletin 81.* London: The Stationery Office. ISBN 0112709176.
(To be revised)

Department for Education and Skills (2004) Briefing Framework for Secondary Schools, Building Bulletin 98, and Briefing Framework for Primary Schools, Building Bulletin 99.
(Revision of Building Bulletin 82: Area Guidelines for Schools.) London:
The Stationery Office.
www.teachernet.gov.uk/sbareaguidelines

Department for Education and Employment (1997). *Music accommodation in secondary schools, Building Bulletin 86.* London: The Stationery Office. ISBN 0112710026.

Department for Education and Skills (2003). *Guidelines for environmental design in schools, Building Bulletin 87.* 2nd ed. London: The Stationery Office. ISBN 0112710131.
www.teachernet.gov.uk/energy

Department for Education and Employment (1998). *Fume cupboards in schools, Building Bulletin 88.* London: The Stationery Office. ISBN 0112710271.

Department for Education and Employment (1998). *Art accommodation in secondary schools, Building Bulletin 89.* London: The Stationery Office. ISBN 0112710298.

Department for Education and Employment (2000). *Modern foreign languages accommodation, Building Bulletin 92.* London: The Stationery Office. ISBN 011271093X.

Department for Education and Skills (2002). *Schools for the future, designs for learning communities, Building Bulletin 95.* London: The Stationery Office. ISBN 0112711286.
www.teachernet.gov.uk/schoolbuildings

Department for Education and Skills (2001). *Guidance on the constructional standards for schools, DfES/0142/2001.* www.teachernet.gov.uk/sbconstand

Special Educational Needs

The Disability Discrimination Act 1995 Part 3 & 4: as amended by The Special Educational Needs & Disability Act 2001.

The Special Educational Needs & Disability Act 2001, MSO1997 ISBN 0101378521.

Department for Education and Employment (1999). *What the Disability Discrimination Act (DDA) 1995 means for schools and LEAs: DfEE Circular 20/99.*

Department for Education and Skills (2001). *Special educational needs code of practice, ref DfES/581/2001.* London: The Stationery Office. ISBN 1841855294.

Department for Education and Skills (2002). *Accessible schools: summary schools, ref: DfES/0462/2002.* www.dfes.gov.uk/sen

Department for Education and Skills (2002). *Accessible schools: planning to increase access to schools for disabled pupils, ref: LEA/0168/2002.* www.dfes.gov.uk/sen

Department for Education and Skills (1998). *Meeting special educational needs, A programme of action.* London: The Stationery Office. ISBN 0855229063.

Department for Education and Employment (1992). *Designing for pupils with special educational needs: special schools, Building Bulletin 77.* London: The Stationery Office. ISBN 0112707963.

Department for Education and Employment (1999). *Access for disabled people to school buildings, Building Bulletin 91.* London: The Stationery Office. ISBN 011271062X.

Department for Education and Skills (2001). *Inclusive school design: accommodating pupils with special educational needs and disabilities in mainstream schools, Building Bulletin 94.* London: The Stationery Office. ISBN 011271109X.

National Grid for Learning. *Inclusion website.* inclusion.ngfl.gov.uk

NDCS (2001). *Deaf Friendly Schools – a Guide for Teachers and Governors.* NDCS.

RNID (2001). *Guidelines for mainstream teachers with deaf pupils in their class.* RNID.

Classroom Acoustics

Finitzo-Hieber, T. and Tillman, T.W. (1978). *Room acoustics effects on monosyllabic word discrimination ability for normal and hearing-impaired children.* J. Speech Hear. Res. 21(3), 440-458.

Miller, J. (1993). *Design standards for the sound insulation of music practice rooms.* Acoustics Bulletin of the Institute of Acoustics 18(6), 54-58.

Mills, J. (1993). *Music in the primary school.* 2nd ed. Cambridge: Cambridge University Press. ISBN 0521448255.

Picard, M. and Bradley, J.S. (2001). *Revisiting speech interference in classrooms.* Audiology 40(5), 221-244.

Shield, B. and Dockrell, J. (2003). *The effects of noise on children at school:* a review. Building Acoustics 10(2), 97-116.

General Acoustics

Association of Noise Consultants (2001). *ANC guidelines: Measurement and assessment of groundborne noise and vibration.* Fresco. ISBN 0953951618.

Association of Noise Consultants (1997). *ANC guidelines: Noise measurement in buildings, Part 1: Noise from building services.* Association of Noise Consultants. www.association-of-noise-consultants.co.uk

Association of Noise Consultants (1998). *ANC guidelines: Noise measurement in buildings, Part 2: Noise from external sources within buildings*. Association of Noise Consultants. www.association-of-noise-consultants.co.uk

Barron, M. (1993). *Auditorium acoustics and architectural design*. London: Spon Press. ISBN 0419177108.

BBC Engineering (1990). *BBC guide to acoustic practice*. 2nd ed. London: BBC. ISBN 0563360798.

BRE Digest 337, *Sound insulation: basic principles*, October 1988. ISBN 0851253652. CRC Ltd.

BRE Report BR 358, *Quiet homes, A guide to good practice and reducing the risk of poor sound insulation between dwellings*, 1998. CRC Ltd. ISBN 1860812562.

BRE Report BR 406, *Specifying dwellings with enhanced sound insulation: a guide*, 2000. CRC Ltd. ISBN 1860814344.

BRE information paper IP6/94, The sound insulation provided by windows, 1994. CRC Ltd.

BRE and CIRIA (1993). *Sound control for homes*. BRE report 238, CIRIA report 127. BRE ISBN 08512555900, CIRIA ISBN 0860173623, CIRIA ISBN 0305408X.

Chu, W.T. and Warnock, A.C.C. (2002). *Detailed directivity of sound fields around human talkers*. National Research Council Canada RR-104. http://irc.nrc-cnrc.gc.ca/fulltext/rr104/

CIBSE (2002). Guide B5: *Noise and Vibration Control for HVCA*. CIBSE. ISBN 1903287251.

Craik, R.J.M. (1996). *Sound transmission through buildings using statistical energy analysis*. Aldershot: Gower. ISBN 0566075725.

Egan, M.D. (1988). *Architectural acoustics*. New York: McGraw-Hill. ISBN 0070191115.

Fry, A. ed. (1988). *Noise control in building services*. Oxford: Pergamon. ISBN 0080340679.

Harris, C.M. (1993). *Noise control in buildings*. New York: McGraw-Hill. ISBN 0070268878.

Lord, P and Templeton, D. (1996). *Detailing for acoustics*. 3rd ed. London: Spon Press. ISBN 0419202102.

McLoughlin, J., Saunders, D.J. and Ford, R.D. (1994). *Noise generated by simulated rainfall on profiled steel roof structures*. Applied Acoustics 42, 239-255.

Morfey, C.L. (2001). *Dictionary of acoustics*. London: Academic Press. ISBN 0125069405.

Templeton, D. ed. (1993). *Acoustics in the built environment*. Oxford: Butterworth Architecture. ISBN 0750605383.

Standards

BS EN ISO 140-1:1998 *Acoustics - Measurement of sound insulation in buildings and of building elements. Part 1. Requirements for laboratory test facilities with suppressed flanking transmission*. British Standards Institution.

BS EN 20140-2:1993 *Acoustics - Measurement of sound insulation in buildings and of building elements. Part 2. Determination, verification and application of precision data*. British Standards Institution.

BS EN ISO 140-3:1995 *Acoustics - Measurement of sound insulation in buildings and of building elements. Part 3. Laboratory measurement of airborne sound insulation of building elements*. British Standards Institution.

BS EN ISO 140-4:1998 *Acoustics - Measurement of sound insulation in buildings and of building elements. Part 4. Field measurements of airborne sound insulation between rooms*. British Standards Institution.

BS EN ISO 140-5:1998 *Acoustics - Measurement of sound insulation in buildings and of building elements. Part 5. Field measurements of airborne sound insulation of façade elements and facades*. British Standards Institution.
BS EN ISO 140-6: 1998 *Acoustics -*

Measurement of sound insulation in buildings and of building elements. Part 6. Laboratory measurement of impact sound insulation of floors. British Standards Institution.

BS EN ISO 140-7: 1998 *Acoustics - Measurement of sound insulation in buildings and of building elements. Part 7. Field measurement of impact sound insulation of floors*. British Standards Institution.

BS EN ISO 140-8: 1998 *Acoustics - Measurement of sound insulation in buildings and of building elements. Part 8. Laboratory measurements of the reduction of transmitted impact noise by floor coverings on a heavyweight standard floor*. British Standards Institution.

BS EN 20140-9:1994 *Acoustics - Measurement of sound insulation in buildings and of building elements. Part 9. Laboratory measurement of room-to-room airborne sound insulation of a suspended ceiling with a plenum above it*. British Standards Institution.

BS EN 20140-10:1992 *Acoustics - Measurement of sound insulation in buildings and of building elements. Part 10. Laboratory measurement of airborne sound insulation of small building elements*. British Standards Institution.

BS EN ISO 140-12: 2000 *Acoustics - Measurement of sound insulation in buildings and of building elements. Part 12. Laboratory measurement of room-to-room airborne and impact sound insulation of an access floor*. British Standards Institution.

BS EN ISO 717-1:1997 *Acoustics - Rating of sound insulation in buildings and of building elements. Part 1. Airborne sound insulation*. British Standards Institution.

BS EN ISO 717-2:1997 *Acoustics - Rating of sound insulation in buildings and of building elements. Part 2. Impact sound insulation*. British Standards Institution.

BS EN ISO 3382:2000 *Acoustics - Measurement of the reverberation time of rooms with reference to other acoustical parameters*. British Standards Institution.

BS 8233: 1999 *Sound insulation and noise reduction for buildings - Code of Practice*. British Standards Institution.

BS EN ISO 11654:1997 *Acoustics - Sound absorbers for use in buildings. Rating of sound absorption*. British Standards Institution.

BS EN 12354-1:2000 *Building Acoustics - Estimation of acoustic performance in buildings from the performance of elements. Part 1. Airborne sound insulation between rooms*. British Standards Institution.

BS EN 12354-2:2000 *Building Acoustics - Estimation of acoustic performance in buildings from the performance of elements. Part 2. Impact sound insulation between rooms*. British Standards Institution.

BS EN 12354-3:2000 *Building Acoustics - Estimation of acoustic performance in buildings from the performance of elements. Part 3. Airborne sound insulation against outdoor sound*. British Standards Institution.

BS EN 20354:1993 *Acoustics - Measurement of sound absorption in a reverberation room*. British Standards Institution.

BS EN 60268-16:1998 *Sound system equipment – Part 16: Objective rating of speech intelligibility by speech transmission index*. British Standards Institution.

BS EN 60804:2001 *Integrating-averaging sound level meters*. British Standards Institution.

ANSI S12.60-2002 *American national standard acoustical performance criteria, design requirements and guidelines for schools*. American National Standards Institute.

ANSI S3.5-1997 (R2002) *Methods for the calculation of articulation index*. American National Standards Institute.

List of organisations

Department for Education and Skills (DfES)
School Buildings & Design Unit
7th Floor, Caxton House, 6-12 Tothill Street, London, SW1H 9NA
Website: www.teachernet.gov.uk/acoustics

Office of the Deputy Prime Minister (ODPM)
Building Regulations Division
3/D1, Eland House, Bressenden Place, London, SW1E 5DU
Website: www.odpm.gov.uk

Institute of Acoustics
77A St Peter's Street, St Albans, Herts, AL1 3BN
Tel: 01727 848195 Fax: 01727 850553
Email: ioa@ioa.org.uk
Website: www.ioa.org.uk

Association of Noise Consultants
6 Trap Road, Guilden Morden, Royston, Herts, SG8 0JE
Tel: 01763 852958 Fax: 01763 853252
Email: mail@association-of-noise-consultants.co.uk
Website: www.association-of-noise-consultants.co.uk

Voice Care Network UK
29 Southbank Road, Kenilworth, Warwickshire, CV8 1LA
Tel: 01926 864000 Fax: 01926 864000
Email: vcnuk@btconnect.com
Website: www.voicecare.org.uk

National Deaf Children's Society
15 Dufferin Street, London, EC1Y 8UR
Tel: 020 7490 8656 (voice & text) Fax: 020 7251 5020
Email: ndcs@ndcs.org.uk
Website: www.ndcs.org.uk

British Association of Teachers of the Deaf (BATOD)
175 Dashwood Avenue, High Wycombe, Buckinghamshire, HP12 3DB
Email: secretary@batod.org.uk
Website: www.batod.org.uk

Royal National Institute for the Deaf (RNID)
19-23 Featherstone Street, London, EC1Y 8SL
Website: www.rnid.org.uk

Association of Building Engineers (ABE)
Lutyens House, Billing Brook Road, Weston Favell, Northampton, NN3 8NW
Tel: 01604 404121
Email: building.engineers@abe.org.uk
Website: www.abe.org.uk

Chartered Institution of Building Services Engineers (CIBSE)
222 Balham High Road, Balham, London, SW12 9BS
Tel: 020 8675 5211
Fax: 020 8675 5449
Website: www.cibse.org

Published by TSO (The Stationery Office) and available from:

Online
www.tso.co.uk/bookshop

Mail, Telephone, Fax & E-mail
TSO
PO Box 29, Norwich, NR3 1GN
Telephone orders/General enquiries: 0870 600 5522
Fax orders: 0870 600 5533
E-mail: book.orders@tso.co.uk
Textphone 0870 240 3701

TSO Shops
123 Kingsway, London, WC2B 6PQ
020 7242 6393 Fax 020 7242 6394
68-69 Bull Street, Birmingham B4 6AD
0121 236 9696 Fax 0121 236 9699
9-21 Princess Street, Manchester M60 8AS
0161 834 7201 Fax 0161 833 0634
16 Arthur Street, Belfast BT1 4GD
028 9023 8451 Fax 028 9023 5401
18-19 High Street, Cardiff CF10 1PT
029 2039 5548 Fax 029 2038 4347
71 Lothian Road, Edinburgh EH3 9AZ
0870 606 5566 Fax 0870 606 5588

TSO Accredited Agents
(see Yellow Pages)

and through good booksellers

ISBN 0 11 271105 7